Innovative Solutions for Implementing Global Supply Chains in Emerging Markets

Ashish Dwivedi
University of Hull Business School, UK

A volume in the Advances in Logistics, Operations, and Management Science (ALOMS) Book Series

BUSINESS SCIENCE Reference
An Imprint of IGI Global

Published in the United States of America by
Business Science Reference (an imprint of IGI Global)
701 E. Chocolate Avenue
Hershey PA, USA 17033
Tel: 717-533-8845
Fax: 717-533-8661
E-mail: cust@igi-global.com
Web site: http://www.igi-global.com

Library of Congress Cataloging-in-Publication Data

Names: Dwivedi, Ashish N., editor.
Title: Innovative solutions for implementing global supply chains in emerging markets / Ashish Dwivedi, editor.
Description: Hershey, PA : Business Science Reference, [2016] | Includes bibliographical references and index.
Identifiers: LCCN 2015042001| ISBN 9781466697959 (hardcover) | ISBN 9781466697966 (ebook)
Subjects: LCSH: Business logistics--Developing countries.
Classification: LCC HD38.5 .I5485 2016 | DDC 658.7--dc23 LC record available at http://lccn.loc.gov/2015042001

This book is published in the IGI Global book series Advances in Logistics, Operations, and Management Science (ALOMS) (ISSN: 2327-350X; eISSN: 2327-3518)

British Cataloguing in Publication Data
A Cataloguing in Publication record for this book is available from the British Library.

All work contributed to this book is new, previously-unpublished material. The views expressed in this book are those of the authors, but not necessarily of the publisher.

For electronic access to this publication, please contact: eresources@igi-global.com.

Advances in Logistics, Operations, and Management Science (ALOMS) Book Series

John Wang
Montclair State University, USA

ISSN: 2327-350X
EISSN: 2327-3518

Mission

Operations research and management science continue to influence business processes, administration, and management information systems, particularly in covering the application methods for decision-making processes. New case studies and applications on management science, operations management, social sciences, and other behavioral sciences have been incorporated into business and organizations real-world objectives.

The **Advances in Logistics, Operations, and Management Science** (ALOMS) Book Series provides a collection of reference publications on the current trends, applications, theories, and practices in the management science field. Providing relevant and current research, this series and its individual publications would be useful for academics, researchers, scholars, and practitioners interested in improving decision making models and business functions.

Coverage

- Finance
- Risk Management
- Marketing engineering
- Decision analysis and decision support
- Organizational Behavior
- Production management
- Services management
- Information management
- Networks
- Computing and information technologies

IGI Global is currently accepting manuscripts for publication within this series. To submit a proposal for a volume in this series, please contact our Acquisition Editors at Acquisitions@igi-global.com or visit: http://www.igi-global.com/publish/.

The Advances in Logistics, Operations, and Management Science (ALOMS) Book Series (ISSN 2327-350X) is published by IGI Global, 701 E. Chocolate Avenue, Hershey, PA 17033-1240, USA, www.igi-global.com. This series is composed of titles available for purchase individually; each title is edited to be contextually exclusive from any other title within the series. For pricing and ordering information please visit http://www.igi-global.com/book-series/advances-logistics-operations-management-science/37170. Postmaster: Send all address changes to above address.

Titles in this Series

For a list of additional titles in this series, please visit: www.igi-global.com

Managerial Strategies and Practice in the Asian Business Sector
U Zeyar Myo Aung (Tourism Promotion and International Relations Department, Myanmar) and Patricia Ordoñez de Pablos (The University of Oviedo, Spain)
Business Science Reference • copyright 2016 • 395pp • H/C (ISBN: 9781466697584) • US $195.00 (our price)

Designing and Implementing Global Supply Chain Management
Sudhanshu Joshi (School of Management, Doon University, India) and Rohit Joshi (Indian Institute of Management IIM, Shillong, India)
Business Science Reference • copyright 2016 • 342pp • H/C (ISBN: 9781466697201) • US $195.00 (our price)

Innovative Management Education Pedagogies for Preparing Next-Generation Leaders
Shalini Rahul Tiwari (Institute of Management Technology, India) and Lubna Nafees (Institute of Management Technology, India)
Business Science Reference • copyright 2016 • 313pp • H/C (ISBN: 9781466696914) • US $195.00 (our price)

Handbook of Research on Global Supply Chain Management
Bryan Christiansen (PryMarke, LLC, USA)
Business Science Reference • copyright 2016 • 689pp • H/C (ISBN: 9781466696396) • US $310.00 (our price)

Analyzing Risk through Probabilistic Modeling in Operations Research
Dariusz Jacek Jakóbczak (Technical University of Koszalin, Poland)
Business Science Reference • copyright 2016 • 441pp • H/C (ISBN: 9781466694583) • US $225.00 (our price)

Organizational Change Management Strategies in Modern Business
Aslı Goksoy (American University in Bulgaria, Bulgaria)
Business Science Reference • copyright 2016 • 460pp • H/C (ISBN: 9781466695337) • US $215.00 (our price)

Maximizing Management Performance and Quality with Service Analytics
Yixin Diao (IBM T. J. Watson Research Center, USA) and Daniela Rosu (IBM T. J. Watson Research Center, USA)
Business Science Reference • copyright 2015 • 465pp • H/C (ISBN: 9781466684966) • US $255.00 (our price)

Optimization of Supply Chain Management in Contemporary Organizations
Ehap Sabri (JDA Software, USA & University of Texas at Dallas, USA)
Business Science Reference • copyright 2015 • 337pp • H/C (ISBN: 9781466682283) • US $205.00 (our price)

In the memory of my parents – Dr S.N. Dwivedi and Mrs Basanti Dwivedi.

Table of Contents

Detailed Table of Contents

supply chain design. This chapter explains the practical application of theories, concepts and frameworks in the area of Supply Chain Design, Risk Mitigation and Social Networks. The practical application is based on a case study of ABC, one of the world's leading agribusiness companies with global operations that made a successful foray in emerging markets facilitated by sound decisions pertaining to the design of its global supply chain network. Certain names and other identifying information were disguised to protect confidentiality.

The core object of this chapter is to develop efficiency measurement technique which links SC processes' efficiency (SCPE) to a firm's financial policy (FFP) through demonstrating and utilising the connection between SCPE and a firm's financial efficiency (FFE). The Dempster Shafer/Analytical Hierarchy Processes (DS/AHP) model is employed to link SCPE to the firm's financial efficiency through defining the relevant importance weights of SCPE measures regarding the preferences of FFE. The chapter also introduces a SC Financial Link Index (SCFLI) to examine the degree to which SCPE is linked to the FFE objectives. This index offers an effective supply chain management (SCM) tool to provide continuous feedback on SCPE and recognise the suitable employed remedial actions. Analysing this index offers opportunities for detailed evaluation of SCPE and empowers firms to trace SC processes that need improvement, causing in more control on daily SC operations. The proposed method permits the evaluation, monitoring, and control of SCPE to enhance SC policy for better alignment with the FFP. Linking SCPE to the FFP objectives empowers firms to gain competitive advantages and formulate strategies for improved SCM through linking such strategies to the focus area of improving the FFE.

Supply chains today are, primarily, measured by Key Performance Indicators (KPIs) such as order-fulfillment, availability to the consumer (percent in-stock) and cost reduction, as well as financial KPIs such as return on investment (ROI), return on brand equity and inventory. These KPIs measure the performance of supply chain as a whole. A supply chain is a network of nodes. The performances of individual nodes are measured with KPIs such as production rate, shipment rate, inventory and the like. These metrics may indicate the performance but may not indicate the cause of such performance. For example, a node whose production rate is below the desired level may be because of poor supply of inputs of production by its supplier node.Thus mere identification of KPIs and their evaluation will not enable to identify the root cause of a problem in a supply chain. Therefore, we need a business intelligence framework that will satisfy the objectives, namely, identification of outcome of each node of the supply chain and its cause. The existing Supply Chain Intelligence (SCI) frameworks aims at identifying metrics that reflect the performance of individual nodes and the total supply chain, but fail to identify the cause of such outcomes. It implies that the linkages or association between the KPIs of individual nodes are required

Existing research work has established that many of today's manufacturing organizations have failed to develop a comprehensive supply chain performance measures. In this chapter, the authors intend to empirically assess the effects of supplier buyer relations and human metrics on supply chain performance in the context of Indian manufacturing organizations. After rigorous literature review, total 18 variables have been identified which are later on reduced in number by factor analysis. As a pilot study, primary data is collected from 100 manufacturing organizations by means of a questionnaire and a scale is developed. On a sample size of 100, the proposed hypotheses are tested by applying two-tailed tests. t-test and factor analysis resulted in 5 factors, 2 related to supplier-buyer relations and 3 related to human metrics. The overall reliability of the scale comes out to be 0.697. The research work provides a new approach to the manufacturing organizations to understand the factors affecting supply chain performance. The present study is limited to Indian manufacturing organizations.

The automotive industry is truly a global industry. Manufacturing plant of the company is set up in one corner of the earth, sourcing the automobile parts from many companies located all over the globes, and selling the final product in another country. This leads to complex supply chain. Production strategies like just-in-time (JIT) and just-in-sequence (JIS) creates testing times but presents opportunities for the logistics management. But currently, the industry is going through a rough phase. Manufacturing costs are increasing, rise in fuel prices is having its own affect on the industry, competitive landscape is changing, and globalization presents new political and economic challenges. Currently, the company needs to manage its supply chains efficiently, reduce cost and improve the accuracy of processes delivery. Today, the automobile industry is facing significant other challenges due to complex logistics. Logistics is about co-coordinating various activities like materials handling, storage, inventory management, freight transport and the related information flow and processing in a way that meets customer requirements at minimum cost. This paper will talk about the current challenges, opportunities, growth drivers and future strategies of the companies' logistics management. It will also cover how companies are adapting to the changing economic environment. The study is based on the qualitative analysis of secondary data. The challenges and opportunities for the companies exist from the current operations management and risks of globalization. Authors have looked at the challenges and opportunities in detail in this chapter. Improving the organization's performance and securing competitive advantage over others greatly depends on the supply chain management as it is seen that the competition is no longer between the organizations, rather they are amongst supply chains. From the research, it has been found that companies believe that for sustainable transportation and logistics, green supply chain is a must for long term sustainability of the company. Companies finds export compliance to the emerging markets as another major challenge.

and also making it available throughout the year. Two main parts of cold chain are transportation and storage systems. The key Indian industries where cold chain logistics play very important role are fruits and vegetables, milk and milk products (ice cream), Poultry and processed meat, marine products, pharmaceutical (mainly vaccines) and chemicals. An efficient cold chain industry ensures availability of food products as well as prevents spoilage of medicines. Country like India, where infrastructure is one of the major challenges, cold chain plays a critical role. Analysis for this study shows that cold supply chain network does not differ significantly from products to products at least in Indian scenario. Some of the challenges to the growth of sector in India are high energy cost, power deficit, rising real estate cost, lack of logistical support and uneven distribution of capacity. All these challenges bring down the operating margin of a company and makes it not so attractive business sector. But during last couple of years there is a positive environment being created for this sector in India. The growth in organized retail, growing interest in horticultural crops, demand for cold chain logistics from Pharmaceutical industry and various initiatives by government are some of the reasons why there is a renewed interest in this sector especially by private sector players. This study, which is focused on Indian cold chain logistics, analyzes the industry on PEST (Political, Economy, Social and Technology) model and presents top 3 factors on each of these 4 parameters.

In manufacturing industries or the supply chain, very few attempts have been made in the past to measure service quality or to test the linkages between service quality with supplier's satisfaction and loyalty. Thus, to fill the existing void in the literature, an attempt has been made to conceptualise a model proposing linkages of service quality with supplier's satisfaction and loyalty in the supply chain of automotive industry. The research design for this paper includes a combination of literature review and a survey of 220 and 170 practitioners from four and two automobile manufacturing companies located in Northern India and Western India, respectively. Structural Equation Modelling Techniques (SEM) has been used to examine the relationships among the variables and to test the goodness fit of the proposed model. Set of propositions related to model was empirically tested at supplier-manufacturer dyad and with an optimum fit of model, found that the data fits the model.

Food cold chain management (FCCM) is a specialized kind of supply chain supported by the two different technologies first, refrigeration and second information technology. The refrigeration equipments like cold storages and cold storing devices. They can keep the food quality cum safety intact and enhance food items shelf life. The information technology enables the tracking and tracing through data loggers etc. Despite the huge fresh produce availability in India, food inflation and food security issues are major concerns for policy makers. The food wastage in India is close to US$ 8309 million annually (US$ 60.17)

from the Australian Bureau of Statistics (ABS 2006), analysis has been conducted on employment data containing information about where people work and what industry they work within the close vicinity of case study seaports. The results show that the spatial extent Australian port centric logistics clusters, which tend to vary both in size and shape. Overall size of employment in port-related industries in Australia has grown substantially. Road freight is dominating industry in the port-centric logistics cluster with a contribution of 23.02% and 37.54% in 2001 and 2006 respectively. The range of port-centric logistics cluster in Melbourne is towards western suburbs and in Botany Bay the port-centric logistics cluster is discerned mainly around eastern suburbs and in Brisbane the spread of port-centred logistics cluster is towards southeast inner Brisbane and Northwest Inner Brisbane suburbs. This shows the impact of land use consolidation by the State Government in their effort to boost transport and warehousing employment closer to Australian container ports. The establishment of port-centric logistics clusters, we argue, could mean the opportunities for organisations to achieve agglomeration economies, increase rivalry among organizations to promote competition, access to greater pool of customers, availability of skilled labour force, closer proximity between customers and supplier, sharing of public infrastructure and resources, increased inter-firm interactions, and knowledge spill-over.

Preface

Advances in Information Technologies have transformed the way organisations interact with each other, and with their customers. Customers and organisations have become more demanding, desiring customised products and services in-reaction to a constantly changing business environment wherein revolutionary technologies are resulting in creation of innovative products with shorter product life cycles, whilst being constant pressure to reduce lead times.

Organisations have recognised that they cannot compete alone. Today supply chains compete. Furthermore, the exploitation of knowledge across the supply chain is fundamental to business optimization. Organisations have accepted and recognised that in the dynamic modern day business environment, knowledge is the prime resource for providing an organisation with a sustainable competitive advantage.

In this text, we seek to explore the role of Knowledge Management and Information Technology in creating innovative solutions for implementing global best practices in supply chain management in emerging markets. We explore the current trends in Supply Chain Management, the efficacy of current Knowledge creation/acquisition, and transfer mechanisms among Supply Chain Partners, and also focus on new approaches and skills are for solutions for implementing global best practices in supply chain management in emerging markets.

The purpose of this book to: (a) contribute to building bridges between Supply Chain Management and Knowledge Management paradigms and (b) to facilitate critical thinking in Supply Chain Management and Knowledge Management paradigms in emerging markets.

Prof. Anthony Giddens in his definition of Globalisation says that "Globalisation link distant localities in such a way that local happenings are shaped by distant events" (Giddens, 1990) and vice versa. He goes on to state that advances in information and communication technologies are the main force behind globalisation. This according to him is because it allows information to flow more effectively, which in turn helps in building of a global perspective. Prof. Giddens goes on to state that we would have to create new institutions or modify the existing ones for the new global age. The parallel existence and mutual interaction of 'Information and Communication Technologies' (ICTs) and 'Industry' in the 20th century has resulted in the creation of the 'knowledge economy'.

In the recent years, the impact of these 'knowledge based organisational entities' has become a global phenomena. These 'knowledge based organisational entities' such as "Microsoft, Oracle" have fundamentally altered the way we transform data into information. The use of generic software applications, such as spreadsheets or databases have affected each and every corporation in the world, irrespective of its country of origin. The question I would like to raise is "Are institutions such as Microsoft or Oracle examples of model institution's that shall define our new global age?"

Neef (1999) contends that knowledge management cannot be seen or defined in terms of precise advantages that would accrue to a corporation that adopts it (Neef, 1999). At best, knowledge management can be understood and valued optimally, if seen in context of the dramatic changes occurring in the global economy.

Today everyone speaks about how the ICT revolution is affecting us all. World-wide ICT stocks are fund managers favorites and must buy items on their lists. Today every analyst while recommending companies for investment, looks at how a company is looking at the Information Technology (IT) revolution and its strategy to take advantage of the same (e.g. Data Analytics). A couple of decades ago, this was not happening. The Arbitrage Pricing Theory (APT) was and continues to be the base to all models to measure risk and returns.

In the week (commencing 21/04/2000) NASDAQ tumbled 160 points mainly due to rumours that Microsoft might be split up into three entities. This just shows how the influence of an IT corporation, which epitomises the knowledge management revolution can affect financial markets world-wide.

Malhotra states that ongoing synthesis of knowledge management and E-business would continue to transform organisations, its work practices in the "digital economy" (Malhotra, 2015). The statement by him "Everything is up for grabs including all tried and tested thumb rules, best practices and business models!" reflects how deeply this management trend has affected all those connected with organisational strategy and decision making.

According to Drucker (1999), the most significant contribution of management in the 20th century was "to increase manual worker productivity fifty-fold". He goes on to state that "the most important contribution of management in the 21st century will be to increase knowledge worker productivity - hopefully by the same percentage" (Drucker, 1999). Drucker further states that till date, this issue has not been addressed. He adds that it is also known that to increase knowledge worker productivity to desired levels "the methods, however, are totally different from those that increased the productivity of manual workers" (Drucker, 1999).

Information technology has evolved from a mere method to achieve functional leverage to a critical component in a firm corporate strategy, which is essential for its survival. About six to seven years ago, opening / hosting a website was a luxury. This has drastically changed today it is an essentiality for survival. All the major high street retailers whether it be TESCO or even the conservative M & S have started to sell products via their website and this is expected to grow exponentially.

This observation is also supported by Botten and McManus (1998) who state "The advancement of communications and computing technologies has increased the stature of IT from an operational resource unrelated to strategic goals, to a component in strategy formulation and implementation leading to competitive advantage. The need to integrate technology into the strategic plans of a firm is widely recognised as vital to the health and longevity of the company."

With regard to the IT strategy in the U.K. and E.U., Botten and McManus (1998) have mentioned an Information technology strategy is one "in whom lies the greatest wealth creating potential across the European Union." Neef (1999) asserts that "there is a broad "knowledge-based revolution" taking place and it comes in a matching set: the knowledge-based economy for nations, and knowledge management for organizations themselves". This statement is interesting if we look how the third world countries, particularly the East Asian bloc have dramatically increased their share in world trade.

What would happen in the next years for them would be a focal theme of the research? The west, particularly the U.S. has made the shift into the knowledge economy, wherein it has a huge leap in comparison to Japan. What would happen in the next few years, with regard to the U.S. - Japan trade scenario? The research from here would focus on whether would Japan would catch up with the U.S.? Or would the U.S. have continued exponential growth in applications of information and communication technologies, so as to make it the driver of the Knowledge economy of the world, thereby maintaining its current position in the world economy.

Powell (2000) states that "One of the most important challenges in effectively applying competitive intelligence, as with other kinds of corporate data, is that of knowledge management - how to extract business value from the information". As a corporation adopts information and communication technologies, its success would be determined in the way the management views it and react by altering its management structure. Hansen, Nohria, and Tierney (1999) has looked effect of linkages in the form, the type of associations between business units and performance. He contends that knowledge can be either explicit or tacit & it is only tacit knowledge requires interpretation, which therefore calls for higher associations to ensure its success. Hansen states that "A company's knowledge management strategy should reflect its competitive strategy". With reference to a strategy for knowledge management Hansen further says that a corporate will adopt either Codification (Everything is in form of Relational databases) or the Personalisation strategy (All knowledge is closely linked with the developers and is best disseminated on one to one basis). He gives the example of McKinsey (Personalisation strategy) whose daily charges in 1997 were about $2,000 charges in comparison to Andersen Consulting (Codification Strategy) at $600 daily.

Neef (1999) comments that "In this new world, entire industries may spring up, thrive and be eliminated in a decade, as knowledge-based growth continues to shorten product life cycles, compress development times, drive new product prices downward, and increase the competition for better technical standards". He gives the case of the computing industry where about 70 per cent of revenue is earned from products which were created within two years. Neef (1999) implies that that this has happened due to the revolution in the global telecommunications infrastructure.

An interesting perspective on the above issue would be to look at the evolution of the finance function in the management structure of an organisation from the early 1900s, mid 1950s and now in 2000. Earlier, Finance function was represented at a generic level as Accounts, Marketing as Sales, Operations and Logistics as Production and well Personnel was the promoters individual managerial philosophy. There has been only one addition i.e. Information Systems / Computing or by whatever epithet, we choose to call it. It is interesting, when we look in hindsight, we see that the 20th century witnessed two of the most horrific wars in the history of mankind. Simultaneously, for the first time, in the history of civilisation, freedom in its fullest scope was experienced by a majority of the world. The 20th century also saw attainment of extra ordinary standards of living for the world as a whole. What this implies is that despite all the technological breakthroughs, there has been only one function that has added value over an extended period of time. When we talk of Information Systems/ Computing and attempts to relate it to one individual, one name comes to mind – Bill Gates. Well, it is because, for the first time in the history of mankind, we see that the richest person (Bill Gates) in the world is a knowledge worker.

The rise of the new knowledge based society is very interesting, in the sense that we are forced to ponder on the implications of the same on our future. Traditionally, researchers, often look for new linkages in the existing body of knowledge. In a way, we look at past research done in a particular field of knowledge and try to build it up further. The French saying, "He who looks deepest into the past, looks farthest into the future", sums up on one of the perspectives towards research. It tells us how crucial it is to learn from the past, so as to alter future. The early 20th century saw the advent of management philosophies, as a prime source of bringing out productivity in the work place. This process of shifting towards a 'knowledge economy' which is global in its existence will, challenge every industrialist, manager and worker to be efficient and competitive on an international - not necessary level playing field.

I have managed to solicit chapters, which discusses how organisations can create innovative solutions for implementing global best practices in supply chain management in emerging markets.

Ashish N. Dwivedi
University of Hull Business School, UK
30 May 2015

REFERENCES

Botten, N., & McManus, J. (1998). Competitive strategies for service organisations - the role of information technology in business. *Management Services*, *42*(10), 16–24.

Drucker, P. F. (1999). Knowledge-worker productivity: The biggest challenge. *California Management Review*, *41*(2), 79–94. doi:10.2307/41165987

Giddens, A. (1990). *The consequences of modernity*. Stanford, CA: Stanford University Press.

Hansen, M. T., Nohria, N., & Tierney, T. (1999). What's your strategy for managing knowledge? *Harvard Business Review*, *77*(2), 106–116. PMID:10387767

Malhotra, Y. (2015). Knowledge management and e-business in the new millennium. *Global Risk Management Network*. Retrieved 20 December, 2015, from www.brint.com/advisor/a011700.htm

Neef, D. (1999). Making the case for knowledge management: The bigger picture. *MCB Management Decision*, *37*(1), 72–78. doi:10.1108/00251749910252049

Powell, T. (2000). Disinformation about knowledge management. *Competitive Intelligence Magazine*, *3*(1), 51–52.

Chapter 1
B-School Selection by Fuzzy TOPSIS and AHP

Vivek Agrawal
GLA University, India

Vikas Tripathi
GLA University, India

Nitin Seth
Indian Institute of Foreign Trade, India

ABSTRACT

Rapid liberalization of education sector in India has resulted in increased competition. As a result, we have witnessed rapid rise in number of management institutes. The student's evaluation about an institute/college is based on multiple criteria. Realizing the need a focused review on the literature was made to understand the subject. The review highlighted that conventional methods for B-school evaluation are inadequate for dealing with the imprecise, uncertain or vague nature of linguistic assessment. To overcome this difficulty, due to MCDM problem, Fuzzy multi-criteria decision-making methods are proposed. The aim of this study is to use fuzzy technique for order preference by similarity to ideal solution (TOPSIS) and Analytical Hierarchal process (AHP) methods for the selection of better B-school. The proposed methods have been applied to a B-School selection problem of the students of NCR and results are presented. This chapter contributes to previous researches by adding a new avenue, where the MCDM technique can be useful. The selection of an institution for getting a professional degree is a very tough task for the students and as well as for their guardians. This method can help them to find a better solution by providing a quantitative framework.

INTRODUCTION

In this competitive environment parents are very much concerned about B school selection after the completion of graduation of their wards. This study is intense on the best business school selection in NCR region in India. Liberalization and development of NCR region become the primary factors for the rapid growth of institutions (government and private) providing business education. This rapid growth

DOI: 10.4018/978-1-4666-9795-9.ch001

of institution had opened an opportunity for students/parents to get B-education but also gave birth to a problem of selection of best B school. It has also been seen and observed that these institutions are running various courses without establishing a proper infrastructure and arranging other facilities. At these instant many candidates seeking admission in business school are puzzled for making decision of their admission to best business school. For the sake of this students and parents should concentrate on some points while taking decision of school selection resembling – where they are having good campus interview, library facilities, management, quality of teachers, research facility, discipline etc. Multi criteria decision making (MCDM) approach provides a better idea to students as well as their parents for the selection of business school. Analytical hierarchal process (AHP) is the method for pair wise comparison developed by Saaty (1980 and 1983) is very useful to compare the different criterions and the alternative for selection of best Business School. Hwang and Yoon (1981) developed Technique for order preference by similarity to the ideal solution. TOPSIS is based on the idea that the chosen alternative should have the shortest distance from the positive ideal solution and on the other side the farthest distance of the negative ideal solution. In this work Fuzzy TOPSIS and AHP Methodology is used which is a combined methodology of AHP and TOPSIS for defining the different weights of different criteria and calculate the overall ranking of the business institutions. To handle the vagueness and uncertainty of human judgment, above said two methods of MCDM has been used to produce the better result.

BACKGROUND

In this paper researcher used two step method consists of Fuzzy TOPSIS as well as AHP. At the first step AHP is used for weights calculations of the characteristics as well as the overall weights of the business school in each characteristic. After calculating the weights with AHP method, at the second step Fuzzy TOPSIS method applied to check the result of AHP and Fuzzy TOPSIS. The following steps are used, with the help of a case.

ANALYTICAL HIERARCHY PROCESS (AHP)

The Analytical Hierarchy Process is most widely used and one of the best multi-criteria decision making MCDM approach. The first reference for AHP has been found from 1972 (Saaty, 1972). After that method has been precisely described in 1977 (Saaty, 1977).

The method of AHP has been evolved from various past studies. The use of paired (pair-wise) comparison, the core of AHP, instead of direct allocation of weights has been used by several researchers (Thurstone, 1927, Yokoyama, 1921). A major feature of AHP, the formulation of criteria hierarchy was first proposed and applied by Miller (1966, 1969 & 1970). The 1-9 point scale based on psychological observations (Fechner 1860, Stevens, 1957). The number of items in each level is inspired by (Miller, 1956). AHP has been widely used technique since its introduction shown in table 1.

Several researchers have compiled the AHP success stories (Forman & Gass, 2001, Golden *et* al., 1989, Ho, 2008, Kumar & Vaidya, 2006, Liberatore & Nydick, 2008, Omkarprasad & Sushil, 2006, Saaty & Forman, 1992, Shim, 1989, Sipahi & Timor, 2010, Vargas, 1990, Zahedi, 1986).

Table 1. Review on AHP

Area	Authors
Banks	Haghighi *et* al., 2010, Seçme *et* al. 2009
Manufacturing Systems	Iç & Yurdakul, 2009, Li & Huang, 2009, Yang et al., 2009
Site selection	Önüt *et* al., 2009
Drugs selection	Vidal *et* al., 2010)
Operators evaluation	Sen & ÇInar, 2010
Software evaluation	Cebeci, 2009, Chang *et* al., 2009
Evaluation of website performance	Liu & Chen, 2009
Selection of recycling technology	Hsu *et* al., 2010
Weapon selection	Dagdeviren *et* al., 2009
Underground mining method selection	Naghadehi *et* al., 2009
Software design	Hsu *et* al, 2009
Firms competence evaluation	Amiri *et* al., 2009
Organisational performance evaluation	Tseng & Lee, 2009
Strategy selection	Chen & Wang, 2010, Li & Li, 2009, Mansar *et* al., 2009, Wu *et* al., 2009
Staff recruitment	Celik *et* al., 2009, Khosla *et* al., 2009
Construction method selection	Pan, 2009
Supplier selection	Chamodrakas *et* al., 2009, Labib, 2011, Wang *et* al., 2010, Wang & Yang, 2009
University evaluation	Lee, 2010
Customer requirement rating	Li *et* al., 2010, Lin *et* al., 2010
Energy selection	Kahraman & Kaya, 2010
Project selection	Amiri, 2010
Route planning	Niaraki & Kim, 2009
Technology evaluation	Lai & Tsai, 2009

FUZZY THEORY APPROACH

When the fuzziness exist in defining variables, fuzzy theory has been used for assisting in decision making (Eldukair & Ayyub, 1992, Feng & Xu, 1999, Tah & Carr, 2000, Seo *et* al., 2004). The specification of decision attributes and decision makers' opinion involve fuzziness, and so it is intended to apply fuzzy theory in improving the effectiveness of decision making for School selection. So many definitions of fuzzy principles and some definitions are following:

1. If X is a collection of objects denoted generically by x, then a fuzzy set $\tilde{A}$ in X is a set of ordered pairs: $\tilde{A} = \{(x, \mu_{\tilde{A}}(x)), x \in X\}$ $\mu_{\tilde{A}}$ is a value assigned to represent the membership of x in $\tilde{A}$ (Zimmermann, 2001).

2. A fuzzy set Ã is convex (Zimmermann, 2001) if:

$$\mu_{\tilde{A}}\left(\lambda x_1+(1-\lambda)x_2\right)\geq \min\left\{\mu_{\tilde{A}}(x_1),\mu_{\tilde{A}}(x_2)\right\}, x_1,x_2\in X, \lambda\in[0,1]$$

where $\mu_{\tilde{A}}(x_1), \mu_{\tilde{A}}(x_2)$ are the membership values of x_1, x_2 belonging to fuzzy set $\tilde{A}$ and λ is a real number $\lambda\in[0,1]$

3. For a triangular fuzzy number with member p denoted by $\tilde{N}(p_1,p_2,p_3)$, the membership function of the member p is defined as (Zimmermann, 2001):

$$\mu_{\tilde{N}}(p)=\begin{cases}0, & p\leq p^1,\\ (p-p^1)/(p^2-p^1), & p^1\leq p\leq p^2,\\ (p^3-p)/(p^3-p^2) & p^2\leq p\leq p^3,\\ 0, & p\geq p_3,\end{cases}$$

where p_1, p_2, p_3 stand for the left bound value, mean value and right bound value, respectively, in the distribution of a triangular fuzzy number.

4. Let $\tilde{a}=[a_1,a_2,a_3]$ and $\tilde{n}=[n_1,n_2,n_3]$ be two triangular fuzzy numbers (Laarhoven, 1983), then:
   a. $\tilde{a}+\tilde{n}=[a_l+n_l, a_m+n_m, a_u+n_u]$
   b. $\tilde{a}\times\tilde{n}=[a_l\times n_l, a_m\times n_m, a_u\times n_u]$
   c. $\mu\tilde{a}=[\mu x_1,\mu x_2,\mu x_3], \mu_X>0$
5. Let $\tilde{a}=[a_1,a_2,a_3]$ and $\tilde{n}=[n_1,n_2,n_3]$ be two triangular fuzzy numbers the distance between ã and ñ is defined as:

$$d(\tilde{a},\tilde{n})=\begin{cases}(1/3\,(|a_l-n_l|^p+|a_m-n_m|^p+|a_u-n_u|^p))^{1/p}, & 1\leq p<\infty\\ max\,(|a_l-n_l|,|a_m-n_m|,|a_u-n_u|), & p=\infty\end{cases}$$

LINGUISTIC VARIABLE SET

Generally, the decision making problem is made under uncertainties, vagueness, fuzziness, risk, time pressure and some information is incomplete or missing. For example, it is difficult for decision makers to give an exact value to express their opinion on a company's capability. They prefer to describe their feeling in the fuzzy term of "good", "fair", or "poor", etc. For attribute weighting, decision makers can use fuzzy terms such as "very important", "important", "low", "very low", to express their opinion. The triangular fuzzy number (TFN) is the simplest form of and is mostly used in expressing the linguistic terms in research (Chen, 2000 & Deng, 2006).

An appropriate linguistic variable set can help decision makers to give right judgments on decisions. The linguistic terms and corresponding membership functions can be elicited from expert assessment and past data, and can be modified to incorporate individual situations. In this paper the linguistic terms are defined for demonstration based on previous research (Chen, 2000, Lin & Chen, 2004, Li *et* al., 2007).

Selection of a B-School is usually based on the factors, which affect the image of any institute. Since the selection problem is associated with uncertainty and complexity, fuzzy multiple criteria decision analysis is suitable for making decision under this situation and hence decision making model should be developed. Out of many famous MCDM methods, technique for order performance by similarity to ideal solution (TOPSIS) method is a practical and useful technique for ranking and selection of a number of possible alternatives through measuring distances.

Many studies have proposed extensions of fuzzy for the TOPSIS method to grasp the uncertainty that is inherent in the corresponding evaluation problems (Chen, 2000, Chen *et* al., 2006, Yong, 2006). In this paper, we adopt the extension of TOPSIS method proposed by Chen (2000) and Chen *et* al. (2006) to achieve the final ranking results in the fuzzy environment.

FUZZY TOPSIS

Technique for order preference by similarity to an ideal solution (TOPSIS), known as a traditional MCDM method, which has been developed by Hwang & Yoon (1981) for solving the MCDM problem. It is based on the concept that the selected alternative should have the shortest distance from the positive ideal solution (PIS), and, on the other side, the farthest distance from the negative ideal solution (NIS). If the assessment values are known to have various types of vagueness or subjectiveness, then the general decision making techniques are not useful for this type of problems. In the last few years, numerous researches have been carried out to apply fuzzy set theory to multiple criteria evaluation methods (Buckley, 1985, Carlsson, 1982 and Zimmermann, 1996). Chen and Tzeng (2004) transform a fuzzy MCDM problem into a non fuzzy MCDM using fuzzy integral. Tsaur *et* al. (2002) convert a fuzzy MCDM problem into non fuzzy MCDM problem using the TOPSIS approach. Instead of using distance, they employ a grey relation grade to define the relative closeness of each alternative. Chu (2002) also changes a fuzzy MCDM problem into a crisp one and solves the problem using the TOPSIS approach. Opposed from the others, he first derives the membership functions of all the weighted ratings in a weighted normalization decision matrix using interval arithmetic of fuzzy numbers and then de-fuzzifies them into crisp values using the ranking method of mean of removals. Chen (2000) extends the TOPSIS approach to fuzzy group decision making situations by defining a crisp Euclidean distance between any two fuzzy numbers. Triantaphyllou and Lin (1996) develop a fuzzy version of the TOPSIS approach based on fuzzy arithmetic operations, resulting in a fuzzy relative closeness for each alternative.

Hsu and Chen (1996) discuss an aggregation of fuzzy opinions under group decision making. Li (2005) proposes a simple and efficient fuzzy model to deal with multi-judges/MCDM problems in a fuzzy environment. Li (1999) proposes several linear programming models and methods for multi attribute decision making under "intuitionist fuzziness", where the concept of intuitionist fuzzy sets is a generalization of the concept of fuzzy sets. Liang (1999) incorporates fuzzy set theory and the basic concepts of positive ideal and negative ideal points, and extends MADM to a fuzzy environment. Ölçer and Odabasi (2005), while dealing with problems of ranking and selection, propose a new fuzzy multi-attribute decision making method for multiple attributive group decision making problems in a fuzzy environment. Yeh *et* al. (2000) propose a fuzzy MCDM method based on the concepts of positive ideal and negative ideal points to evaluate performance of bus companies. Despite the applicability of these methods to many decision making problems, typical fuzzy multi criteria analyses require the comparison of fuzzy numbers. However, the comparison process can be quite complex and produce unreliable results (Bortolan & Degani, 1985, Chen & Klein, 1997, Yeh & Deng, 2004). It is also widely used technique in different areas as shown in table 2.

Table 2. Review on TOPSIS

Area	Authors
Aircraft	Wang & Chang, 2007
Weapon selection	Dağdeviren *et* al., 2009
Waste management	Gumus, 2009, Kim *et* al. 2013
Transshipment site selection	Onut & Soner 2008
TOM consultant selection	Saremi *et* al.. 2008
Supplier selection	Yayla *et* al., 2012, Zouggari & Benoucef, 2012, Bai & Cui, 2006, Dalalah *et* al., 2011, Feng *et* al., 2011, Kara, S. S. 2011
Selection of reverse logistics	Kannan *et* al., 2009
Selection among logistics information technologies	Kahraman *et* al., 2007
Robot selection	Chu & Lin, 2003
Programming problems	Abo- Sinna *et* al.,2005
Location selection	Chu, 2002, Yong, 2006
Operating system selection	Ballı & Korukoğlu, 2009
Performance evaluation mode	Secme *et* al., 2009
Project selection	Amiri, 2010
Machine tool alternatives	Zeki & Rifat, 2012.
Outsourcing of logistics services,	Bottani & Rizzi, 2006
Risk assessment	Zayed & Chang, 2002, Wang & Elhag, 2006, Zeng *et* al., (2007)
Selecting an expatriate host country	Chen & Tzeng, 2004
Large scale multi-objective non-linear	Abo- Sinna *et* al., 2005
Selection of computer-integrated	Ic, 2012
Airline industry	Torlak *et* al., 2010
Automobile market	Kim *et* al., 2011
Bank	Secme *et* al., 2009, Shaverdi *et* al., 201, Che *et* al., 2010
Competitive advantages of shopping websites	Sun & Lin, 2009
Customer evaluation for order acceptance	Chamodrakas *et* al., 2009
Customer-driven product design process.	Lin *et* al., 2008
Decision making	Shih *et* al., 2007

Here the researchers extended the concept and approach of TOPSIS to develop a methodology for solving multi criterion decision making problems in fuzzy environments. Considering the fuzziness in the decision data, linguistic variables are used to assess the weight of each criterion and the rating of each alternative with respect to each criterion. We convert the decision matrix into a fuzzy decision matrix and construct a weighted normalized fuzzy decision matrix once the decision makers' fuzzy ratings have been pooled. The lower bound value of alternatives has been designed to obtain the distance value of the corresponding alternatives for detecting the fuzzy positive ideal solution (FPIS) and the fuzzy negative ideal solution (FNIS). Then, we calculate the fuzzy similarity degree of each alternative from FPIS and FNIS, respectively. Finally, a closeness coefficient is defined for each alternative to determine the rankings of all alternatives. The higher value of closeness coefficient indicates that an alternative is closer to FPIS and farther from FNIS simultaneously.

MAIN FOCUS OF THE CHAPTER

As we know that nowadays due to IT revolution and reach of internet to the farthest areas have opened a window for those who are information seekers. But it has also created a problem in selection of information. Before utilizing this information the user has to test it on different criterions. This testing had given birth to the need of selection tools which could be used as helping hand to decision maker. The MCDM methods had emerged as tools helping in selection of an option among available alternatives. In this chapter researchers had tried to implement an already available tool of MCDM in different area of institution selection providing B-education.

Issues

The researcher had tried to deal with the problem of selection of a B-School among available various B-Schools. This selection issues in nowadays gained importance among parents/students because of mushrooming of institutions providing B-education. The parents/students are in search of those techniques which can provide them an appropriate comparison about the B-schools.

Problems

Selection of criterions and the finding out of different weights of those criterions was the basic problem in implementing AHP and TOPSIS method. The criterions may be different from institutions to institutions due to their different orientation.

Solutions and Recommendations

This section focus on the analysis with two methods of MCDM which are discussed above i.e AHP followed by Fuzzy TOPSIS. For the same three students opinion are considered, who are seeking admission for business education as expert opinion for calculation of weights by MCDM technique to selecting a better B-School and given in table 3.

Table 3. Expert opinion

Criterion	Alternative Expert	B_1	B_2	B_3	B_4	B5
A_1	D_1	Vg	Vg	E	Vg	M
	D_2	G	M	Vg	G	G
	D_3	M	G	E	G	G
A_2	D1	Vg	G	Vg	Vg	M
	D2	Vg	G	Vg	Vg	G
	D3	Vg	G	E	G	M
A_3	D1	G	G	Vg	M	G
	D2	G	G	Vg	Vg	M
	D3	M	M	Vg	E	M
A_4	D1	Vg	Vg	E	Vg	G
	D2	G	Vg	E	G	G
	D3	Vg	M	E	Vg	G
A_5	D1	Vg	G	Vg	Vg	G
	D2	Vg	G	E	G	M
	D3	G	G	Vg	M	G
A_6	D1	E	M	Vg	Vg	E
	D2	G	V	E	V	G
	D3	G	G	E	V	G
A_7	D1	V	V	E	V	G
	D2	G	G	V	M	G
	D3	G	V	E	V	M

AHP involves the structuring multiple criterion into a hierarchy than take a decision for the selection of a business school. After assessing the relative importance of each criterion an overall ranking (Douligeris and Pereira, 1994) will determine.

AHP technique has at least three levels. All the three levels which are required in this problem are as follows.

Level 1: (goal) – Best B-School selection
Level 2: (Criterion) – 7 major criterions and those are:

- **A_1:** Infrastructure
- **A_2:** Library facility
- **A_3:** Good Management
- **A_4:** Co-curricular activities
- **A_5:** Placement
- **A_6:** Faculty and staff
- **A_7:** Research facility

Level 3: (Alternatives) – 5 B-school in NCR have been considered those are:

- **B_1:** MBA Institute with UPTU
- **B_2:** Standalone University Department
- **B_3:** PGDM Institute
- **B_4:** MBA Institute with UPTU as part of Engineering College
- **B_5:** PGDM Institute with other Colleges

Step 1: Take the pair wise comparison for all the alternatives. For the pair wise comparison a matrix given by Saaty's scale given in Tables 4 and 5 for the criterion infrastructure as follows.

Figure 1. Hierarchy for best B-School selection

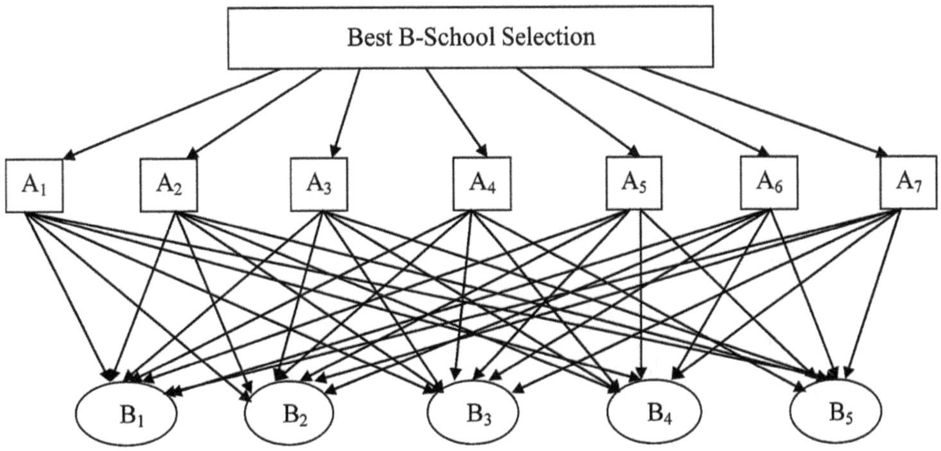

Table 4. Saaty's 9-point scale

Scale	1	3	5	7	9	2,4,6,8
Compare Factor	Equally Important	Inadequately Important	Strongly Important	Very Strongly important	Exceptionally Important	Middle Value

Table 5. Pair wise comparison for infrastructure according to first expert

Experts	B_1	B_2	B_3	B_4	B_5
B_1	1	1	0.5	1	4
B_2	1	1	0.5	1	4
B_3	2	2	1	2	5
B_4	1	1	0.5	1	4
B_5	0.25	0.25	0.2	0.25	1

Step 2: Calculate the sum for each column $\sum_i C_{ij}$

Step 3: Standardized each cell by $X_{ij} = \frac{C_{ij}}{\sum_i C_{ij}}$

Step 4: Calculate the row sum by using the formula

$$R_i = \sum_i X_{ij}$$

and

Weights by $W_i = \frac{R_i}{N}$,

where n is the number of colleges

Step 5: Calculate vector for priority by using

$V_i = A.W_i$ where $i = 1, 2, 3, ..., N$

Step 6: Calculate $\lambda_i = \frac{V_i}{W_i}$ and calculate λ_{max} by averaging the λ_i's shown in Table 6.

Table 6. Calculation for λ_{max}

Experts	B_1	B_2	B_3	B_4	B_5	Column Sum
B_1	1	1	2	1	0.25	5.25
B_2	1	1	2	1	0.25	5.25
B_3	0.5	0.5	1	0.5	0.2	2.7
B_4	1	1	2	1	0.25	5.25
B_5	4	4	5	4	1	18
Xij	0.19048	0.19048	0.38095	0.19048	0.04762	
	0.19048	0.19048	0.38095	0.19048	0.04762	
	0.18519	0.18519	0.37037	0.18519	0.07407	
	0.19048	0.19048	0.38095	0.19048	0.04762	
	0.22222	0.22222	0.27778	0.22222	0.05556	
R	0.97884	0.97884	1.79101	0.97884	0.27249	
W_1	0.19577	0.19577	0.3582	0.19577	0.0545	
V_1	0.9847	0.9846	1.8052	0.9845	0.2732	5.0322
Λ	5.0300	5.0294	5.0396	5.0289	5.0131	N = 5
λmax	5.0282					

Step 7: Calculate consistency index and consistency ratio

$$C.I. = \frac{\lambda_{\max} - n}{N - 1} = 0.00705$$

$$C.R. = \frac{C.I.}{R.I.} = 0.00629 < 0.10,$$

where R.I. means random Index shown in Table 7

From the above the value of C.R. is less than 0.10, which is an allowable value. Hence, matrix is found to be acceptable due the consistency in the judgments which is in the acceptable range. But if in any case the value of C. R. is greater than 0.10 then the researcher need to revise the subjective judgment (Rana *et* al., 2012)

Step 8: Repeat step 1 to step 7 for each expert and check the consistency ratio for each case.

Step 9: Calculate Geometric Mean of each cell for expert 1, expert 2 and expert 3 for determining the overall weight for each B- School for the criteria infrastructure, shown in table 8.

Table 7. R. I. (Saaty, 1980)

Matrix	1,2	3	4	5	6	7	8	9	10	11
Random Index	0	0.52	0.89	1.12	1.26	1.36	1.41	1.46	1.49	1.52

Table 8. Overall weight for criteria infrastructure

B-School	B_1	B_2	B_3	B_4	B_5
A_1	0.1645	0.1643	0.3126	0.2016	0.1556

Table 9. Overall weights of each B-School

Weight	School	B_1	B_2	B_3	B_4	B5
0.1699	A_1	0.1645	0.1643	0.3126	0.2016	0.1556
0.0844	A_2	0.2016	0.1785	0.2853	0.2016	0.1324
0.0423	A_3	0.1766	0.1766	0.2907	0.2266	0.1291
0.1698	A_4	0.1906	0.1644	0.3039	0.1906	0.1501
0.0843	A_5	0.2293	0.1786	0.2858	0.1673	0.1386
0.0251	A_6	0.1923	0.1222	0.2737	0.2192	0.1923
0.2973	A_7	0.1901	0.2124	0.3041	0.151	0.1419

Step 10: Now steps 1-7 have to be repeated for each remaining criterions and also calculate weights for each school according each criterion and calculate the weight of each criterion by the step 1 to 7 as shown in Table 9.

By taking the averages of all the determinants/ criterions for each business school find the result and then rank the Business School, shown in table 10.

Second Step Methodology: TOPSIS

In second step the Fuzzy TOPSIS method is used. In this method only first three institute's result or closeness coefficient have been considered. The process of this method has been described with the help of following steps.

Step 1: Defining fuzzy decision matrix.

Three experts are selected as the experts for assessing the selection of business school. Since Saaty and Varges (1994) suggest that three to seven experts are suitable.

The linguistic weights are taken on judgment of decision makers on attribute as shown in Table 12 (a). (Chen & Hwang, 1992, Liang, 1999, Tsaur *et* al., 2002, Chu, 2002, Chu & Lin, 2003, Chen & Tzeng, 2004). The ratings of alternatives formulated based on decision makers judgments as shown in table 13(a).

Table 10. Performance of each school with rank

B-School	Result	Rank
B_3	0.2503	1
B_1	0.193	2
B_4	0.193	2
B_2	0.171	3
B5	0.128	4

Table 11. Linguistic terms

Linguistic Terms		Triangular Fuzzy Numbers (TFN)
Weightings	Ratings	
Very low (VL)	Very Poor (VP)	(0.0,0.0,0.1)
Low (L)	Poor (P)	(0.0,0.1,0.3)
Medium Low (ML)	Medium poor (MP)	(0.1,0.3,0.5)
Medium (M)	Fair (F)	(0.3,0.5,0.7)
Medium high (MH)	Medium good (MG)	(0.05,0.7,0.9)
High (H)	Good (G)	(0.7,0.9,1.0)
Very high (VH)	Very Good (VG)	(0.9,1.0,1.0)

Table 12. Attributes Weights

Attributes	Decision Makers		
	D_1	D_2	D_3
12 (a). Decision Makers' judgment on attribute weightings			
A_1	H	H	VH
A_2	M	MH	M
A_3	H	H	MH
A_4	H	MH	M
A_5	VH	H	H
A_6	M	MH	MH
A_7	H	H	H
12 (b). TFN for decision makers' judgment on attribute (criterion) weightings			
A_1	(0.7,0.9,1.0)	(0.7,0.9,1.0)	(0.9,1.0,1.0)
A_2	(0.3,0.5,0.7)	(0.05,0.7,0.9)	(0.3,0.5,0.7)
A_3	(0.7,0.9,1.0)	(0.7,0.9,1.0)	(0.05,0.7,0.9)
A_4	(0.7,0.9,1.0)	(0.05,0.7,0.9)	(0.3,0.5,0.7)
A_5	(0.9,1.0,1.0)	(0.7,0.9,1.0)	(0.7,0.9,1.0)
A_6	(0.3,0.5,0.7)	(0.05,0.7,0.9)	(0.05,0.7,0.9)
A_7	(0.7,0.9,1.0)	(0.7,0.9,1.0)	(0.7,0.9,1.0)

Table 13. Ratings by Decision Makers

Attributes	Decision Makers								
	D_1			D_2			D_3		
	B_1	B_2	B_3	B_1	B_2	B_3	B_1	B_2	B_3
13 (a). Decision Makers' judgment on ratings									
A_1	VG	F	F	VG	F	MG	G	MG	F
A_2	MG	F	MG	F	F	F	MG	MG	F
A_3	F	F	F	F	F	F	F	MP	MG
A_4	F	G	MG	F	MG	MG	MG	G	MG
A_5	G	F	MG	MG	F	MG	MG	F	MG
A_6	G	MG	F	F	MG	F	MG	MG	F
A_7	G	G	F	MG	MG	F	MG	F	F
Table 13 (b). TFN for decision makers' judgment on ratings									
A_1	(0.9,1.0,1.0)	(0.3,0.5,0.7)	(0.3,0.5,0.7)	(0.9,1.0,1.0)	(0.3,0.5,0.7)	(0.5,0.7,0.9)	(0.7,0.9,1.0)	(0.5,0.7,0.9)	(0.3,0.5,0.7)
A_2	(0.5,0.7,0.9)	(0.3,0.5,0.7)	(0.5,0.7,0.9)	(0.3,0.5,0.7)	(0.3,0.5,0.7)	(0.3,0.5,0.7)	(0.5,0.7,0.9)	(0.5,0.7,0.9)	(0.3,0.5,0.7)
A_3	(0.3,0.5,0.7)	(0.3,0.5,0.7)	(0.3,0.5,0.7)	(0.3,0.5,0.7)	(0.3,0.5,0.7)	(0.3,0.5,0.7)	(0.3,0.5,0.7)	(0.1,0.3,0.5)	(0.5,0.7,0.9)
A_4	(0.3,0.5,0.7)	(0.7,0.9,1.0)	(0.5,0.7,0.9)	(0.3,0.5,0.7)	(0.5,0.7,0.9)	(0.5,0.7,0.9)	(0.5,0.7,0.9)	(0.7,0.9,1.0)	(0.5,0.7,0.9)
A_5	(0.7,0.9,1.0)	(0.3,0.5,0.7)	(0.5,0.7,0.9)	(0.5,0.7,0.9)	(0.3,0.5,0.7)	(0.5,0.7,0.9)	(0.5,0.7,0.9)	(0.3,0.5,0.7)	(0.05,0.7,0.9)
A_6	(0.7,0.9,1.0)	(0.5,0.7,0.9)	(0.3,0.5,0.7)	(0.3,0.5,0.7)	(0.5,0.7,0.9)	(0.3,0.5,0.7)	(0.5,0.7,0.9)	(0.5,0.7,0.9)	(0.3,0.5,0.7)
A_7	(0.7,0.9,1.0)	(0.7,0.9,1.0)	(0.3,0.5,0.7)	(0.5,0.7,0.9)	(0.5,0.7,0.9)	(0.3,0.5,0.7)	(0.5,0.7,0.9)	(0.3,0.5,0.7)	(0.3,0.5,0.7)

Step 2: Formulating the complex fuzzy decision matrix.

The fuzzy attribute weightings and fuzzy decision matrix are formulated by the converting the linguistic terms into TFN [table 12(b) and table 13(b)]. After TFN, convert this into complex decision matrix (table 14) by using following formulas:

$\tilde{a}_{ij} = 1/t\ [\tilde{a}_{ij}^{1} + \tilde{a}_{ij}^{2} + \ldots\ldots + \tilde{a}_{ij}^{t}],\ i = 1,2,\ldots\ldots,s,\ j = 1,2,\ldots$

$\tilde{w} = 1/t\ [\tilde{w}_i^{1} + \tilde{w}_i^{2} + \ldots\ldots + \tilde{w}_i^{t}],\ i = 1,2,\ldots,s$

Step 3: Normalizing the complex fuzzy decision matrix.

The fuzzy decision matrix now normalized (table 15) by using the formulas

$$\tilde{r}_{ij} = \left(\frac{a_{lij}}{a^{*}_{uij}}, \frac{a_{mij}}{a^{*}_{uij}}, \frac{a_{uij}}{a^{*}_{uij}}\right)\ i \in B$$

$$\tilde{r}_{ij} = \left(\frac{a^{-}_{li}}{a_{uij}}, \frac{a^{-}_{li}}{a_{mij}}, \frac{a^{-}_{li}}{a_{lij}}\right) i \in C$$

Table 14. Averaged frequency weightings and ratings of three Schools

Attributes	Weights	B_1	B_2	B_3
A_1	(0.77,0.93,1.0)	(0.83,0.97,1.0)	(0.37,0.57,0.77)	(0.37,0.57,0.77)
A_2	(0.37,0.57,0.77)	(0.43,0.63,0.83)	(0.37,0.57,0.77)	(0.37,0.57,0.77)
A_3	(0.63,0.87,0.97)	(0.30,0.50,0.70)	(0.23,0.43,0.63)	(0.37,0.57,0.77)
A_4	(0.50,0.70,0.87)	(0.37,0.57,0.77)	(0.63,0.83,0.97)	(0.5,0.7,0.9)
A_5	(0.77,0.93,1.0)	(0.57,0.77,0.93)	(0.30,0.50,0.70)	(0.5,0.7,0.9)
A_6	(0.43,0.63,0.83)	(0.50,0.70,0.87)	(0.50,0.70,0.90)	(0.3,0.5,0.7)
A_7	(0.70,0.90,1.0)	(0.57,0.77,0.93)	(0.5,0.7,0.87)	(0.3,0.5,0.7)

Table 15. Normalized fuzzy decision matrix

Attributes	B_1	B_2	B_3
A1	(o.83,0.97,1.0)	(0.37,0.57,0.77)	(0.37,0.57,0.77)
A2	(0.44,0.58,0.85)	(0.48,0.65,1)	(0.48,0.65,1)
A3	(0.38,0.65, 0.91)	(0.3,0.57,0.83)	(0.48,0.74,1)
A4	(0.38,0.59,0.79)	(0.66,0.86,1)	(0.52,0.72,0.93)
A5	(0.61,0.82,1)	(0.32,0.54,0.75)	(0.54,0.75,0.96)
A6	(0.35,0.43,0.6)	(0.33,0.43,0.6)	(0.43,0.6,1)
A7	(0.32,0.39,0.53)	(0.35,0.43,0.6)	(0.43,0.43,1)

Step 4: Construction of weighted normalized fuzzy decision matrix.

With the normalized fuzzy numbers now construct the weighted normalized fuzzy decision matrix (table 16) by using the formula

$\tilde{v}_{ij} = \tilde{w}_{I} * r_{ij}$, i = 1,2,3 -----s, j = 1,2,3,..............,n

Step 5: Calculate the FPIS and FNIS.

A* = [(1,1,1), (1,1,1), (1,1,1), (1,1,1), (1,1,1), (1,1,1), (1,1,1)]

A^- = [(0,0,0), (0,0,0), (0,0,0), (0,0,0), (0,0,0), (0,0,0), (0,0,0)]

Step 6: Calculate the distance.

The distance for each alternative form FNIS and FPIS can be calculated according to the following formulas (table 17):

$$dj^* = \sum_{i=1}^{s} d(\tilde{v}_{ij}\tilde{v}_i^*) \text{ j} = 1,2,\ldots\ldots s$$

$$dj^- = \sum_{i=1}^{s} d(\tilde{v}_{ij}\tilde{v}_i^-) \text{ j} = 1,2,\ldots\ldots s$$

Table 16. Weighted normalized fuzzy decision matrix

Attributes	B_1	B_2	B_3
A1	(0.64,0.9,1.0)	(0.28,0.53,0.77)	(0.28,0.53,0.77)
A2	(0.16,0.33,0.65)	(0.18,0.37,0.77)	(0.18,0.37,0.77)
A3	(0.25,0.54,0.88)	(0.19,0.47,0.8)	(0.30,0.62,0.97)
A4	(0.19,0.41,0.69)	(0.33,0.60,0.87)	(0.26,0.51,0.81)
A5	(0.47,0.77,1.0)	(0.25,0.50,0.75)	(0.41,0.70,0.96)
A6	(0.15,0.27,0.50)	(0.14,0.27,0.50)	(0.19,0.38,0.83)
A7	(0.23,0.35,0.56)	(0.24,0.39,0.60)	(0.30,0.54,1.0)

Table 17. Distance measurement of alternatives

	d*	d^-
B_1	3.677	3.887
B_2	4.013	3.567
B_3	3.57	4.269

Table 18. Closeness coefficient of each alternative

B-School	CC_i	Ranks
B_1	0.5138	2
B_2	0.4705	3
B_3	0.5446	1

Step 7: Calculate the closeness coefficient of each alternative using the following formula (table 18).

$$CC_i = \frac{d_j^-}{d_j^* + d_j^-} \quad j = 1,2,\ldots\ldots n$$

According to the closeness coefficient the three institute can be ranked as $B_3 > B_1 > B_2$, from highest to the lowest. The results indicate that B_3 is the favorite choice of students for taking admission. B_2 perform worse than B_1 in terms of A_1, A_2, A_3 and A_5. So B_2 has some more gap to improve these four attributes to improve the standard of their institute.

FUTURE RESEARCH DIRECTION

Further research may focus on studying the criticality of attributes or criterion for selecting for other professionals institute. The study can also be replicated in other states of India and also other economies. This research further could be deepening by exploring sub criterions of attributes found. The Institutes can also go for the comparative analysis of factors/attributes provided by them vis a vis their competition. The increased competition had demanded better facility and to improve their features and standard to provide better education and other facility to the students. The future research can provide more areas (dimensions) for improvement and comparison.

CONCLUSION

From table 18, it can be concluded that B_3 is the best school for the education of business management in NCR followed by B_1 and B_2. Thus this project submits a good idea for selecting the best business school in NCR for the fresh students depending on various criterions rather than, depending on a particular criterion. This methodology or model can be extended for the selection of best business school all over India.

The proposed model provides a great opportunity to the students or decision makers for selection of best business school in India/World. Finally this article introduces a multi criteria decision making approach that integrates Fuzzy, AHP and TOPSIS to support Best business school Selection decisions.

REFERENCES

Abo- Sinna, M., & Abou-El- Enien, T. H. M. (2005). An interactive algorithm for large scale multiple objective programming problems with fuzzy parameters through TOPSIS approach. *Applied Mathematics and Computation*, *177*(2), 515–527.

Abo-Sinna, M.-A., Amer, A.-H., & Ibrahim, A.-S. (2008). Extensions of TOPSIS for large scale multi-objective non-linear programming problems with block angular structure. *Applied Mathematical Modelling*, *32*(3), 292–302. doi:10.1016/j.apm.2006.12.001

Amiri, M. (2010). Project selection for oil-fields development by using the AHP and fuzzy TOPSIS methods. *Expert Systems with Applications*, *37*(9), 6218–6224. doi:10.1016/j.eswa.2010.02.103

Amiri, M., Zandieh, M., Soltani, R., & Vahdani, B. (2009). A hybrid multi-criteria decision-making model for firms competence evaluation. *Expert Systems with Applications*, *36*(10), 12314–12322. doi:10.1016/j.eswa.2009.04.045

Bai, R., & Cui, B. (2006). Application in supplier selection using TOPSIS method. *Railway Transport Ecoomies.*, *28*(9), 14–17.

Ballı, S., & Korukoğlu, S. (2009). Operating system selection using fuzzy AHP and TOPSIS methods. *Mathematical and Computational Applications*, *14*(2), 119–130.

Boran, F. E., Genc, S., Kurt, M., & Akay, D. (2009). A multi-criteria intuitionistic fuzzy group decision making for supplier selection with TOPSIS method. *Expert Systems with Applications*, *36*(8), 11363–11368. doi:10.1016/j.eswa.2009.03.039

Bortolan, G., & Degani, R. (1985). A review of some methods for ranking fuzzy subsets. *Fuzzy Sets and Systems*, *15*(1), 1–19. doi:10.1016/0165-0114(85)90012-0

Bottani, E., & Rizzi, A. (2006). A fuzzy TOPSIS methodology to support outsourcing of logistics services. *Supply Chain Management: An International Journal*, *11*(4), 294–308. doi:10.1108/13598540610671743

Braglia, M., Frosolini, M., & Montanari, R. (2003). Fuzzy TOPSIS approach for failure mode, effects and criticality analysis. *Quality and Reliability Engineering International*, *19*(5), 425–443. doi:10.1002/qre.528

Buckley, J. J. (1985). Fuzzy hierarchical analysis. *Fuzzy Sets and Systems*, *17*(3), 233–247. doi:10.1016/0165-0114(85)90090-9

Büyüközkan, G., & Çifçi, G. (2012). A novel hybrid MCDM approach based on fuzzy DEMATEL, fuzzy ANP and fuzzy TOPSIS to evaluate green suppliers. *Expert Systems with Applications*, *39*(3), 3000–3011. doi:10.1016/j.eswa.2011.08.162

Cano, A., & Cruz, M. P. (2002). Integrated methodology for project risk management. *Journal of Construction Engineering and Management*, *128*(6), 473–485. doi:10.1061/(ASCE)0733-9364(2002)128:6(473)

Carlsson, C. (1982). Tackling an MCDM-problem with the help of some results from fuzzy set theory. *European Journal of Operational Research*, *10*(3), 270–281. doi:10.1016/0377-2217(82)90226-0

Cebeci, U. (2009). Fuzzy AHP-based decision support system for selecting ERP systems in textile industry by using balanced scorecard. *Expert Systems with Applications*, *36*(5), 8900–8909. doi:10.1016/j.eswa.2008.11.046

Celik, M., Kandakoglu, A., & Er, D. (2009). Structuring fuzzy integrated multi-stages evaluation model on academic personnel recruitment in MET institutions. *Expert Systems with Applications*, *36*(3), 6918–6927. doi:10.1016/j.eswa.2008.08.057

Chamodrakas, A., Alexopoulou, N., & Martakos, D. (2009). Customer evaluation for order acceptance using a novel class of fuzzy methods based on TOPSIS. *Expert System Application*, *36*(4), 7409–7415. doi:10.1016/j.eswa.2008.09.050

Chang, C.-W., Wu, C.-R., & Lin, H.-L. (2009). Applying fuzzy hierarchy multiple attributes to construct an expert decision making process. *Expert Systems with Applications*, *36*(4), 7363–7368. doi:10.1016/j.eswa.2008.09.026

Che, Z. H., Wang, H. S., & Chuang, C. L. (2010). A Fuzzy AHP and DEA Approach for Marketing Bank Loan Decisions for Small and Medium Enterprises in Taiwan. *Expert Systems with Applications*, *37*(10), 7189–7199. doi:10.1016/j.eswa.2010.04.010

Chen, C. B., & Klein, C. M. (1997). An efficient approach to solving fuzzy MADM problems. *Fuzzy Sets and Systems*, *88*(1), 51–67. doi:10.1016/S0165-0114(96)00048-6

Chen, C. T. (2000). Extensions of the TOPSIS for group decision making under fuzzy environment. *Fuzzy Sets and Systems*, *114*(1), 1–9. doi:10.1016/S0165-0114(97)00377-1

Chen, C. T. (2000). Extension of the TOPSIS for group decision-making under fuzzy environment. *Fuzzy Sets and Systems*, *114*(1), 1–9. doi:10.1016/S0165-0114(97)00377-1

Chen, C. T., Lin, C. T., & Huang, S. F. (2006). A fuzzy approach for supplier evaluation and selection in supply chain management. *International Journal of Production Economics*, *102*(2), 289–301. doi:10.1016/j.ijpe.2005.03.009

Chen, M. F., & Tzeng, G. H. (2004). Combining grey relation and TOPSIS concepts for selecting an expatriate host country. *Mathematical and Computer Modelling*, *40*(13), 1473–1490. doi:10.1016/j.mcm.2005.01.006

Chen, S. J., & Hwang, C. L. (1992). *Fuzzy Multiple Attribute Decision Making: Methods and Applications*. Berlin: Springer. doi:10.1007/978-3-642-46768-4

Chu, T. (2002). Selecting plant location via a fuzzy TOPSIS approach. *International Journal of Advanced Manufacturing Technology*, *20*(11), 859–864. doi:10.1007/s001700200227

Chu, T., & Lin, Y. (2002). Improved extensions of the TOPSIS for group decision making under fuzzy environment. *Journal of Information and Optimization Sciences*, *23*(2), 273–286. doi:10.1080/02522667.2002.10698996

Chu, T. C. (2002). Facility location selection using fuzzy TOPSIS under group decisions. *International Journal of Uncertainty, Fuzziness and Knowledge-based Systems*, *10*(6), 687–701. doi:10.1142/S0218488502001739

Chu, T. C. (2002). Selecting plant location via a fuzzy TOPSIS approach. *International Journal of Advanced Manufacturing Technology, 20*(11), 859–864. doi:10.1007/s001700200227

Chu, T. C., & Lin, Y. C. (2003). A fuzzy TOPSIS method for robot selection. *International Journal of Advanced Manufacturing Technology, 21*(4), 284–290. doi:10.1007/s001700300033

Dağdeviren, M., Yavuz, S., & Kılınc, N. (2009). Weapon selection using the AHP and TOPSIS methods under fuzzy environment. *Expert Systems with Applications, 36*(4), 8143–8151. doi:10.1016/j.eswa.2008.10.016

Dagdeviren, M., Yavuz, S., & Kılınç, N. (2009). Weapon selection using the AHP and TOPSIS methods under fuzzy environment. *Expert Systems with Applications, 36*(4), 8143–8151. doi:10.1016/j.eswa.2008.10.016

Dalalah, D., Hayajneh, M., & Batieha, F. (2011). A fuzzy multi-criteria decision making model for supplier selection. *Expert Systems with Applications, 38*(7), 8384–8391. doi:10.1016/j.eswa.2011.01.031

Deng, H., Yeh, C. H., & Willis, R. J. (2000). Inter-company comparison using modified TOPSIS with objective weights. *Computers & Operations Research, 27*(10), 963–973. doi:10.1016/S0305-0548(99)00069-6

Deng, Y. (2006). Plant location selection based on fuzzy TOPSIS. *International Journal of Advanced Manufacturing Technology, 28*(7-8), 839–844. doi:10.1007/s00170-004-2436-5

Douligeris, C., & Pereira, I. J. (1994). A Telecommunications Quality Study Using the Analytic Hierarchy Process. *IEEE Journal on Selected Areas in Communications, 12*(2), 241–250. doi:10.1109/49.272873

Eldukair, Z. A., & Ayyub, B. M. (1992). Multi-attribute fuzzy decisions in construction strategies. *Fuzzy Sets and Systems, 46*(2), 155–165. doi:10.1016/0165-0114(92)90128-Q

Ertuğrul, İ., & Karakaşoğlu, N. (2008). Comparison of fuzzy AHP and fuzzy TOPSIS methods for facility location selection. *International Journal of Advanced Manufacturing Technology, 39*(7-8), 783–795. doi:10.1007/s00170-007-1249-8

Feng, B., Fan, Z., & Li, Y. (2011). A decision method for supplier selection in multi-service outsourcing. *International Journal of Production Economics, 132*(2), 240–250. doi:10.1016/j.ijpe.2011.04.014

Feng, S., & Xu, L. D. (1999). Decision support for fuzzy comprehensive evaluation of urban development. *Fuzzy Sets and Systems, 105*(1), 1–12. doi:10.1016/S0165-0114(97)00229-7

Forman, E., & Gass, S. (2001). The Analytic Hierarchy Process – An Exposition. *Operations Research, 49*(4), 469–486. doi:10.1287/opre.49.4.469.11231

Golden, B., Wasil, E., & Harker, P. (1989). *The Analytic Hierarchy Process: Applications and Studies.* Heidelberg, Germany: Springer-Verlag. doi:10.1007/978-3-642-50244-6

Gumus, A.-T. (2009). Evaluation of hazardous waste transportation firms by using a two step fuzzy-AHP and TOPSIS methodology. *Expert Systems with Applications, 36*(2), 4067–4074. doi:10.1016/j.eswa.2008.03.013

Haghighi, M., Divandari, A., & Keimasi, M. (2010). The impact of 3D e-readiness on e-banking development in Iran: A fuzzy AHP analysis. *Expert Systems with Applications*, *37*(6), 4084–4093. doi:10.1016/j.eswa.2009.11.024

Ho, W. (2008). Integrated analytic hierarchy process and its applications - A literature review. *European Journal of Operational Research*, *186*(1), 211–228. doi:10.1016/j.ejor.2007.01.004

Hsu, H. M., & Chen, C. T. (1996). Aggregation of fuzzy opinions under group decision making. *Fuzzy Sets and Systems*, *79*(3), 279–285. doi:10.1016/0165-0114(95)00185-9

Hsu, S. H., Kao, C.-H., & Wu, M.-C. (2009). Design facial appearance for roles in video games. *Expert Systems with Applications*, *36*(3), 4929–4934. doi:10.1016/j.eswa.2008.05.049

Hsu, Y.-L., Lee, C.-H., & Kreng, V. B. (2010). The application of Fuzzy Delphi Method and Fuzzy AHP in lubricant regenerative technology selection. *Expert Systems with Applications*, *37*(1), 419–425. doi:10.1016/j.eswa.2009.05.068

Hwang, C. L., & Yoon, K. (1981). *Multiple Attribute Decision Making: Methods and Applications*. Berlin: Springer. doi:10.1007/978-3-642-48318-9

Ic, Y. (2012). An experimental design approach using TOPSIS method for the selection of computer-integrated manufacturing technologies. *Robotics and Computer-integrated Manufacturing*, *28*(2), 245–256. doi:10.1016/j.rcim.2011.09.005

Iç, Y. T., & Yurdakul, M. (2009). Development of a decision support system for machining center selection. *Expert Systems with Applications*, *36*(2), 3505–3513. doi:10.1016/j.eswa.2008.02.022

Kahraman, C., Ates, N. Y., Çevik, S., Gülbay, M., & Erdogan, S. A. (2007). Hierarchical fuzzy TOPSIS model for selection among logistics information technologies. *Journal of Enterprise Information Management*, *20*(2), 143–168. doi:10.1108/17410390710725742

Kahraman, C., & Kaya, I. (2010In Press). *A fuzzy multicriteria methodology for selection among energy alternatives. Expert Systems with Applications*. Corrected Proof; doi:10.1016/j.eswa.2010.1002.1095

Kannan, G., Pokharel, S., & Kumar, S. P. (2009). A hybrid approach using ISM and fuzzy TOPSIS for the selection of reverse logistics provider. *Resources, Conservation and Recycling*, *54*(1), 28–36. doi:10.1016/j.resconrec.2009.06.004

Kara, S. S. (2011). Supplier selection with an integrated methodology in unknown environment. *Expert Systems with Applications*, *38*(3), 2133–2139. doi:10.1016/j.eswa.2010.07.154

Kaya, T., & Kahraman, C. (2011). Multicriteria decision making in energy planning using a modified fuzzy TOPSIS methodology. *Expert Systems with Applications*, *38*(6), 6577–6585. doi:10.1016/j.eswa.2010.11.081

Khosla, R., Goonesekera, T., & Chu, M.-T. (2009). Separating the wheat from the chaff: An intelligent sales recruitment and benchmarking system. *Expert Systems with Applications*, *36*(2), 3017–3027. doi:10.1016/j.eswa.2008.01.090

Kim, S., Lee, K., Cho, J. K., & Kim, C. O. (2011). Agent-based diffusion model for an automobile market with fuzzy TOPSIS-based product adoption process. *Expert Systems with Applications*, *38*(6), 7270–7276. doi:10.1016/j.eswa.2010.12.024

Kim, Y., Chung, E., Jun, S.-M., & Kim, S. U. (2013). Prioritizing the best sites for treated waste water in stream use in an urban watershed using fuzzy TOPSIS. *Resources, Conservation and Recycling*, *73*, 23–32. doi:10.1016/j.resconrec.2012.12.009

Kumar, S., & Vaidya, O. (2006). Analytic hierarchy process: An overview of applications. *European Journal of Operational Research*, *169*(1), 1–29. doi:10.1016/j.ejor.2004.04.028

Labib, A., & Shah, J. (2001). Management decisions for a continuous improvement process in industry using the Analytical Hierarchy Process. *Journal of Work Study*, *50*(5), 189–193. doi:10.1108/EUM0000000005678

Lai, W.-H., & Tsai, C.-T. (2009). Fuzzy rule-based analysis of firm's technology transfer in Taiwan's machinery industry. *Expert Systems with Applications*, *36*(10), 12012–12022. doi:10.1016/j.eswa.2009.03.054

Lee, E., Park, Y., & Shin, J. G. (2009). Large engineering project risk management using a Bayesian belief network. *Expert Systems with Applications*, *36*(3), 5880–5887. doi:10.1016/j.eswa.2008.07.057

Lee, S.-H. (2010In Press). *Using fuzzy AHP to develop intellectual capital evaluation model for assessing their performance contribution in a university. Expert Systems with Applications.* Corrected Proof; doi:10.1016/j.eswa.2009.1012.1020

Li, D. F. (2005). Multi attribute decision making models and methods using intuitionistic fuzzy sets. *Journal of Computer and System Sciences*, *70*(1), 73–85. doi:10.1016/j.jcss.2004.06.002

Li, R. J. (1999). Fuzzy method in group decision making. *Computers & Mathematics with Applications (Oxford, England)*, *38*(1), 91–101. doi:10.1016/S0898-1221(99)00172-8

Li, S., & Li, J. Z. (2009). Hybridising human judgment, AHP, simulation and a fuzzy expert system for strategy formulation under uncertainty. *Expert Systems with Applications*, *36*(3), 5557–5564. doi:10.1016/j.eswa.2008.06.095

Li, T.-S., & Huang, H.-H. (2009). Applying TRIZ and Fuzzy AHP to develop innovative design for automated manufacturing systems. *Expert Systems with Applications*, *36*(4), 8302–8312. doi:10.1016/j.eswa.2008.10.025

Li, Y., Tang, J., & Luo, X. (2010In Press). *An ECI-based methodology for determining the final importance ratings of customer requirements in MP product improvement. Expert Systems with Applications.* Uncorrected Proof. doi:10.1016/j.eswa.2010.1002.1100

Li, Y. W., Nie, X. T., & Chen, S. Y. (2007). Fuzzy approach to prequalifying construction contractors. *Journal of Construction Engineering and Management*, *133*(1), 40–49. doi:10.1061/(ASCE)0733-9364(2007)133:1(40)

Liang, G. S. (1999). Fuzzy MCDM based on ideal and anti-ideal concepts. *European Journal of Operational Research*, *112*(3), 682–691. doi:10.1016/S0377-2217(97)00410-4

Liberatore, M., & Nydick, R. (2008). The analytic hierarchy process in medical and health care decision making: A literature review. *European Journal of Operational Research, 189*(1), 194–207. doi:10.1016/j.ejor.2007.05.001

Lin, C.-L., Chen, C.-W., & Tzeng, G.-H. (2010In Press). *Planning the development strategy for the mobile communication package based on consumers' choice preferences. Expert Systems with Applications.* Corrected Proof. doi:10.1016/j.eswa.2009.1011.1009

Lin, C. T., & Chen, Y. T. (2004). Bid/no-bid decision making – a fuzzy linguistic approach. *International Journal of Project Management, 22*(7), 585–593. doi:10.1016/j.ijproman.2004.01.005

Lin, M.-C., Wang, C.-C., Chen, M.-S., & Chang, C.-A. (2008). Using AHP and TOPSIS approaches in customer-driven product design process. *Computers in Industry, 59*(1), 17–31. doi:10.1016/j.compind.2007.05.013

Liu, C.-C., & Chen, S.-Y. (2009). Prioritization of digital capital measures in recruiting website for the national armed forces. *Expert Systems with Applications, 36*(5), 9415–9421. doi:10.1016/j.eswa.2008.12.051

Miller, J. (1966). *The assessment of Worth: a systematic procedure and its experimental validation.* MIT.

Miller, J. (1969). Assessing alternative transportation systems. In *Memorandum RM-5865-DOR*. The RAND Corporation.

Miller, J. (1970). *Professional decision-making: a procedure for evaluation complex alternatives.* New York: Praeger Publishers.

Motawa, I. A., Anumba, C. J., & El-Hamalawi, A. (2006). A fuzzy system for evaluating the risk of change in construction projects. *Advances in Engineering Software, 37*(9), 583–591. doi:10.1016/j.advengsoft.2006.01.006

Muralidhar, P., Ravindranath, K., & Srihari, V. (2013). The influence of GRA and TOPSIS for assortment of green supply chain management strategies in Cement industry. *International Journal of Supply Chain Management, 2*(1), 49–55.

Naghadehi, M. Z., Mikaeil, R., & Ataei, M. (2009). The application of fuzzy analytic hierarchy process (FAHP) approach to selection of optimum underground mining method for Jajarm Bauxite Mine, Iran. *Expert Systems with Applications, 36*(4), 8218–8226. doi:10.1016/j.eswa.2008.10.006

Niaraki, A. S., & Kim, K. (2009). Ontology based personalized route planning system using a multi-criteria decision making approach. *Expert Systems with Applications, 36*(2), 2250–2259. doi:10.1016/j.eswa.2007.12.053

Ölçer, A. I., & Odabasi, A. Y. (2005). A new fuzzy multiple attributive group decision making methodology and its application to propulsion/maneuvering system selection problem. *European Journal of Operational Research, 166*(1), 93–114. doi:10.1016/j.ejor.2004.02.010

Omkarprasad, V., & Sushil, K. (2006). Analytic hierarchy process: An overview of applications. *European Journal of Operational Research, 169*(1), 1–29. doi:10.1016/j.ejor.2004.04.028

Önüt, S., Efendigil, T., & Soner Kara, S. (2009). A combined fuzzy MCDM approach for selecting shopping center site: An example from Istanbul, Turkey. *Expert Systems with Applications*. doi:.10.1016/j.eswa.2009.1006.1080

Onut, S., & Soner, S. (2008). Transshipment site selection using the AHP and TOPSIS approaches under fuzzy environment. *Waste Management (New York, N.Y.)*, *28*(9), 1552–1559. doi:10.1016/j.wasman.2007.05.019 PMID:17768038

Paksoy, T., Pehlivan, N. Y., & Kahraman, C. (2012). Organizational strategy development in distribution channel management using fuzzy AHP and hierarchical fuzzy TOPSIS. *Expert Systems with Applications*, *39*(3), 2822–2841. doi:10.1016/j.eswa.2011.08.142

Pan, N. (2009). Selecting an appropriate excavation construction method based on qualitative assessments. *Expert Systems with Applications*, *36*(3), 5481–5490. doi:10.1016/j.eswa.2008.06.097

Saaty, T. (1972). *An eigenvalue allocation model for prioritization and planning*. Working Paper. Energy Management and Policy Center, University of Pennsylvania.

Saaty, T. (1977). A scaling method for priorities in hierarchical structures. *Journal of Mathematical Psychology*, *15*(3), 234–281. doi:10.1016/0022-2496(77)90033-5

Saaty, T. (1994). Highlights and critical points in the theory and application of the Analytic Hierarchy Process. *European Journal of Operational Research*, *74*(3), 426–447. doi:10.1016/0377-2217(94)90222-4

Saaty, T., & Forman, E. (1992). *The Hierarchon: A Dictionary of Hierarchies* (Vol. 5). Pittsburgh, PA: RWS Publications.

Saaty, T. L. (1980). *The Analytic Hierarchy Process*. New York: McGraw-Hill.

Saaty, T. L. (1983). Priority Setting in Complex Problems. *IEEE Transactions on Engineering Management*, *30*(3), 140–155. doi:10.1109/TEM.1983.6448606

Saremi, M. S., Mousavi, F., & Sanayei, A. (2008). TQM consultant selection in SME with TOPSIS under fuzzy environment. *Expert Systems with Applications*, 36.

Secme, N. Y., Bayrakdarogˇlu, A., & Kahraman, C. (2009). Fuzzy performance evaluation in Turkish Banking Sector using Analytic Hierarchy Process and TOPSIS. *Expert Systems with Applications*, *36*(9), 11699–11709. doi:10.1016/j.eswa.2009.03.013

Sen, C. G., & Çınar, G. (2010). Evaluation and pre-allocation of operators with multiple skills: A combined fuzzy AHP and max-min approach. *Expert Systems with Applications*, *37*(3), 2043–2053. doi:10.1016/j.eswa.2009.06.075

Seo, S., Aramaki, T., Hwang, Y. W., & Hanaki, K. (2004). Fuzzy decision-making tool for environmental sustainable buildings. *Journal of Construction Engineering and Management*, *130*(3), 415–423. doi:10.1061/(ASCE)0733-9364(2004)130:3(415)

Shaverdi, M., Akbari, M., & Tafti, S. F. (2011). Combining Fuzzy MCDM with BSC Approach in Performance Evaluation of Iranian Private Banking Sector. *Advances in Fuzzy Systems*, *2011*, 1–12. doi:10.1155/2011/148712

Shih, H. S., Syur, H. J., & Lee, E. S. (2007). An extension of TOPSIS for group decision making. *Mathematical and Computer Modelling*, *45*(7-8), 801–813. doi:10.1016/j.mcm.2006.03.023

Shim, J. (1989). Bibliography research on the analytic hierarchy process (AHP). *Socio-Economic Planning Sciences*, *23*, 161–167. doi:10.1016/0038-0121(89)90013-X

Sipahi, S., & Timor, M. (2010). The analytic hierarchy process and analytic network process: An overview of applications. *Management Decision*, *48*(5), 775–808. doi:10.1108/00251741011043920

Stevens, S. (1957). On the psychophysical law. *Psychological Review*, *64*(3), 153–181. doi:10.1037/h0046162 PMID:13441853

Sun, C. C. (2010). A performance evaluation model by integrating fuzzy AHP and fuzzy TOPSIS methods. *Expert Systems with Applications*, *37*(12), 7745–7754. doi:10.1016/j.eswa.2010.04.066

Sun, C. C., & Lin, G. T. R. (2009). Using fuzzy TOPSIS method for evaluating the competitive advantages of shopping websites. *Expert Systems with Applications*, *36*(9), 11764–11771. doi:10.1016/j.eswa.2009.04.017

Tah, J. H. M., & Carr, V. (2000). A proposal for construction project risk assessment using fuzzy logic. *Construction Management and Economics*, *18*(4), 491–500. doi:10.1080/014461900050024905

Thurstone, L. (1927). A law of comparative judgments. *Psychological Review*, *34*(4), 273–286. doi:10.1037/h0070288

Torlak, G., Sevkli, M., Sanal, M., & Zaim, S. (2010). Analyzing business competition by using fuzzy TOPSIS method: An example of Turkish domestic airline industry. *Expert Systems with Applications*, *38*(4), 3396–3406. doi:10.1016/j.eswa.2010.08.125

Triantaphyllou, E., & Lin, C. T. (1996). Development and evaluation of five fuzzy multi attribute decision-making methods. *International Journal of Approximate Reasoning*, *14*(4), 281–310. doi:10.1016/0888-613X(95)00119-2

Tsaur, S. H., Chang, T. Y., & Yen, C. H. (2002). The evaluation of airline service quality by fuzzy MCDM. *Tourism Management*, *23*(2), 107–115. doi:10.1016/S0261-5177(01)00050-4

Tseng, Y.-F., & Lee, T.-Z. (2009). Comparing appropriate decision support of human resource practices on organizational performance with DEA/AHP model. *Expert Systems with Applications*, *36*(3), 6548–6558. doi:10.1016/j.eswa.2008.07.066

Vargas, L. (1990). An overview of the analytic hierarchy process and its applications. *European Journal of Operational Research*, *48*(1), 2–8. doi:10.1016/0377-2217(90)90056-H

Vidal, L.-A., Sahin, E., Martelli, N., Berhoune, M., & Bonan, B. (2010). Applying AHP to select drugs to be produced by anticipation in a chemotherapy compounding unit. *Expert Systems with Applications*, *37*(2), 1528–1534. doi:10.1016/j.eswa.2009.06.067

Wang, H. S., Che, Z. H., & Wu, C. (2010). Using analytic hierarchy process and particle swarm optimization algorithm for evaluating product plans. *Expert Systems with Applications*, *37*(2), 1023–1034. doi:10.1016/j.eswa.2009.05.073

Wang, T. C., & Chang, T. H. (2007). Application of TOPSIS in evaluating initial training aircraft under a fuzzy environment. *Expert Systems with Applications*, *33*(4), 870–880. doi:10.1016/j.eswa.2006.07.003

Wang, T.-Y., & Yang, Y.-H. (2009). A fuzzy model for supplier selection in quantity discount environments. *Expert Systems with Applications*, *36*(10), 12179–12187. doi:10.1016/j.eswa.2009.03.018

Wang, Y.-M., & Elhag, T.-M.-S. (2006). Fuzzy TOPSIS method based on alpha level sets with an application to bridge risk assessment. *Expert Systems with Applications*, *31*(2), 309–319. doi:10.1016/j.eswa.2005.09.040

Wu, C.-R., Lin, C.-T., & Lin, Y.-F. (2009). Selecting the preferable bancassurance alliance strategic by using expert group decision technique. *Expert Systems with Applications*, *36*(2), 3623–3629. doi:10.1016/j.eswa.2008.02.016

Yang, C.-L., Chuang, S.-P., & Huang, R.-H. (2009). Manufacturing evaluation system based on AHP/ANP approach for wafer fabricating industry. *Expert Systems with Applications*, *36*(8), 11369–11377. doi:10.1016/j.eswa.2009.03.023

Yayla, A. Y., Özbek, A., & Yildiz, A. (2012). Fuzzy TOPSIS Method in Supplier Selection and Application in the Garment Industry. *FIBRES & TEXTILES in Eastern Europe*, *4*(93), 20–23.

Yeh, C. H., & Deng, H. (2004). A practical approach to fuzzy utilities comparison in fuzzy multi criteria analysis. *International Journal of Approximate Reasoning*, *35*(2), 179–194. doi:10.1016/j.ijar.2003.09.002

Yeh, C. H., Deng, H., & Chang, Y.-H. (2000). Fuzzy multi criteria analysis for performance evaluation of bus companies. *European Journal of Operational Research*, *126*(3), 459–473. doi:10.1016/S0377-2217(99)00315-X

Yokoyama, M. (1921). The Nature of the affective judgment in the method of paired comparison. *The American Journal of Psychology*, *32*(3), 357–369. doi:10.2307/1414000

Yong, D. (2006). Plant location selection based on fuzzy TOPSIS. *International Journal of Advanced Manufacturing Technology*, *28*(7-8), 839–844. doi:10.1007/s00170-004-2436-5

Zahedi, F. (1986). The analytic hierarchy process: A survey of the method and its applications. *Interface*, *16*(4), 96–108. doi:10.1287/inte.16.4.96

Zayed, T. M., & Chang, L. M. (2002). Prototype model for build-operate-transfer risk assessment. *Journal of Management Engineering*, *18*(1), 7–16. doi:10.1061/(ASCE)0742-597X(2002)18:1(7)

Zeng, J., An, M., & Smith, N. J. (2007). Application of a fuzzy based decision making methodology to construct project risk assessment. *International Journal of Project Management*, *25*(6), 589–600. doi:10.1016/j.ijproman.2007.02.006

Zimmermann, H. J. (1996). *Fuzzy Set Theory and its Applications*. Boston: Kluwer Academic Publishers. doi:10.1007/978-94-015-8702-0

Zou, P. X. W., Zhang, G., & Wang, J. (2007). Understanding the key risks in construction projects in china. *International Journal of Project Management*, *25*(6), 601–614. doi:10.1016/j.ijproman.2007.03.001

ADDITIONAL READING

Saaty, T. (1972). An eigenvalue allocation model for prioritization and planning. In *Working paper, Energy Management and Policy Center*: University of Pennsylvania.

Saaty, T. (1977). A scaling method for priorities in hierarchical structures. *Journal of Mathematical Psychology*, *15*(3), 234–281. doi:10.1016/0022-2496(77)90033-5

Saaty, T. (1986). Axiomatic Foundation of the Analytic Hierarchy Process. *Management Science*, *32*(7), 841–855. doi:10.1287/mnsc.32.7.841

Saaty, T. (1990). An Exposition of the AHP in Reply to the Paper —Remarks on the Analytic Hierarchy ProcessP. *Management Science*, *36*(3), 259–268. doi:10.1287/mnsc.36.3.259

Saaty, T. (1991). Response to Holder's Comments on the Analytic Hierarchy Process. *The Journal of the Operational Research Society*, *42*(10), 909–929. doi:10.1057/jors.1991.176

Saaty, T. (2003). Decision-making with the AHP: Why is the Principal Eigenvector necessary? *European Journal of Operational Research*, *145*(1), 85–91. doi:10.1016/S0377-2217(02)00227-8

Saaty, T. (2006). Rank from Comparisons and from Ratings in the Analytic Hierarchy/Network Processes. *European Journal of Operational Research*, *168*(2), 557–570. doi:10.1016/j.ejor.2004.04.032

Saaty, T., & Hu, G. (1998). Ranking by Eigenvector Versus other Methods in the Analytic Hierarchy Process. *Applied Mathematics Letters*, *11*(4), 121–125. doi:10.1016/S0893-9659(98)00068-8

Saaty, T., & Ozdemir, M. (2003). Negative Priorities in the Analytic Hierarchy Process. *Mathematical and Computer Modelling*, *37*(9-10), 1063–1075. doi:10.1016/S0895-7177(03)00118-3

Saaty, T., & Takizawa, M. (1986). Dependence and Independence: From Linear Hierarchies to Nonlinear Networks. *European Journal of Operational Research*, *26*(2), 229–237. doi:10.1016/0377-2217(86)90184-0

Saaty, T., & Vargas, L. (1984a). Comparison of Eigenvalue, Logarithmic Least Squares and Least Squares Methods in Estimating Ratios. *Mathematical Modelling*, *5*(5), 309–324. doi:10.1016/0270-0255(84)90008-3

Saaty, T., & Vargas, L. (1984b). Inconsistency and Rank Preservation. *Journal of Mathematical Psychology*, *28*(2), 205–214. doi:10.1016/0022-2496(84)90027-0

Saaty, T., & Vargas, L. (1984c). The legitimacy of rank reversal. *Omega*, *12*(5), 513–516. doi:10.1016/0305-0483(84)90052-5

Saaty, T., & Vargas, L. (2007). Dispersion of group judgments. *Mathematical and Computer Modelling*, *46*(7-8), 918–925. doi:10.1016/j.mcm.2007.03.004

Saaty, T. L., & Vargas, L. G. (2005). The possibility of group welfare functions. *International Journal of Information Technology & Decision Making*, *4*(02), 167–176. doi:10.1142/S0219622005001453

KEY TERMS AND DEFINITIONS

AHP: Thomas Saaty developed the Analytic Hierarchy Process (AHP) under the multi criterion decision methods to find the solution in a complex problem. In this decision maker allows to make a hierarchy model on the basis of goal, criterions and alternatives and then select the best alternative.

B-School: The educational institute or entity which provide the teaching and knowledge about business or business services.

Criterions: These are the measures, standards or expectations used in evaluation and used as the basis for comparison against which other things can be compare and judge or evaluated.

Fuzzy: It means the concept, ideas, responses etc. are not fixed, can vary according to conditions in which they exist. Generally it means that the concept in condition of uncertainty, lacking, vague concepts etc.

Linguistic Variable: It is an important concept of Fuzzy logic. It represents crisp information in a form and precision for the problem. These are the variables which are expressed in plain language words and statements in place of numeric numbers.

MCDM: It is a sub discipline of Operations Research where one is having number of decision making criterion in decision making environment.

TOPSIS: Hwang and Yoon developed this technique in 1981. It is Technique for order or preference by similarity to ideal solution (TOPSIS) for the multi criterion decision making.

Chapter 2
Innovative Solutions for Implementing Global Supply Chains in Emerging Markets

Raj V. Amonkar
Goa Institute of Management, India

ABSTRACT

In a supply chain network, facilities are the primary components where a product is manufactured or stored. During the phase of design of a supply chain network, a company decides how to configure the supply chain over the next several years. Key driver of supply chain performance in terms of responsiveness and efficiency will be taken into consideration in the decision process and the decisions include the role assigned to each facility, its capacity to perform the assigned role, the number and location of the facilities. Since supply chain design decisions pertaining to facilities are typically made for the long term and are very expensive to alter on short notice, the decisions must take into account uncertainty in anticipated market conditions over the next few years. Decisions regarding facilities are therefore a crucial part of supply chain design. This chapter explains the practical application of theories, concepts and frameworks in the area of Supply Chain Design, Risk Mitigation and Social Networks. The practical application is based on a case study of ABC, one of the world's leading agribusiness companies with global operations that made a successful foray in emerging markets facilitated by sound decisions pertaining to the design of its global supply chain network. Certain names and other identifying information were disguised to protect confidentiality.

INTRODUCTION

Globalization is not a new phenomenon. The networking of the world's economy has been evolving for centuries, with companies gradually expanding beyond their national borders. What is new is the dramatic acceleration of the process. Global production provides an unparalleled opportunity for companies to grow into new markets while at the same time boosting their competitiveness. However, most of today's networks are legacy structures-only a fraction was strategically planned. As a result, there is

DOI: 10.4018/978-1-4666-9795-9.ch002

huge potential to be captured from rethinking traditional structures, approaches, and supply relationships (Abele, et al., 2008). Supply chain design decisions are typically made for the long term and are very expensive to alter on short notice. These decisions must take into account uncertainty in anticipated market conditions over the next few years. As a result, managers must account for both opportunities and uncertainties over the long term when designing a global supply chain network (Chopra, et al., 2013). Melnyk, et al., 2013 say that global supply chain design presents managers and researchers with its own set of issues, concerns and obstacles since this concept is relatively new and the salient issues that define its content, scope and boundaries are still emerging.

In this chapter, the author explains how theories, concepts and frameworks in the area of Supply Chain Design, Risk Mitigation and Social Networks could be practically applied. The research is based on a case study of ABC, one of the world's leading agribusiness companies with global operations that made a successful foray in emerging markets facilitated by sound decisions pertaining to the design of its global supply chain network. Certain names and other identifying information were disguised to protect confidentiality.

The chapter is divided into four sections. In the next section, a literature review summarizes the theories, concepts and frameworks based on published research in supply chain design, risk mitigation and social networks. Third section presents the discussion and findings. Fourth section elucidates the conclusions and recommendations.

LITERATURE REVIEW

Supply Chain Design

Chopra, et al., 2013 state that a company's competitive strategy has a significant impact on network design decisions within the supply chain.

Melnyk, et al., 2014 propose a framework of supply chain design that comprehends three key levels of factors critical in understanding supply chain design: influencers, design decisions and building blocks. Influencers are higher-level considerations such as the business and political environment, the business model employed, the firm's desired outcomes and the supply chain life cycle. Design decisions include the social, behavioral and physical/structural design elements that define a supply chain. Building blocks include the inventory, transportation, capacity and technology decisions that are used to implement the supply chain. Supply chain design needs to comprehend these three levels of analysis.

Badenhorst-Weiss, et al., 2011 have developed a conceptual framework with which organisations can analyse their supply chain designs. They tested the framework empirically and found that framework is in fact a workable instrument to analyse supply chain designs. The framework encompasses the three basic phases of supply chain design (SCD): understanding end customer's needs and determining how to meet these needs, selecting a supply chain strategy and structuring the supply chain. The framework constitutes thirteen supply chain design elements as summarized in Figure 1.

Chopra et.al. 2013 cite supporting examples of companies that focus on cost leadership find the lowest cost location for their manufacturing facilities, even if that means locating far from the markets they serve and companies that focus on responsiveness tend to locate facilities closer to the market and may select a high-cost location if this choice allows the company to react quickly to changing market needs. Chopra et.al. 2013, further state that the global supply chain networks can best support their strategic objectives

Figure 1. Framework for supply chain design elements

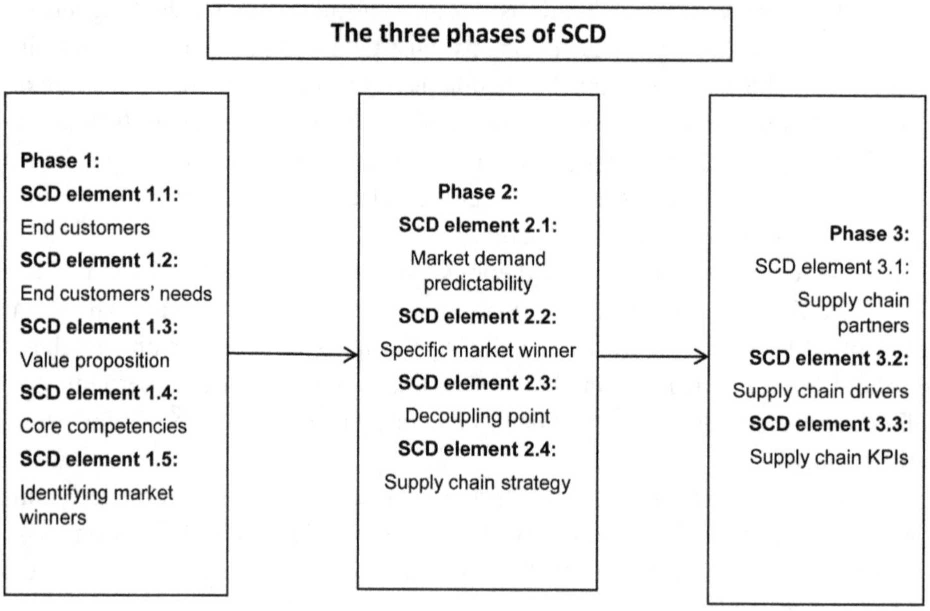

Kruse, 2004 offers solutions for migration of manufacturing to low cost countries. One such solution offered is the use of the Puttick Grid. The Puttick Grid positions a company's products against a product complexity/demand uncertainty matrix and identifies four key quadrants as show in Figure 2.

Products in different quadrants essentially have different levels of supply chain challenges and migration potential, and mapping a company's products to the matrix is a good indicator on which product groups are most suitable for migration (accepting all the time that one company might well have products

Figure 2. Puttick Grid

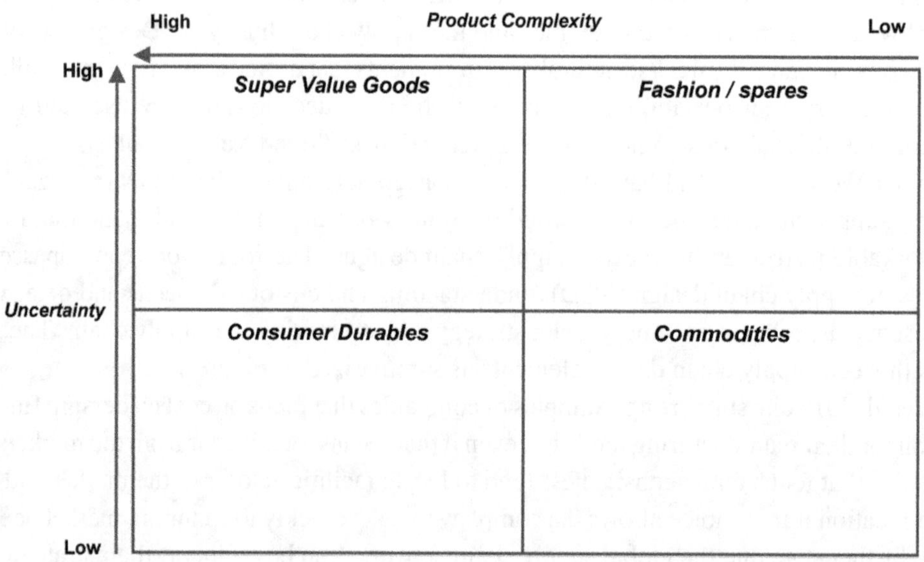

in more than one quadrant).Fashion goods are inherently not good for migration though their product simplicity has in the past made them popular for such projects. Commodities are a 'natural' for migration. Consumer durables are more complex but tend to be fairly labour intensive so that the labour cost benefits can be high. Super value goods are not generally suitable for migration but their components are, especially since many components fall into the commodity quadrant.

Turner, 2002 states that the Puttick Grid could also be used to evaluate how the market will view a company's product-market arena today and tomorrow. Thus the company is able to prepare its strategies and plans in the light of the how the customer views and appreciates the products it wishes to acquire and the characteristics of the acquisition process, rather than how the organisation views its own products, strengths and weaknesses.

Chopra, et al., 2013 state that the global supply chain networks can best support their strategic objectives with facilities in different countries playing different roles. They explain that it is therefore important for a company to identify the strategic role of each facility when designing its global network and to configure the resources of the facility accordingly. Chopra et al., 2013 also explain the use of Network Optimization Models to decide how markets are allocated to facilities. The demand allocation decision can be altered on a regular basis as costs change and markets evolve.

Risk Mitigation

Lassar, et al., 2010 confirm based on a qualitative assessment of the sample firms— multinational buyers along with suppliers—that a symbiotic triad of resources, networks, and performances are key determinants of strategic risk management in supply chains in one important emerging market—Mexico and that a framework comprised of the symbiotic triad for analyzing strategic risk management in supply chains is a viable paradigm for studying supply chains in emerging markets.

Chopra, et al., 2013 have emphasised the fact that global supply chains today are subject to more risk factors than localized supply chains of the past. They state that it is thus critical for global supply chains to be aware of the relevant risk factors and build in suitable mitigation strategies. They have categorized supply chain risks and their drivers for consideration during network design. The categorization is reproduced in Table 1.

Chopra, et al., 2013 also state that global supply chains should generally use a combination of mitigation strategies designed into the supply chain along with financial strategies to hedge uncovered risks. They have suggested some tailored mitigation strategies as outlined in Table 2.

Singh, et al., 2012 propose an integrated model based on a detailed mathematical programming formulation to evaluate the location of the plants and warehouses (strategic decisions of global supply chain) and number of the quantities to flow between SC echelons (tactical decisions of global supply chain). Simultaneously, the model incorporates a set of risk factors, their expected values, and probability of their occurrence. Production, distribution risk, demand risk, supply risk, and interaction risk have been considered in the model which incorporates the late shipments, exchange rates, customs delays, quality control problems, and logistics and transport breakdowns.

Goetschalckx et al., 2012 have developed a comprehensive and integrated model of the structure and the behavior of the supply network of a manufacturing enterprise. In this strategic model the inherent uncertainty of the future environmental conditions in which the supply network will have to function is captured in a large number of scenarios. The variability of the profits of the scenarios is used as a proxy for the supply network risk. The model allows for the explicit trade off of the systems life cycle profit

Table 1. Categorization of risks

Category	Risk Drivers
Disruptions	Natural disaster, war, terrorism
	Labor disputes
	Supplier bankruptcy
Delays	High capacity utilization at supply source
	Inflexibility of supply source
	Poor quality or yield at supply source
Systems risk	Information infrastructure breakdown
	System integration or extent of systems being networked
Intellectual property risk	Vertical integration of supply chain
	Global outsourcing and markets
Procurement risk	Exchange-rate risk
	Price of inputs
	Fraction purchased from single source
	Industry-wide capacity utilization
Receivable risk	Number of customers
Inventory risk	Rate of product obsolescence
	Inventory holding cost
	Product value
	Demand and supply uncertainty
Capacity risk	Cost of capacity
	Capacity flexibility

Table 2. Tailored risk mitigation strategies

Risk Mitigation Strategy	Tailored Strategies
Increase capacity	Focus on low-cost decentralized capacity for predictable demand
	Build centralized capacity for unpredictable demand; increase decentralization as cost of capacity drops
Get redundant suppliers	More redundant supply for high-volume products,; less redundancy for low-volume products in a few flexible suppliers
Increase responsiveness	Favour cost over responsiveness for commodity products; favour responsiveness over cost for short-life cycle products
Increase inventory	Decentralize inventory of predictable, low value products; centralize inventory of less predictable, higher value products
Increase flexibility	Favour cost over flexibility for predictable, high-volume products; favour flexibility for less predictable, low-volume products; centralize flexibility in a few locations if it is expensive
Pool or aggregate demand	Increase aggregation as unpredictability grows
Increase source capability	Prefer capability over cost for high-value, high-risk products
	Favour cost over capability for low-value commodity products
	Centralize high capability in flexible source if possible

with the variability of this profit. An efficient solution algorithm eliminates the vast majority of feasible supply network configurations from consideration and identifies all the optimal network configurations. However, the final selection of the supply network configuration depends on the risk characteristics of the designer and the enterprise. Selecting the supply network configuration with the appropriate robustness becomes one of the principal mitigation policies for supply chain risks.

Dorian, 2011 examines the most common risk mitigation strategies and how they can be used to effectively manage risk. Three steps are accordingly listed as follows:

1. Set the level of risk the board and management is willing to take.
2. Prioritize, or rank, each risk for significance and likelihood based on Risk prioritization.
3. Determine appropriate risk mitigation strategies. The four most common mitigation strategies are avoidance, acceptance, transference, and control.

Manufacturing Technology Committee-Risk Management Working Group in their training guide on Risk Ranking and Filtering present a simple two-Criterion evaluation model. The two-criterion model lends itself to a classical "Risk Matrix" as shown in Table 3. The "Risk Matrix" leads to a two-dimensional diagram of probability of occurrence measured against the severity of the consequences if it did occur. The technique works by assigning values to probability of occurrence and the severity of the outcome to give a two-dimensional view. A risk that is present but highly unlikely to occur has a low probability with a score of 1 assigned. Account should also be taken of what the consequences would be if the risk did become reality i.e. severity. If the consequences were severe in effect then this would be assigned a severity of high with a score of 3. This translates as a risk score 1 x 3 = 3 (medium).For each identified risk the probability and severity are multiplied to give a risk score, with 1 as the lowest and 9 as the highest score.

Risk Avoidance/Acceptance is applicable in case of risks with Scores of 4 and below whereas Risk Transference/Control is applicable in case of risks with Scores in excess of 4.

Social Networks

Muelle, et al., 2007 introduce the concept of Supply Networks based on the judgment that by its currency, the supply chain is one of the more successful metaphors in economics. They explain how the metaphor borrows from mechanics the idea of the chain, that is something that consists of elements that are linked to each of their two immediate neighbors and which jointly provide a strong but flexible connection and then transplants the chain-idea into the sphere of economics where, before the introduction of supply chains, chains were, for most economists, things best left in the care of ironmongers. They argue that

Table 3. Risk matrix

Increasing Probability	High	3	6	9
	Medium	2	4	6
	Low	1	2	3
		Low	Medium	High
	Increasing Severity			

the network metaphors, like the chain metaphor, are likely to fade into oblivion, unless they are backed up by network models of supply and therefore suggest that network models based on graph theory, such as those that have been developed and successfully applied in social network analysis (SNA), also hold promise when applied to the analysis of supply networks in agribusiness (SNA is a branch of sociology which studies collections of individuals and the linkages among them. For modeling social networks SNA employs ideas and methods from graph theory, algebra, and statistics). Their argument is based on the fact that supply networks are emergent, stable patterns in the relationships among specialized economic and public agents that are involved in the manifold coordinated activities which culminate in providing end users with a product or service. The relationships can be of many kinds but four are essential in food supply networks in Europe. The first relationship involves the exchange of goods or services and of money and the second involves the transport of goods. The third relationship is concerned with the exchange of information about a food product's quality characteristics, including traceability information. Finally, agents need to exchange meta-information that allows them to coordinate the first three essential relationships. In contrast to organizations, supply networks are rarely if ever under the control of a single manager. And in contrast to self-organizing markets, supply networks require some concerted coordination efforts by their members. Because a relationship always involves more than one agent, productive supply network relationships are unlikely to come about easily when the agents of a potential supply network do not share a common perception of their positions in the network. Supply networks are, however, mental constructs and there may be as many perceptions of a given network as there are minds perceiving it. Like a shared map that allows two travelers to meet at some agreed upon point, a shared model of the supply network showing agents and the linkages among them, greatly facilitates managers' task to define their position in the network, to identify agents with whom to link processes, and to define the nature of the links to specific agents. Graphical models of supply networks abound.

Elements of SNA and Their Uses

Muelle, et.al. 2007 lucidly explain the elements of SNA, their uses, network data potential for SNA in supply chain management as follows:

1. A set N of actors: The actors N in a social network can be anything that can be represented by nodes in a graph. In a supply network, for instance, the actors could be the producers, processors, transporters, retailers, regulatory agencies, or certification providers.
2. A collection L of links or ties that represent relationships between ordered pairs of actors: The set L of links or ties in a social network represents whatever connects pairs of actors. In a supply network this could be flows of material, flows of money, flows of product information, information required to carry out transactions, or of meta-information required for the network coordination. Beyond representing the essential flows in a supply network, links may also reflect relationships that are unrelated to the core functions of the network, such as financial partnerships. Relationships between the actors in a network may be of many forms, but the ability to model them is limited in graph theory. Links may be simple and only indicate that a connection of a certain kind exists between actors. Directed links indicate the direction of a relationship, e.g. the direction of an information or product flow. Valued or weighted links also indicate the intensity or frequency of a relationship. Finally, actors may be connected by several relationships, each requiring a separate link.

3. A "sociograph" Gd consisting of nodes, which represent actors, and directed or undirected lines between the nodes, which represent the relations among actors: The natural representation of a graph is a drawing of it. Convention suggests that network actors are represented as vertices or dots whereas relationships are shown as the edges or lines of a network drawing. In some pictures of networks actor characteristics are encoded by the size, color, or shading of the dots. Directed links are represented by arrows and valued links are represented by lines with numbers attached. Drawings of networks are without dimensions. The positions where the nodes are placed in a network drawing are therefore unconstrained by coordinates and the distances between the dots and angles between the lines of a network drawing are meaningless. How a network is actually drawn is, however, not totally arbitrary and several conventions for drawing aesthetically pleasing networks have evolved.
4. A sociomatrix or adjacency matrix A, which has as many rows and columns as there are actors and where the elements ai,j of this matrix represent the links or ties between the actors: Conventionally, all elements on the main diagonal of an adjacency matrix are set at aii = 0.Networks with undirected and unvalued links yield symmetric adjacency matrices with elements aij = {0,1}. Typical extensions of the most basic network models are models with directed edges or arcs which represent ties that run in certain directions, such as flows of merchandise, or information. Such networks are known as digraphs an adjacency matrices of digraphs usually are not symmetric because they may contain elements for which aij ≠ aji holds. In networks with weighted or valued links the elements may be some real number: aij $\in \Re$. Networks with several types of links cannot be represented with a single adjacency matrix but require a separate matrix for each type of link.
5. A characteristics matrix C, which has as many rows as there are actors and as many columns as there are attributes of interest.
6. A social network then is defined by: $S = \{ N, L\, G_d, \mathbf{A}, \mathbf{C} \}$. Social network analysis is concerned with the relationships among actors and actor attributes other than those related to the network are a secondary concern. Nevertheless, network managers and analysts may be interested in correlating network attributes of actors, which may be derived from the adjacency matrix, with other actor characteristics. Data on these characteristics may then be assembled in an actor characteristics matrix C which has as many rows as there are actors in the network and a column for each actor characteristic. The values of the entries c_{ij} in this matrix depend on the scale with which the characteristics are measured; they may be binary $c_{ij} = \{0,1\}$ or real valued $c_{ij} \in \Re$.

Potential for SNA in Supply Chain Management

SNA provides a workable model of networks which relates network attributes to measures of network outcomes, and it provides a framework for collecting and analyzing data useful for planning and monitoring changes in network operations. Essential tools that network models based on graph theory can provide for network managers and analysts are:

1. Network diagrams: Managers, when they describe their organization, usually use an organization chart, which is a specific graph, a tree, that shows the actors of the organization and the relationships between them. Using drawings for navigating through a supply network is therefore natural for most or all managers. Like all graphics, network drawings reduce search, facilitate recognition, and they may suggest interesting new perspectives and inferences about a supply network.

2. Quantitative measure describing network-related attributes of individual network members: The degree of an actor is defined by the number of its links with other actors in the network. For the purposes of comparison, the "actor degree" may be standardized by the number of actors in the network. When the relationship data are binary, the degree is equal to the row sum or the column sum of the adjacency matrix. The degree of an agent in a supply network indicates, for example, how intensively that agent is connected with his peers. The degree says, however, nothing about the intensity of the relationships. In directed networks a distinction is made between the *indegree* and *outdegree* of an actor. The *indegree* of an actor i is the number of directional links that terminate at that actor, and the *outdegree* is number of directional links ties that originate from that actor. On the basis of the number of their in- and out-degrees actors may be classified as "transmitters" when their *indegree* is zero and their *outdegree* is larger than zero; "receivers" have a zero *outdegree* and a non-zero *indegree*, and the *indegrees* and *outdegrees* are both larger than zero for "carriers". Degree is a measure of local centrality because only the direct connections to other actors are taken into account. Closeness centrality, in contrast, is a measure of global centrality. Closeness centrality is the inverse of the sum of the distances of an actor to all other actors in the network. In an undirected and unweighted network, distance is measured as the geodesic between the actors, which is the path with the smallest number of actors along the way. The actor with the smallest sum of his geodesics to all other agents in the network then is the most globally central actor in the network. The measure can be extended to directed and weighted networks. Betweenness Centrality, finally, is a measure based on the idea that an actor who lies on the paths connecting many other actors exerts control over the flow of resources or information that travel along these paths. Betweenness Centrality of actor some actor j therefore counts the geodesics between all other actors i and k and which pass through actor j.
3. Quantitative and qualitative measures of the whole network: The most basic characteristics of a network are its diameter and its completeness. The diameter of a network is the length of the largest geodesic between any pair of actors in the network. A network is complete when each actor is connected to all other actors. A complete network has the highest possible density of a network, where the density is defined as the proportion of possible links that are actually present in the network. Knowing the density of a network may be useful for the manager of a network to assess whether agents need to be encouraged to link up with each other. Cut points, cut sets, and bridges are elements of strategic importance for a network. A cut point is an actor whose removal from the network causes the network to disintegrate into two or more sub networks that are unconnected with each other; cut sets are similarly defined. A bridge is a link of equivalent strategic importance: if removed, a bridge-link causes a network to disintegrate. The importance for managers of knowing the cut points, cut sets, and bridges in a supply network is obvious. A k-core is defined as a maximal sub network in which each agent is adjacent to at least k other agents; in other words, all agents that belong to a k-core have a degree equal to k or higher. Comparison of k-cores of a network for different levels of k provides some insight into the strength of the connectedness of the agents.

Ego Network

Borgatti, et al., 2009 introduce the concept of *ego network* based on the fact that in practice many studies that draw on the network theoretical perspective do not actually take into account the full network. Instead, they utilize only a part of the network what is called the *ego network*. The Ego Network comprises of a focal actor known as ego, the set of actors with any kind of tie to ego (known as alters) and all ties among the alters and between the alters and the ego. The main reason to restrict attention to the *ego network* is, in principle, the belief that more distal connections are not relevant to the specific mechanisms at hand. Of course, it is also much more convenient to collect ego network data than full network data, and if the ego network provides a reasonable proxy for position in the larger structure, there is little reason to collect the whole network. Borgatti, et.al. 2009 also throw light on, a useful ego-network property referred to as the 'quality' of ego's alters. This measure is simply the average (or sum, or maximum value, as needed) of the attribute values for a given ego's alters. They extend this logic further by means of two measures, u and v. A firm gets a high score on u to the extent that it supplies firms that have many suppliers. A firm gets a high score on v to the extent it is supplied by firms with many customers. In the context of web sites linking to other web sites, nodes with high u scores are referred to as *hubs*, while nodes with high v scores are called *authorities* In the supply chain context, firms that are both hubs (sellers to those with many alternative suppliers) and authorities (buyers from those with alternative customers) are in very competitive markets. These are firms that must be highly *agile* to survive. Firms that are hubs but not authorities are firms that face serious competition in selling their goods, but have a good situation with respect to their suppliers. These firms have been labeled as *sales-oriented*. Firms that are authorities but not hubs have to be nimble procurers, but face relatively simple sales environments. These firms are labelled as *procurement-oriented*. Finally, firms that are neither hubs nor authorities are in relatively calm waters on both fronts and are labelled as *comfortable*.

Borgatti, et al., 2009 state in the case of a network defined by supply ties, the closeness measure remains interesting even when there are unreachable pairs. Let us define in-closeness as the average valid graph-theoretic distance from all firms to the focal firm. If we discard as invalid the supply chains that do not reach the focal firm, what we have left is all chains from the "bottom" of the economy (e.g., natural resource processors) to the firm. The average length of these chains indicates the number of steps that a focal firm's raw materials had to go through to get to the focal firm. All else being equal, we might hypothesize that longer chains provide greater opportunities for disruption and cost increases. Borgatti, et al., 2009 also state that a node has high betweenness to the extent that it lies along many shortest paths between pairs of others. A node gets a high score to the extent that it exclusively controls the short paths between all pairs of others – i.e., there are no other equally short paths that bypass the node. Thus, given a network of supply ties, a firm will have high betweenness to the extent that all of the shortest (and, let us assume, best) chains from first makers to end consumers pass through that firm. Such firms are structurally important to the economy itself, because if they disappear or slow production, they will affect more other firms than if they had lower betweenness. These are "key players" that need to be healthy for the rest of the network to be healthy.

Beamon, 1998 concludes that supply chain systems are inherently complex. Thus, the models and methods used to accurately study these systems are, expectedly, also complex. Beamon, 1998 therefore propounds that, if supply chain systems could be classified on the bases of specific characteristics, such as uncertainty or volume of demand, number of echelons, or number of items produced, there may be rule-of-thumb techniques that suggest operational characteristics that may achieve a certain objective (or set of objectives). Thus, research that develops a meaningful classification scheme for supply chain systems that leads to rules-of-thumb associations between decision variables and performance objectives is needed.

Table 4. Summary and classification of literature review

Area	Application	Articles
Supply Chain Design	Impact of competitive strategy on network design decisions	Chopra,et.al.,2013
	Influencers and design decisions	Melnyk, et.al. 2014
	Supply chain design elements	Badenhorst-Weiss,et.al., 2011
	Puttick Grid	Kruse, 2004 Turner, 2002
	Strategic role of facilities	Chopra,et.al.,2013
	Assignment of customers to Distribution Centres	Chopra et al.,2013
Risk Mitigation	Resources, Networks, Performance	Lassar, et. al.,2010
	Risk Drivers	Chopra,et.al.,2013
	Risk Mitigation Strategy	Chopra,et.al.,2013
	Scenarios	Goetschalckx,et.al., 2012
	Risk Factors	Singh, et.al., 2012
	Steps to mitigate Risks	Dorian, 2011
	Risk Ranking and Filtering	Manufacturing Technology Committee-Risk Management Working Group training guide on Risk Ranking and Filtering
Social Networks	Social Network Analysis	Muelle, et.al. 2007
	Ego Network	Borgatti,et.al.,2009
General	Rule-of-thumb techniques	Beamon,1998

Summary and Classification of Literature Review

The Literature reviewed is summarized and classified in Table 4.

DISCUSSION AND FINDINGS

The Company under Consideration: ABC

ABC (the identity of the company is not revealed for maintaining confidentiality) is one of the world's leading agribusiness companies with a broad range of safe and efficient crop protection products that improve crop yields and health in all stages of plant development. Efficient production and reliable supply are essential to meet customers' requirements for high-quality products in the right amount and at the right time. A typical product constitutes the bulk formulation packed as per the requirements of the local market. The bulk formulation in turn is produced based on active ingredients. The bulk formulations and active ingredients are procured from suppliers (along with other raw materials) or produced at ABC's manufacturing sites. Active ingredients are made at eight major manufacturing sites and a further 18 facilities around the world produce and package the finished products. Major suppliers play a critical role in ABC's ability to meet demand for crop protection products.

One of the unique things about ABC is that its actions and products can help address one of the planet's most challenging dilemmas: how to grow more crops from lesser resources. That is why ABC's ambition is to bring greater food security in an environmentally sustainable way to an increasingly populous world by creating a worldwide step-change in farm productivity. Through deploying world-class science, ABC aims to transform the way crops are grown and look beyond yield. This includes novel go-to-market models and reaching new customers in emerging markets.

Issues and Concerns

ABC launched Product 1 in the 1990's, as a selective herbicide for rice, with an objective to protect the crop against grasses and broad leaved weeds. The key product attribute at the time of launch was its excellent action against the grasses and broad leaved weeds whereby the product is taken up by the shoots and to a lesser extent by the roots of the germinating grasses and weeds which die shortly after emergence. This provides early elimination of the weeds and clean start to the crop. This attribute was well received by the Farmer. The major market for the product was in the emerging markets of the Asian region comprising of India, Pakistan, China and Bangladesh. The technology for manufacture of Product 1 was based on in-company research and development and was patented. Plant 1 located in Europe, tailor-made for manufacture of new products and well equipped with research and development facilities, was earmarked for manufacture of Product 1. The plant manufactured the active ingredient, formulation as well as the finished packed product. The key raw materials were procured from reputed suppliers in Germany and USA and other materials were procured from suppliers located in close proximity to Plant1. The product was sold to the farmers through local distributors in the consuming countries. The product maintained high growth rates during the initial phase and has developed into a reasonably strong brand mainly in the South Asian markets. However with the expiry of the patent and consequent pressure from competition and generics, the demand levelled off with sales in excess of 70% of expected sales peak and is almost on the threshold of a decline.

Challenges and Uncertainties

1. Performance of supply chain partners.

ABC has experienced some critical challenges during the course of sourcing from low cost countries. In case of some of the countries, the supplier base is fragmented, with huge differences between suppliers in terms of capabilities. The first top tier suppliers can easily compete with their peers in developed markets by following world-class practices and adopting latest management techniques. Tier 2 and tier 3 suppliers are often rural and lag far on the top tier. The major issues of concern with these suppliers are reliability of supply, product quality and the caliber of day-to-day management. These suppliers have no significant process or quality management posing risk of sourcing.

2. Logistics complexity.

The major logistics related hurdles experienced by ABC can be attributed to the key infrastructural challenges involving the active role of roads and the road freight industry, ports and shipping. A major quantum of roads is structurally unsuitable for loaded trucks. A very small portion of the roads constitutes the express highways. Off the highways, transportation links become very scarce to non-existent. Firms can only use trucks that are smaller than 6 meters (2-axle rigid trucks and smaller). Access to the interior of the mainland is limited. The ports experience infrastructure constraints such as lack of storage space and the outdated handling equipment. This in turn results in high pre-berthing time, demurrage charges and turnaround time. Shipping to these ports involves a large amount of paperwork, which is one of the reasons why customs clearance takes 6 days on average. The situation is compounded with frequent labor issues and strikes at the ports. The logistics complexity therefore causes risk of supplies.

3. Challenges with transport regulations.

Another key challenge encountered by ABC pertains to lack of transport regulations in some countries or lack of a proper enforcement mechanism in countries where a formal transport regulation exists. This results in unsafe transportation practices, use of untrained drivers and unfit vehicles/containers. This assumes importance in light of the fact that most of the products and raw materials involved are hazardous substances posing risk of transportation.

SOLUTIONS AND RECOMMENDATIONS

Supply Chain Design

- **Impact of competitive strategy on network design decisions (Chopra et al., 2013):** In recognition of the impact of the competitive strategy on network design decisions, ABC constituted the Product Leadership Team (dedicated to Product 1) that also included the Product Supply Chain Manager. The Product Leadership Team was responsible to formulate the competitive strategy and supply chain strategy aligned with the competitive strategy.
- **Influencers and Design Decisions (Melnyk, et al., 2014):** The Product Leadership team considered Influencers such as broad-based environmental factors (competition), life cycle considerations (generics), desired supply chain outcomes (reduction of cost and complexity).The key Design Decision was the redesign of the supply chain which encompassed the physical network design (location and geographic dispersion of sites, distribution network) and sourcing strategies.
- **Supply Chain Design Elements (Badenhorst-Weiss, et al., 2011):** Supply Chain Design Elements taken into consideration were End customers (farmers), End customer needs (protect crops against grasses and broad leaves),Value proposition (grow more crops from lesser resources),Core competencies (deploying world class science),Identifying market winners (step change in farm productivity),Market demand predictability (identifying the launch and growth phases for the product)Specific market winner (price),Decoupling point (packing units in consuming countries supplying tailor made packages),Supply chain strategy (reducing cost and complexity of the supply chain),Supply chain partners (manufacturing plants, contract manufacturers, distributors, transporters, suppliers), Supply chain drivers (facilities, sourcing, transportation, information),Supply chain KPIs (balanced score card).
- **Puttick Grid (Kruse, 2004; Turner, 2002:** The position of Product 1 was ABC assessed based on per the Puttick Grid. The product was in the Super value quadrant of the grid at the time of launch and then moved to the Consumer durable quadrant with growth. This is illustrated by means of the following.

Table 5 shows the key to the Product Complexity and Uncertainty scores using Rule-of- thumb techniques (Beamon, 1998).

Table 6 shows the Product Complexity and Uncertainty scores for the product at the time of launch (Product 1a) and during growth (Product 1b).

The data from Table 6 is in turn plotted in the Puttick Grid as shown in Figure 3.

The transition from Product 1a to Product 1b triggered the redesign of the supply chain of the product and the decision to transfer the manufacture of the product from Europe to India was a sound one considering the fact that consumer durable type of products are good candidates for migration to low cost countries.

Resultant Competitive Strategy and Supply Chain Design: The Product Leadership Team formulated core strategies in terms of leveraging brand equity, promoting new brand identity, strategic alliances with select partners, optimizing cost of goods sold, reducing supply chain complexity and supporting generic monitoring activity. In light of the high growth rates of Product 1 and with due regard to its major consumption the Asian Region, the "Go East" philosophy was adopted. The manufacture of active ingredients and formulations was transferred to Plant 2 and a contract manufacturer located in India respectively. The packing facilities were set up in the individual consuming countries in collaboration with contract manufacturers for supply of products in tailor made packages. The downstream distribution network was retained. The key raw materials continued to be procured from the suppliers in Germany

Table 5. Key to the product complexity and uncertainty scores

Product Complexity	Score
Number of Variants of the product	
Not more than 4	1
Between 4 and 10	2
At least 10	3
Number of Layers in bills of materials pertaining to the manufacture of the product	
Not more than 2	1
At least 3	3
Number of Tiers in supply chain pertaining to the manufacture of the product	
Not more than 2	1
Between 2 and 5	2
At least 5	3
Uncertainty	Score
Number of changes in product specifications	
Nil	1
At least one in two years or more	2
At least one per year	3
Demand forecast accuracy for the product	
At least 80%	1
Between 70% and 80%	2
Note more than 70%	3

Table 6. Product complexity and uncertainty scores

Product Complexity Score	Product1a	Product1b
Variants of the product	3	3
Layers in bills of materials pertaining to the manufacture of the product	3	3
Tiers in supply chain pertaining to the manufacture	3	3
Total score	**9**	**9**
Uncertainty Score	**Product1a**	**Product1b**
Product Specifications	3	1
Level of demand	3	2
Total score	**6**	**3**

Figure 3. Puttick Grid for Product 1a and Product 1b

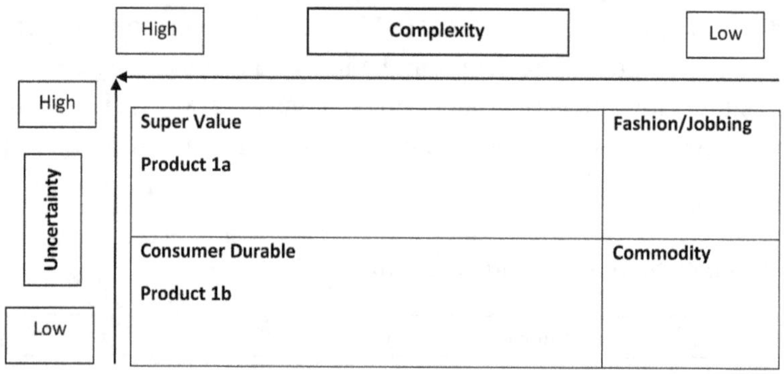

and USA. The other materials were procured from sources in close proximity to the respective plants. Facilities were assigned roles (Chopra et al.,2013) such as the European plant 1 earmarked for manufacture of new products and the Indian plant 2 assigned the role of cost effective manufacture of products with higher volumes. The packing facilities performed the role of a Server facility to supply the market where these were located. The Network Optimization Model (Chopra et al., 2013) was used to assign customers to distributors.

Risk Mitigation

Resources, Networks, Performance (Lassar, et. al., 2010): ABC's ability to manage risks is substantiated by the resource availability (Product Leadership Team for each product). ABC worked with suppliers to assure the suppliers are tailored for individual geographical regions and invested in these relationships to minimize risks and maximize benefits. An important part of ABC's relationship with suppliers was to ensure they meet high standards on health, safety and environment, and labor standards (ABC's Minimum Requirements for Suppliers).

ABC also monitored financial health of suppliers prevent interruptions in supply. ABC conducted regular checks of existing suppliers and all new suppliers must undergo a financial health check before they are contracted. Suppliers were kept informed about ABC's business to build relationships. ABC's aim was to be a preferred customer for its strategic suppliers. Regular meetings, along with open and constructive audits, helped promote long-term relationships of suppliers with ABC. A comprehensive performance management system based on the Balanced Score Card was used by ABC to monitor the operations for all its products.

Risk Drivers (Chopra et al., 2013): Risk Drivers were identified in the category of Disruptions/Delays/Procurement risk/Inventory risk (performance of supply chain partners, logistics complexity and challenges with transport regulations).

Risk Mitigation Strategy (Chopra, et al., 2013; Goetschalckx et al., 2012; Singh et al., 2012; Dorian, 2011; Manufacturing Technology Committee-Risk Management Working Group training guide on Risk Ranking and Filtering): ABC adopted and implemented a robust Risk Management Strategy. This is illustrated as follows:

Table 7 shows the key to the probability and severity scores using Rule-of-thumb techniques (Beamon, 1998).

Table 8 shows the risk factors (probability, severity and risk scores) for a few scenarios (Scenario 1 through Scenario 8) and the classification of the scenarios into "High", "Medium" or "Low" depending upon the risk scores.

Table 7. Key to the probability and severity scores

Value	Probability	Severity
Low	Virtually impossible	Insignificant financial loss and/or Insignificant regulatory/legal implications and/or No loss of company reputation and/or image
Medium	Only remotely possible	Significant financial loss and/or Significant regulatory/legal Implications and/or No loss of company reputation and/or image
High	Almost certain; might well be expected	High financial loss and/or Serious regulatory/legal implications and/or Loss of company reputation and/or image

Table 8. Risk factors (probability, severity and risk scores) for Scenario 1 through Scenario 7

Scenario	Probability	Severity	Risk Score
Scenario 1	Low (1)	High (3)	Medium (3)
Scenario 2	Medium (2)	Low (1)	Low (2)
Scenario 3	Medium (2)	Medium (2)	Medium (4)
Scenario 4	Medium (2)	High (3)	High (6)
Scenario 5	Low (1)	Low (1)	Low (1)
Scenario 6	High (3)	High (3)	High (9)
Scenario 7	Low (1)	Low (1)	Low (1)

Table 9. Risk ranking and filtering tool

Increasing Probability	**High**	3	6	**9** **Scenario 6**
	Medium	**2** **Scenario 2**	**4** **Scenario 3**	**6** **Scenario 4**
	Low	**1** **Scenario 5** **Scenario 7**	2	**3** **Scenario 1**
		Low	**Medium**	**High**
	Increasing Severity			

The data from Table 8 is in turn plotted in the "Risk Ranking and Filtering" tool as shown in Table 9.

Scenario 4 and Scenario 6 are in the "High" category. Risk Reduction is applicable in case of Scenario 4 and Scenario 6 (reducing Probability or Severity in case of Scenario 4 and reducing Probability and Severity in case of Scenario 6). On the other hand, no action need be taken in case of the other five Scenarios. For instance Scenario 6 could be driven by supplier bankruptcy in the Disruptions Category. The Risk Factors and Risk Ranking and Filtering have resulted in classifying Scenario 6 in the High category necessitating Risk Reduction by reducing the Probability and Severity of the risk). ABC therefore resorted to monitoring the financial health of suppliers to prevent interruptions in supply. ABC also conducted regular financial health checks of existing suppliers (all new suppliers went through a financial health check before they are contracted) to mitigate the risk.

Social Networks

Social Network Analysis: The application of Social Network Analysis for the material flow (supply ties) in respect of the supply chain of Product 1 is illustrated as follows:

Figure 4 shows a simple Ego network (only forward logistics for two markets with clustering of similar agents or actors).

Table 10 shows the Degree Centrality for the network (the intensity of links is not considered).

The key to the Ego Network and Degree Centrality are as follows:

- S1-Germany supplier
- S2-US supplier
- S3-local supplier cluster for P1
- S4-local supplier cluster for P2
- P1-plant 1 (active ingredients)
- P2-plant 2 (formulation)
- P3, P4,-packing plants for two country markets
- D1, D2-distributor clusters for two country markets
- C1, C2, C3, C4 –customers assigned to D1
- C5, C6, C7, C8-customers assigned to D2
- 1-material flow taking place; 2-material flow not taking place

Figure 4. Simple ego network for Product 1

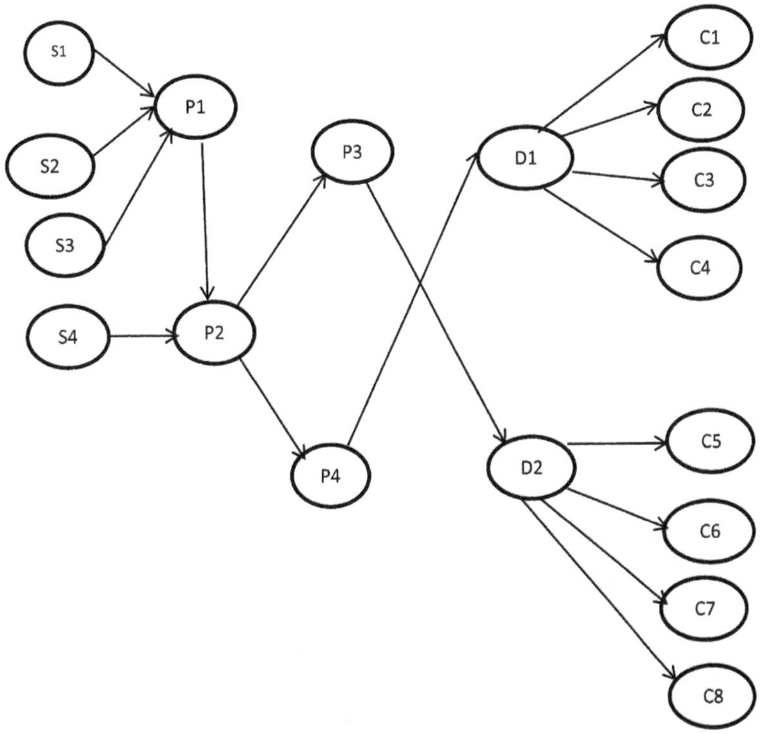

Table 10. Degree centrality for Product 1 supply chain network

	S1	S2	S3	S4	P1	P2	P3	P4	D1	D2	C1	C2	C3	C4	C5	C6	C7	C8	OUT
S1	0	0	0	0	1														1
S2	0	0	0	0	1														1
S3	0	0	0	0	1														1
S4	0	0	0	0	0	1													1
P1	0	0	0	0	0	1													1
P2	0	0	0	0	0	0	1	1											2
P3	0	0	0	0	0	0	0	0	1	1									2
P4	0	0	0	0	0	0	0	0	1	1									2
D1	0	0	0	0	0	0	0	0	0	0	1	1	1	1					4
D2	0	0	0	0	0	0	0	0	0	0	0	0	0	0	1	1	1	1	4
C1	0	0	0	0	0	0	0	0	0	0	0	0	0	0	0	0	0	0	0
C2	0	0	0	0	0	0	0	0	0	0	0	0	0	0	0	0	0	0	0
C3	0	0	0	0	0	0	0	0	0	0	0	0	0	0	0	0	0	0	0
C4	0	0	0	0	0	0	0	0	0	0	0	0	0	0	0	0	0	0	0
C5	0	0	0	0	0	0	0	0	0	0	0	0	0	0	0	0	0	0	0
C6	0	0	0	0	0	0	0	0	0	0	0	0	0	0	0	0	0	0	0
C7	0	0	0	0	0	0	0	0	0	0	0	0	0	0	0	0	0	0	0
C8	0	0	0	0	0	0	0	0	0	0	0	0	0	0	0	0	0	0	0
IN	**0**	**0**	**0**	**0**	**3**	**2**	**1**	**1**	**2**	**2**	**1**	**1**	**1**	**1**	**1**	**1**	**1**	**1**	

D1, D2 and P1 have significant local centrality (D1 and D2 are most significant Transmitters and P1 is the most significant Receiver).

P3 and P4 need to be most Agile in the network since they are supplied by Nodes that have alternate customers and supply to Nodes that have alternate suppliers.

P2 is easily identified as the focal Node with relative high "betweenness" (it falls on most paths from one end of the network to the other).

It is also obvious (without measuring actual geodesic distances) that the path S2-P1-P2 is the longest path to reach P2 and therefore prone to maximum risk of disruption.

ABC is working on ways and means of improving performance of the network based on the above identification of Nodes and paths of strategic importance. For instance the length of the path S2-P1-P2 could be shortened by means of development of supply chain partners in close proximity to one another.

FUTURE RESEARCH DIRECTIONS

The Study carried out is limited to a single product manufactured by a single company in agribusiness with the objective of market expansion. Future studies can be carried out for different companies, businesses, products and objectives. This study will be of considerable interest to industry practitioners as it explains key issues and concerns in the design of global supply chains and suggests means of overcoming the issues and concerns. This study will also help other researchers in the critique of supply chain designs with varying products and businesses. According to the author's knowledge, it is believed that this is the first study carried out to understand the issues and concerns in the design of a global supply chain in the context of one of the world's leading agribusiness companies making a successful foray in emerging markets.

CONCLUSION

Based on the study carried out, it was inferred that the global supply chain network design decisions could be made based on theories, concepts and frameworks in the area of supply chain design, risk mitigation and social networks. The global supply chain networks can best support their strategic objectives with facilities in different countries playing different roles. It is therefore important for a company to identify the strategic role of each facility when designing its global network and to configure the resources of the facility accordingly. The Puttick Grid enables the company to prepare its strategies and plans in the light of the how the customer views and appreciates the products it wishes to acquire and the characteristics of the acquisition process, rather than how the organisation views its own products, strengths and weaknesses. The Grid could therefore be used to identify products suitable for migration to low cost countries. Global supply chains today are subject to more risk factors than localized supply chains of the past. Companies can build in suitable risk mitigation strategies to be aware of the relevant risk factors and build in suitable mitigation strategies.

The Chapter also explains the application of Social Networks in the supply chain context by means of a simple example. Social Network Analysis enables identification of Nodes and Paths of strategic importance in the supply network. For instance the longest paths prone to maximum risk of disruption can be identified and the length of the path could be shortened by development of supply chain partners in close proximity to one another. The Chapter also suggests directions for future research such as critique of supply chain designs with varying products and businesses.

REFERENCES

Abele, E., Meyer, T., Naher, U., Strube, G., & Sykes, R. (Eds.). (2008). *Global Production-A handbook for Strategy and Implementation*. Springer-Verlag Berlin Heidelberg. doi:10.1007/978-3-540-71653-2

Badenhorst-Weiss, J., & Nel, J. (2011). *A conceptual framework to analyze supply chain designs. Acta Commercii*. Special.

Beamon, B. (1998). Supply Chain Design and Analysis: Models and Methods. *International Journal of Production Economics*, *55*(3), 281–294. doi:10.1016/S0925-5273(98)00079-6

Borgati, S. (2009). On Social Network Analysis in a Supply Chain Context. *Journal of Supply Chain Management.*

Chopra, S., Meindl, P., & Kalra, D.V. (2013). *Supply Chain Management Strategy, Planning and Operation*. Dorling Kindersley (India) Pvt. Ltd.

Dorian, L. (2011). *Understanding Risk Mitigation, Industry Insights*. Institute of Chartered Accountants.

Goetschalckx, M., Huangb, E., & Mital, P. (2012). Robust global supply network design. *Information, Knowledge, Systems Management*, *11*, 119–130.

Kruse, G. (2004). *Managing The Extended Supply Chain, Opportunities and Solutions For Manufacturing Migration*. Retrieved April 6, 2014 from http://www.littoralis.info/iom/secure/assets/iom20040901.775512_417681c8304c.pdf

Lassar, W., Haar, J., Montalvo, R., & Hulser, L. (2012). Determinants of Strategic Risk Management in Emerging Markets Supply Chains: The Case of Mexico. *Journal of Economics, Finance and Administrative Science.*

Manufacturing Technology Committee-Risk Management Working Group. (n.d.). *Risk Management Training Guide: Risk Ranking and Filtering*. Retrieved June 14, 2013 from http://www.pqri.org/pdfs/MTC/Risk_Rank_Filter_Training_Guide.pdf

Melnyk, S., Narasimhan, R., & DeCampos, H. (2014). Supply chain design: Issues, challenges, frameworks and solutions. *International Journal of Production Research*, *52*(7), 1887–1896. doi:10.1080/00207543.2013.787175

Mueller, R., Buergelt, D., & Seidel-Lass, L. (2007, February). *Supply Chains And Social Network Analysis*. Paper presented at 1st International European Forum on Innovation and System Dynamics in Food Networks. Innsbruck-Igls, Austria.

Singh, A., Mishra, P., Jain, R., & Khurana, M. (2012). Design of global supply chain network with operational risks. *International Journal of Advanced Manufacturing Technology*, *60*(1-4), 273–290. doi:10.1007/s00170-011-3615-9

Turner, S. (2002). *Tools for Success: A Manager Guide.* Retrieved June 11, 2013 from http://www.google.co.in/url?sa=t&rct=j&q=&esrc=s&source=web&cd=5&ved=0CD8QFjAE&url=http%3A%2F%2Fwww2.warwick.ac.uk%2Ffac%2Fsci%2Fwmg%2Feducation%2Ftranscap%2Fsupportmaterials%2Fhandbooks%2Flvc%2Fgeneral%2Fputtick_grid.ppt&ei=Ltc7U-2iMcWWrgfS-4GwAQ&usg=AFQjCNE2USzvMYgyWc4gw8sGI89zGsXy0g

Chapter 3
Supply Chain Process Efficiency (SCPE) and Firm's Financial Efficiency (FFE):
A Study of Establishing Linkages

Viqar Ali Baig
Lingaya's University, India

Javaid Akhtar
Lingaya's University, India

ABSTRACT

The core object of this chapter is to develop efficiency measurement technique which links SC processes' efficiency (SCPE) to a firm's financial policy (FFP) through demonstrating and utilising the connection between SCPE and a firm's financial efficiency (FFE). The Dempster Shafer/Analytical Hierarchy Processes (DS/AHP) model is employed to link SCPE to the firm's financial efficiency through defining the relevant importance weights of SCPE measures regarding the preferences of FFE. The chapter also introduces a SC Financial Link Index (SCFLI) to examine the degree to which SCPE is linked to the FFE objectives. This index offers an effective supply chain management (SCM) tool to provide continuous feedback on SCPE and recognise the suitable employed remedial actions. Analysing this index offers opportunities for detailed evaluation of SCPE and empowers firms to trace SC processes that need improvement, causing in more control on daily SC operations. The proposed method permits the evaluation, monitoring, and control of SCPE to enhance SC policy for better alignment with the FFP. Linking SCPE to the FFP objectives empowers firms to gain competitive advantages and formulate strategies for improved SCM through linking such strategies to the focus area of improving the FFE.

1. INTRODUCTION

As a result of the strong domestic and international competition that firms currently face, firms will not be able to contend or endure unless they develop strategies to attain cost reduction, quality improvement and enhanced productivity. The real challenge for firms is how to manage the trade-offs between such

DOI: 10.4018/978-1-4666-9795-9.ch003

strategies as they generally work against one another. For example employing a policy to achieve cost reduction could adversely impact quality or result in reduced productivity. The management of material, products, information and time flows through the SC has an immediate effect on the success of these strategies (Chan et al., 2002). A SC is a set of a firm's complete operations directly and indirectly interlinked and interacted to transform inputs into outputs that are supplied to the end customer. Harrison and New (2002) stated the results of a major international survey undertaken in 1998 into the relations between corporate policy, SC policy and SC efficiency management in manufacturing firms across the major industrialised countries. The survey publicized that 90% of the respondents alleged that SCPE was important or very important for attaining competitive advantage in the future (Forslund, 2007). Managers at different levels should be aware of the connection between SCPE and the FFE and how the firm's daily actions can influence the overall FFE. Presutti and Mawhinney (2007) stated that 70% or more of manufacturing firms' expenditures are on SC related activities, which highlights the potential effect of an effectively managed SC in contributing to overall improvement in FFE.

To effectively quantify the impact of SC activities on the FFE, SC efficiency needs to be connected to the firm's financial strategic goals (Kremers, 2010). The challenge for many firms is that the alignment of efficiency measurements between SC and financial roles is still comparatively poor. The main reason for this is that SCPE metrics and FFE metrics are defined in different ways which generates difficulty to translate SC operational measures, with their emphasis on day to day operations, into financial targets (Camerinelli & Cantu, 2006). The chief long-term financial goal of the firm is to maximise profit. To achieve this overall long-term goal, the firm should translate it into meaningful short-term efficiency objectives that can be measured and monitored. These objectives can be realized through recognizing the source of poor efficiency in terms of precise activities and designing short-term strategies for enhancing the efficiency of these activities (Grant, 2005). Understanding the link between SCPE metrics and the overall metrics used to measure the FFE is vital to align SCPE to the firm's financial strategic goal.

The core object of this paper is to develop an efficiency measurement technique which links SC processes' efficiency (SCPE) to a firm's financial policy (FFP) through demonstrating and utilising the connection between SCPE and a firm's financial efficiency. According to the proposed method, SCPE metrics assess the efficiency of SC processes in terms of trustworthiness, responsiveness, agility, cost, and asset management based on SC Operations Reference-model (SCOR) standard efficiency metrics (SEM), while FFE metrics evaluate and analyse the efficiency of the outputs of these processes in terms of efficiency and profitability using Du Pont ratio analysis. To link SCPE to the FFE, the proposed method employs the Dempster Shafer/Analytical Hierarchy Processes (DS/AHP) model developed by Beynon et al. (2000).

According to the DS/AHP model, the significance weight of the evaluation criteria is determined with regard to the preferences of associated decision elements. Applying this model, the significance weights of SCPE measures can be determined with regard to the preferences of the FFP. Consequently, SC policy is formulated based on these preferences through linking SC policy to the focus area of improving the FFE. The developed method introduces a SC Financial Link Index (SCFLI) to test the extent to which SCPE is linked to the firm's short-term financial policy objectives. Analysing this index provides more control over the daily SC operations as it enables firms to trace SC processes that want enhancement and thus identify their connected efficiency pointers for healthier SCM. The remainder of this chapter is organised as follows. Firstly, a review of related literature is presented. In Section 3, the rationale for the proposed method is discussed. Then, the framework for the proposed method is illustrated in Sections 4. Finally, Section 5 discusses conclusion and suggests future work to be built on this research and presents conclusions.

2. LITERATURE REVIEW

Interest in efficiency measurement and management has notably increased in the last 20 years. Firms have acknowledged that monitoring and understanding firms' efficiency have become vital to compete in continuously changing environments (Taticchi et al., 2010). Efficiency measurement is defined as the process of quantifying the effectiveness and efficiency of action (Neely et al., 2005). Effectiveness mentions to the extent to which customer requirements are met, while efficiency assess how economically a firm's resources are exploited when providing a pre-specified level of customer satisfaction (Shepherd and Gunter, 2006). From the fifteenth century until the nineteenth century, firm's efficiency measurement was based on the results of the bookkeeping system. Identifying profit and governing cash flow were the key aspects to dominate firm's efficiency measurement (Morgan, 2004). Although FFE measures have been widely used to measure a firm's efficiency, their ability to capture and reflect the different aspects relating to a firm's efficiency is limited. FFE measures are used to assess inputs and outputs through their codification into financial terms (Neely, 2003).

These measures evaluate how well the firm converts inputs into desired outputs without tracing the way in which the various inputs interacted to produce the outputs. The incompetence to capture the firm's processes that leads to such outputs makes these conventional financial measures incapable to deal with the rapid changes in the business environment (Behn, 2003). In the early 1900s a turning point was achieved in firm's efficiency measurement when William Durant, founder of General Motors, comprehended that profit was not the result of bookkeeping practices, but the consequence of a cost stream that range throughout the SC (Drucker, 1995). SC efficiency measurement provides the tools to monitor SC operations' efficiency and to reveal the effectiveness of a firm's strategies. In addition, it can provide feedback to enable managers to analyse problems and identify success and potential opportunities (Ramaa et al., 2009). Many scholars have proposed various efficiency measurement systems to measure SC efficiency (Neely, 2005; Shepherd & Gunter, 2006; Gunasekaran & Kobu, 2007). However several criticisms were raised against these systems. Amongst the most widely highlighted criticisms of current efficiency measurement systems in SCM are: the lack of a connection with policy (Gunasekaran et al., 2004), the failure to make integration between financial and non-financial measures (Gunasekaran et al., 2001;), and the lack of system thinking (Chan, 2003; Chan & Qi, 2003).

SCM requires standardisation of business processes to be able to link-up processes among the members of the SC and to conduct a meaningful comparison of firms efficiency (Naslund & Williamson, 2010). Current evolution of firms necessitates a shift from the conventional transaction focused measurement to process focused measurement systems (Morgan, 2007). Several authors have recommended implementing business processes in the context of SCM (Srivastava et al., 1999; Bowersox et al., 1999; Lambert & Cooper, 2000; Mentzer, 2001; Huan et al., 2004). Lambert et al. (2005) identified five SCM frameworks that identify the need to implement homogeneous business processes across corporate functions and across firms. The first framework is the SCOR model. The SCOR model was developed in 1996 by the Supply-Chain Council (SCC). This model is based on five core processes (plan, source, make, deliver, and return) and divided into three levels of process detail (top level, configuration level, and process element level). The model attempts to integrate the concepts of business process reengineering, benchmarking, process measurement, and best practice analysis. This permits the top management of a firm to make connections between strategies and measurements and to focus on key processes and measures that have a significant influence on the overall efficiency of a SC (Lockamy & McCormack, 2004; Huang et al., 2005). In this sense, the SCOR delivers standard explanations of SC processes that structure the SC and an outline for defining relationships among these standard processes. In addition, the model includes SEM to measure the efficiency and process evaluation which allow firms to compare and benchmark their processes against those of other firms (Huan et al., 2004).

The second framework was developed in 1996 by the Global SC Forum (GSCF). It consists of three primary related elements: the SC network structure, the SC business processes, and the management (Lambert & Cooper, 2000). The third framework developed by Srivastava et al. (1999) includes three business processes: customer relationship management, product development management, and SCM. The fourth framework was a SCM framework introduced by Bowersox et al. (1999) and focused on three "contexts": operational, planning and control, and behavioural. This framework was further developed by Melnyk et al. (2000) to include eight business processes: plan, acquire, make, deliver, product design/redesign, capacity management, process design/redesign, and measurement. Mentzer et al. (2001) developed the fifth framework which focused on the cross-functional interaction within a firm and on the relationships developed with other SC members. Although these five frameworks suggest the implementation of standard cross-functional business processes, only the GSCF and SCOR frameworks include business processes that could be used by management to achieve cross-functional integration (Lambert et al., 2005). Although the impact of SCM on a firm's efficiency has been discussed by many scholars, few studies have been conducted to find the links between SCM practices and FFE improvements (Gardner, 2004). Toyli et al. (2008) stated that SC efficiency and the firm's FFE have been broadly studied but limited experiential affirmation of their relationship has been presented. A study was conducted from 1997 to 2000 by a research team from Accenture, INSEAD and Stanford University to study the SC-FFE link. The study targeted to test the statistical relationship between firms' financial success and the depth of their SCs. The study statistically analysed publicly available data on 3000 firms, conducted in-depth interviews with more than 75 executives from 60 firms, and analysed 100 responses to a web-based survey designed to capture the SC understandings and experiences of leading officials from firms across North America and Europe. The study's results were published in 2003 showing a statistical correlation between firms' financial success and the depth of their SCs. According to the study's results, SC leaders achieve significantly higher market-capitalisation growth rates than the industry average growth rate. Moreover, the analysis of the study's interviews and surveys revealed that all the winning business strategies of leading firms incorporate SC strategies that provide competitive advantage and dedicate substantial attention to designing integrated SC operating models (D'Avanzo et al., 2003). Deloitte conducted a study of 600 firms in 22 countries which concluded that effectively managing a complex global SC has a positive impact on a firm's FFE (as cited in Colman, 2003). The study revealed that firms which effectively managed their SC realised profit margins 73% higher than other firms with poor SC efficiency and less complex environments. Another survey has been conducted to test the impact of implementing SCM practices on enhancing return on investment. The survey revealed that 76% of responses showed that practices of carefully managed SCs resulted in financial benefits for participating firms (Gunasekaran et al., 2004). Presutti and Mawhinney (2007) demonstrated how SCPE metrics can be connected to FFE metrics to achieve the critical connection between SC productivity and business productivity. The SCPE metrics used were based on SCOR model, while the FFE metrics used were based on the EVA concept. The study concluded that there is a clear and direct link between how effectively SC activities are executed and how well the business performs. The success in making the link between corporate efficiency and SC efficiency results in gratifying two of the firm's most significant stakeholders, its customers and shareholders. Tejas and Srikanth (2007) linked SCPE metrics to financial key efficiency pointers through using scorecards to determine priorities for investments in improving processes and related technology. This linkage helps senior managers to quantify the efficiency of SCPE metrics and understand its impact on the firm's top and bottom lines. Toyli et al. (2008) analysed the relationship between logistics efficiency and FFE in Finnish small and medium- sized enterprises (SMEs). The study covered 424 SMEs that joined in a countrywide Finnish logistics survey in 2006. Logistics efficiency measures were derived from the survey data, and classified into three dimensions: service level describing the service excellence, operational metrics illustrating the time-based logistics efficacy, and logistics costs characterising cost efficiency. The FFE of these firms was then examined in terms of growth and profitability using financial reports based data. The results implied that there was no positive linkage between logistics efficiency and FFE among the surveyed firms, indicating that logistics is just starting to gain more attention among SMEs in Finland and that it might be relatively easy for SMEs to gain competitive advantage by focusing more on logistics efficiency. Woei (2008) administered a study to explore the SCM and financial success relationship. To analyse the relationship between a firm's financial success and its SC efficiency, an experiential study was undertaken based on financial information extracted from public quoted firms in Malaysia during the financial years from 1999 to 2006. Financial accomplishment was calculated by market capitalisation though SC efficiency was determined by four variables specifically revenue, cost of sales as percentage of revenue, cash to cash cycle and return on working capital. The data was statistically analysed to test the correlation between categories of SCM efficiency measures and financial success. The results presented that the correlation was weak for the financial years of 1999–2002, while the correlation became robust for the succeeding financial years of 2003–2006. The results also publicized that firms which implement the complete range of SC measures can find prospects to become financially successful firms in the new business setting. Camerinelli (2009) demonstrated the link between FFE and operational decisions through mapping financial metrics to operational metrics. Since operational metrics assess the operational status of the firm and are linked to operational decisions, a firm's FFE can reflect the quality of the operational decisions taken to achieve it. The scholar identified the operational metrics that can be used to map financial metrics to operational metrics based on the SCOR model SEM through selecting the proper elements from the balance sheet and income statement to be linked to the SCOR level 1 and level 2 metrics. Hutchison et al. (2009) advocated how cash-to-cash (C2C) approaches can be used in a SC setting as active cash management and synergistic tools to comprehend opportunities for enhancing efficiency, productivity, cash-flow management, and communication systems among SC associates. According to this approach, an information sharing atmosphere should be established among trading associates in the SC so as to recognise likely prospects that can eventually improve cash flow and profitability. The C2C calculation contains three financial variables: inventory, accounts receivable, and accounts payable. Since each party in the SC may have an advantage in its weighted average cost of capital (WACC) or inventory carrying cost (ICC), the chain can exploit inventory as well as receivable and payable items to reduce costs that relate to purchases, inventory, and capital so as to benefit all trading partners throughout the SC. Kremers (2010) offered an approach that provides a comprehensive vision of the existing relationship between firms operational and FFE. According to this approach, SCPE can be evaluated in terms of its impact on cash flow, market value, and key internal FFE metrics. This approach tied SCPE to strategic business goals through linking SCOR model efficiency measures to the priorities of FFE drivers (profitability, asset employment, and financial leverage competence). Marquez (2010) developed and evaluated a comprehensive dynamic SCM model to determine operational and financial benefits from various levels of SC integration. The model highlighted the financial implications of different pricing strategies and cost structures when modelling financial aspects of the SC. It explored the operational and financial impacts of various potential problems in SCM, offering a compilation of practical solutions using system dynamics. Wisner (2011) claimed that SCM decisions and resource utilisation could influence the FFE of the firm. To demonstrate such impact, Wisner exemplified how SC functions influence the results shown in the firm's financial statements (income statement, balance sheet, statement of cash flows, and statement of stockholders' equity). She presented a model connecting SCPE metrics to the outcomes of the financial statements. The model identified the SCPE measures relevant to the components of each financial statement in order to ensure that SC actions and decisions are compatible with the firm's financial goals.

The previous review of literature reveals that there is a connection between SC daily practices and FFE. The challenge, however, is how to make that linkage. A well-designed efficiency measurement system is vital for improving business processes, but conventional efficiency measurement systems, which rely on financial measures only, do not fit today's business environment. The new business environment requires efficiency measurement systems to incorporate both financial and non-financial measures.

3. RATIONALE FOR THE PROPOSED METHOD

Connecting SC activities to the firm's financial strategic objectives represents an opportunity for firms to gain competitive advantages by focusing on linking SCPE to the focus area of enhancing the FFE. The link identification problem is handled with regard to the accumulation of multi-criteria expressions in a SC context. Many approaches have been developed to cumulative the efficiency from multi-criteria expressions such as the weighted mean accumulation operator, to handle hierarchical links, the Choquet integral operator, for taking interactions into account, and the AHP technique and the FAHP approach, to quantify the weights and the efficiency elementary expression (Berrah & Clivillé, 2007). The multi-criteria decision-making process (MCDMP) is complex and challenging as the human competence to compare or decide on more than two factors has natural limitations (Abdul Moneim, 2008). In most cases in real life, there are interdependencies among criteria or alternatives while the available data and information are incomplete and the decision environment is uncertain and complex. In these cases, accumulation operators can play a central role to blend information in ways that allow efficiency accumulation. The Choquet integral, as a sophisticated accumulation operator, can be used in many industrial efficiency accumulation applications through ordering the arguments of the accumulation to determine the weights that are assigned to the different argument components (Tan and Chen, 2011).

In the AHP, first, the decision problem is organized in a order of different stages of elements, and then a pair-wise comparison matrix is used to itentify the relative preferences of the decision elements (weights of the criteria). The pair-wise comparisons are acknowledged as linguistic evaluations articulating the respective importance of duos. Finally, the weights of each element in each hierarchical level are cumulated to the next level applying the principle of hierarchic composition (Mikhailov, 2004). Elgazzar et al. (2011a) suggested a technique to link SCPE to the preferences of the firm's financial policy using the FAHP technique. According to this method, SCPE is measured based on the SCOR model metrics, while the firm's FFE preferences are identified using Du Pont model. To link SCPE to the FFE preferences, the relevant importance weights of SCPE measures are identified with respect to the preferences of the firm's financial policy using the FAHP technique, and consequently, the appropriate SC policy is framed based on these preferences.

In this chapter, SC efficiency is framed according to the SCOR model matrix with its five main SCPE measures. The efficiency rates of all measurement traits are cumulated – using the averaging accumulation method – throughout the hierarchy of the SC to determine the efficiency rate of the SCPE measurement traits at the top level. The weighted rates of the five main SCPE measures are then cumulated using the DS/AHP method to determine the firm's SCFLI. These weights quantify the respective contributions of the SCPE measures to the overall FFE. A method resulting from Presutti and Mawhinney (2007) has been developed. To explore the link between SC efficiency and the firm's efficiency, Presutti and Mawhinney (2007) linked the SCOR model metrics for computing SCPE to the Economic Value Added (EVA) components as a comprehensive measure of the firm's profitability in relation to the amount of capital employed. According to this method, SCOR metrics efficiency traits that have a direct impact on the customer were linked to the revenue component of EVA while SCOR metrics efficiency traits that have a direct impact on the firm were linked to the cost and assets components of EVA. Although this method demonstrates how SC metrics can be linked to a firm's financial metrics, it does not specify how this link can be utilised to enhance the firm's overall FFE. The method proposed by Presutti and Mawhinney (2007) assumes that EVA components (revenue, cost, and assets) have the same influence weighting on the firm's FFE. It is considered relevant to set the preferences of these components according to the firm's financial policy in order to highlight the components that need improvement with respect to the focus areas for enhancing the FFE.

Moreover, setting preferences for these components enables the development of SC policy connected to the firm's financial strategic objectives through identifying SC processes and measures that have a significant impact on the focus areas of the firm's financial policy. In addition, EVA metrics measure the value created by the firm through evaluating its profitability in relation to the amount of capital employed. Linking SCPE to the FFE requires FFE metrics which analyse the firm's FFE in terms of operating efficiency as well as profitability. Presutti and Mawhinney method (2007) also addresses SCM as the only factor that can impact FFE. It does not address the impact of the ignorance factors, out of the firm's control, such as the political factors which may impact a firm's FFE and thus should be considered.

To overcome the above obstacles, this paper develops a method to link SCPE to the preferences of FFE in the short-term and evaluates its impact on maximising profit as the firm's primary long-term financial goal. SCPE is evaluated based on SCOR model SEM while FFE is evaluated in terms of efficiency and profitability based on Du Pont ratio analysis. The results of Du Pont analysis allow the preferences of FFE factors (efficiency and profitability) to be determined through evaluating the contribution of each factor and highlighting factors that need improvement in the short-term. Then, the DS/AHP model is used to link SCPE to the preferences of FFE factors through determining the relevant importance weights of the main SCPE measures with respect to these preferences. The developed method illustrates how this link can be utilised to connect SCPE to the firm's short-term financial strategic objectives (STFSO) in order to contribute to improvement in the firm's overall FFE through impacting on its profitability and efficiency. Consequently, SC policy is formulated based on the preferences of FFE factors for better alignment with the firm's STFSO. To test the extent to which SCPE is linked to the firm's STFSO, a SC Financial Link Index (SCFLI) is developed. This index is different from the conventional SCPE index as it takes into consideration the relevant importance weights of SCPE measures. It cumulates the weighted rates of the main SCPE measures to reflect SCPE with respect to the preferences of the firm's

FFE. In addition, the method developed takes into consideration factors outside the firm's control that can impact on a firm's FFE as it permits use of the DS/AHP model to calculate the influence weight of the ignorance factor on the decisions made by the firm.

The SCOR Model

The SCOR model is more applicable to a manufacturing context than the GSCF model (Ellram et al., 2004). It focuses on identifying areas of improvement in order to provide cost reductions and improve asset efficiency making the framework operational efficiency- oriented rather than relationship-oriented. Its processes are developed based on the operations policy as it focuses on engaging partners from the logistics, production and purchasing functions in its five SCM processes (plan, source, make, deliver, and return) which make the model somewhat limited in scope. Ten SEM are provided by the SCOR model such as: perfect order fulfilment, order fulfilment cycle time, upside SC flexibility, upside SC adaptability, downside SC adaptability, SC management cost, cost of goods sold, cash to cash cycle time, return on SC fixed assets and return on working capital. All these fall into five standard efficiency categories: trustworthiness, responsiveness, flexibility, cost and asset metrics. These ten efficiency metrics are designed to provide a view of overall SC efficiency at level 1 (top level) while the SCOR model levels 2 and 3 (configuration level and process element level) supporting metrics are keys to the level 1 metrics (Supply-Chain Council, 2008).

Du Pont Ratio Analysis

Du Pont ratio analysis is a financial ratio commonly used to measure a firm's FFE. The Du Pont ratio analysis evaluates the areas of profitability (P) and operating efficiency (E) through assessing the efficiency of the components contributing to return-on-assets (ROA). These are revenue (sales), cost and total assets. ROA measures how much profit a firm generates compared to the assets employed in the business. It consists of a profitability measure (Net Profit Margin) and an efficiency measure (Total Assets Turnover) which can be expressed in the following formula (Dehning 2002):

1. The Du Pont ratio can also be broken into more components depending upon the needs of the analysis (Nissim & Penman, 2001). DuPont analysis can also be applied based on the return on equity (ROE) ratio. It can be decomposed into the three multiplicative ratios of profit margin, asset turnover, and equity multiplier as follows:
2. The ROE form is not applicable for this research as ROE is affected by changes in the firm's financial structure (Soliman, 2007). Since this research focuses on how the firm performs business operations not on how it decides to finance such operations, the ROA form is more relevant. As presented in Fig. 1, Presutti and Mawhinney (2007) considered the EVA components to link SCPE metrics to the firm's FFE. This is developed further in this paper by incorporating Du Pont analysis in the FFE metrics to illustrate the impact of SCPE on FFE through assessing the contribution of each FFE component (revenue, cost, and assets) to the improvement of the firm's profitability and operating efficiency (figure 1).

Figure 1. Linking SCOR model performance metrics to the financial performance factors
Developed from Prescutti and Mawhinney (2007) and Elgazzar et al. (2011)

	SC performance attributes					Financial performance metrics	
	Customer- Facing			Internal - Facing			
SCOR level 1- stratgic SC metrics	Reliability	Responsiveness	Flexibility	Cost	Assets	EVA components	Du Pont analysis
Perfect Order Fulfilment						Revenue	Profitability & Efficiency factor
Order Fulfilment Cycle Time							
Upside Supply Chain Flexibility						Cost	Profitability factor
Upside Supply Chain Adaptability							
Downside Supply Chain Adaptability							
Supply Chain Management Cost							
Cost of Goods Sold							
Cash-to-Cash Cycle Time						Assets	Efficiency factor
Return on Supply Chain Fixed Assets							
Return on Working Capital							

DS/AHP Model

DS/AHP model is a multi-criteria decision-making model that incorporates Dempster Shafer theory (DST) with the philosophy behind the Analytical Hierarchy Processes (AHP) technique to improve conventional approaches to multi-criteria decision modelling (Beynon, 2005b). DST is a generalisation of the Bayesian theory of subjective probability (Smarandache, 2003). The Bayesian theory quantifies judgements about a question by assigning probabilities to the possible answers to question while DST provides a non-Bayesian way of using mathematical probability to quantify subjective judgements. It permits the derivation of degrees of belief for a question from probabilities for a related question and then considers the implications of these probabilities for the question of interest (Shafer, 2008).

The basic difference between probability theory and DST is DST framework is a broader framework for representing uncertainty than probability. Under the probability structure, the total of probabilities of all likely values of a variable equals one while in the DST, risk is not only allocated to the single elements of the frame but also to all other suitable subsets of the frame and to the whole frame (Bovee et al., 2003). DST gives the ability to assign probability measures (basic probability assignments) to groups of objects rather than in classical probability theory where measures must be given to individual objects.

The utilisation of DST in DS/AHP permits decision makers to make preference judgments on groups of decision choices (D.C.'s) rather than considering all D.C.'s (as in the classical AHP technique) and consequently, the number of comparisons can be reduced (Beynon, 2002). Incorporating DST permits the related measure of ignorance to be calculated on the judgements made by the decision makers. Within DS/AHP decision makers can ignore those D.C.'s they do not have an opinion towards. They only need to give judgments to the D.C.'s they have a level of opinion towards which enables the decision maker to have a greater level of control on their judgements compared to standard AHP methods (Beynon et al., 2001;Beynon, 2005a).

4. THE PROCEDURES FOR THE PROPOSED METHOD

The procedures for the proposed method are illustrated in the following steps.

Step One: Du Pont ratio for the firm is calculated and then compared to the industrial average to reveal the firm's overall FFE relative to the industrial average and highlight FFE factors need improvement. Based on the result of Du Pont ratio analysis, the priorities of FFE factors (profitability and efficiency) are determined using a pair-wise comparison method.

Step Two: To link SCPE to the priorities of the FFE, the respective significance weights of the five main SCPE measures can be determined regarding the priorities of the FFE issues using DS/AHP model. Since the FFE components (revenue, cost, and assets) are categorized into profitability factor and efficiency factor grounded on Du Pont analysis, the five main SCPE measures (trustworthiness, responsiveness, agility, cost, and asset) can drive these FFE components. Fig. 2 illustrates the developed hierarchy framework to link SCPE to the priorities of the FFE. Using DS/AHP model, the firm does not need to consider all decision choices (D.C.'s) (i.e., trustworthiness (RL), responsiveness (RS), agility (AG), cost (CO), and asset management (AM)), instead it considers groups of D.C.'s for each FFE criterion (i.e. profitability (P) and efficiency (E)). The selected group of D.C.'s can drive each FFE criterion is considered based on the Presutti and Mawhinney method (see Fig. 1). As demonstrated in Fig. 1, SCPE measures can drive profitability gears (revenue and cost) are: trustworthiness, responsiveness, agility, and cost though SCPE measures can drive efficacy components (revenue and asset) are: trustworthiness, responsiveness, and asset management.

Step Three: To evaluate the efficiency and the effectiveness of current SC policy, the proposed SCFLI is calculated for the firm in order to replicate the degree to which SCPE is related to the firm's STFSO. To calculate SCFLI, the efficiency rate which is allocated for each of the five main SCPE measures based on the SCOR model's conventional SC efficacy index is adjusted by the respective significance weights of these measures. By multiplying the respective significance weight of each measure by its efficiency rate, the weighted rate of each efficiency measure is determined. The weighted rates of all efficiency measures are then cumulated to determine the firm's SCFLI (figure 2).

Step Four: Having evaluated and analysed its current FFE and SCPE, the firm is now in a position to formulate its new SC policy based on the priorities of FFE with respect to the respective significance weights of the main SCPE measures. According to SCOR model SEM, each SC efficiency measurement characteristic corresponds to specific processes in the SC. Based on the respective significance weights of SCPE measures, the firm can identify the related processes need improvement and their corresponding efficiency indicators to align with SC policy, and consequently with the firm's STFSO.

Step Five: At the end of the accounting period, SCFLI is computed once more to assess the efficacy and the effectiveness of the newly drafted SC policy in contributing to accomplishing the firm's STFSO. Calculating this index at the end of the period reflects SCPE for this period. This index also can be used as an effective SCM instrument as it can be computed at any time during the period hence permitting the firm to get constant feedback on SC policy and take the necessary corrective actions for better results by the end of the period. By examining this index, a firm can seek SC procedures still need enhancement allowing superior control of daily SC operations.

Step Six: Du Pont ratio is calculated by the end of the accounting period to examine the effect of SCPE on enhancing the firm's overall FFE. The complete procedures of linking SCPE to the firm's financial strategic objectives are summarised in Fig. 3. The framework of the proposed research method is presented in Fig. 4. As illustrated, SCPE is measured in terms of agility, cost, trustworthiness, responsiveness, and asset management based on SCOR model SEM. FFE is evaluated in terms of the firm's profitability and operating efficiency through assessing the efficiency of the components contributing to ROA (cost, revenue, and assets). Using Du Pont ratio analysis, the priorities of FFE factors (profitability and efficiency) can be determined according to the assessment of their corresponding components.

Then, SCPE metrics are linked to FFE metrics using the DS/AHP model. This model permits the determination of the importance weights of the five main SCPE measures with respect to FFE priorities. Consequently, SC policy is formulated based on these priorities, resulting in improvement in the overall FFE (figures 3 and 4).

Figure 2. The developed hierarchy framework to link supply chain processes performance to the financial performance using DS/AHP model

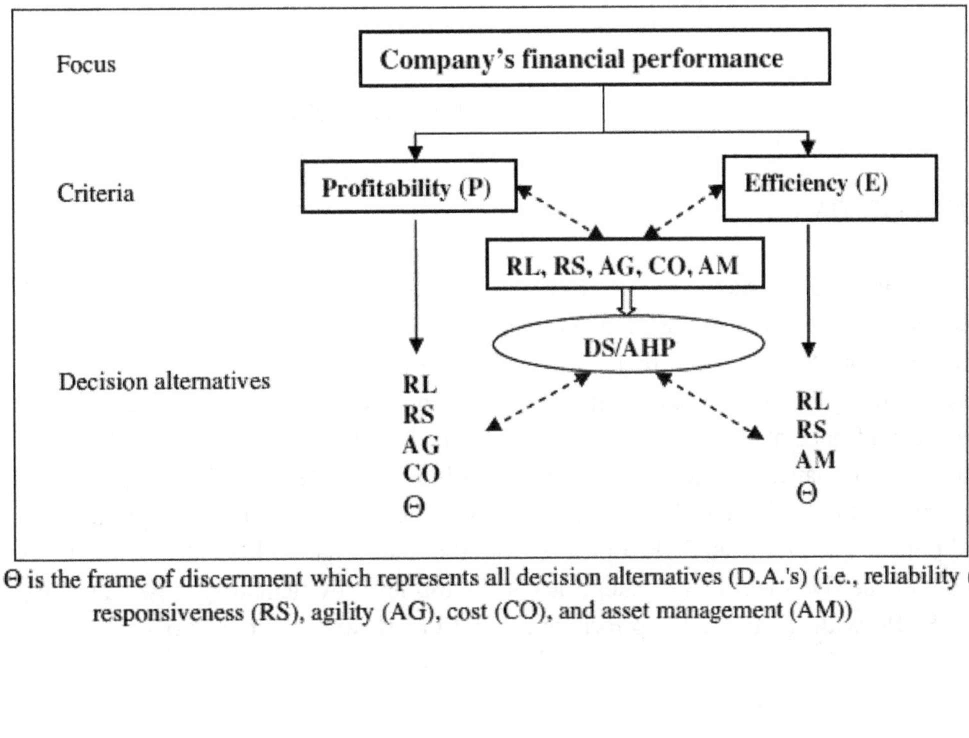

Where: Θ is the frame of discernment which represents all decision alternatives (D.A.'s) (i.e., reliability (RL), responsiveness (RS), agility (AG), cost (CO), and asset management (AM))

Figure 3. Linking SC processes' performance to the company's strategic objectives

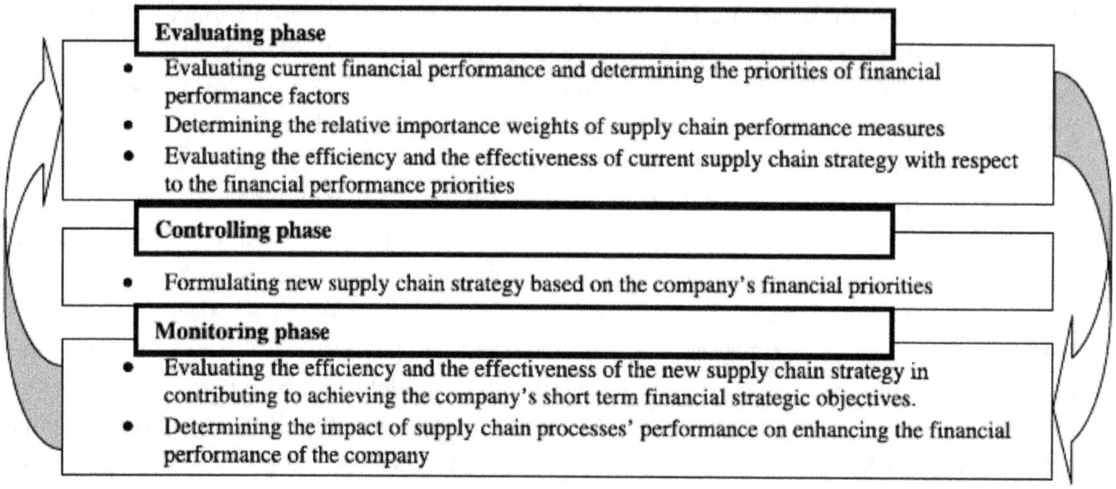

Figure 4. Linking supply chain processes' performance to financial performance
Developed from Elgazzar et al. (2011)

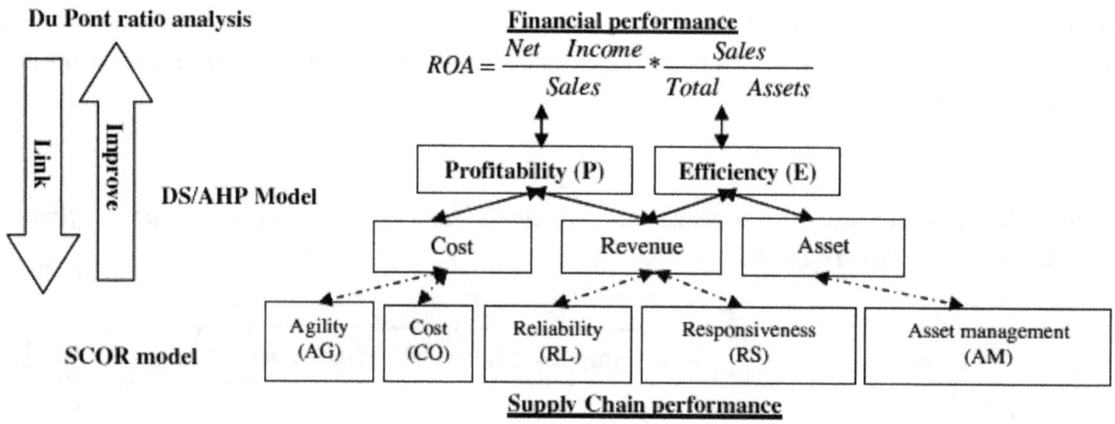

5. CONCLUSION

This research illustrated a method to link SC performance metrics to the priorities of FFE. This method enables companies to connect SCPE to the company's short-term financial strategic objectives through evaluating current SC policy and then formulating the new SC policy based on the priorities of the financial efficiency in the short-term for achieving improvement in the firm's profitability as the primary long-term financial goal. SCFLI was developed to test the extent to which SCPE is linked to the firm's short-term financial strategic objectives. This index provides an effective SCM tool to evaluate, monitor, and control SCPE in order to enhance supply chain policy for better alignment with the firm's financial policy. The proposed SCFLI differs from the traditional SCPE index as it adjusts the performance rate of the five main SCPE categories at the top level by their respective significance weights to reflect SCPE with respect to the priorities of the FFE. However, it does not consider the respective significance weights of SCPE measures at lower levels (configuration level, process element level, and implementation levels) assuming that SCPE measures at lower levels are equally weighted. In reality, determining the relative importance weights of SCPE measures at lower levels enables firms to identify the corresponding processes at each category that need improvement and to assess the performance of SC processes at lower levels in order to identify the core competence processes, identify processes which create value or create waste, and develop strategies to better manage the SC.

To determine the respective significance weights of SCPE measures at lower levels, a technique incorporating FAHP in the SCOR model can be adapted (Elgazzar et al., 2011b).

This chapter demonstrated the developed method within the context of a simple SCM model. The aim of SCM was to enhance the processes to which cost and agility performance measures correspond assuming that all other variables would not change and remain constant. However, it should be noted that in real life, firms' objectives are not mutually exclusive. SCM may include a number of conflicting aims and achieving one of the aims may cause other variables to move into undesirable states. In a further refinement of the proposed method, system dynamics can be deployed to integrate more complex decision variables and evaluate multiple objectives in order to generate more realistic supply chain management dynamics models.

6. FUTURE RESEARCH DIRECTIONS

As a suggestion for future research, the SCOR FAHP technique can be combined with the method proposed in this chapter to determine the respective significance weight of each SCPE measure at lower levels in order to identify SC processes that need improvement at each SCPE category with respect to the priorities of the financial strategy.

REFERENCES

Abdul Moneim, A. F. (2008). Fuzzy genetic prioritization in multi-criteria decision problems. *Jordan Journal of Mechanical and Industrial Engineering*, *2*, 175–182.

Behn, R. D. (2003). Why measure performance? Different purposes require different measure. *Public Administration Review*, *63*(5), 586–606. doi:10.1111/1540-6210.00322

Berrah, L., & Clivillé, V. (2007). Towards an aggregation performance measurement system model in a supply chain context. *Computers in Industry*, *58*(7), 709–719. doi:10.1016/j.compind.2007.05.012

Beynon, M. (2002). DS/AHP method: A mathematical analysis, including an understanding of uncertainty. *European Journal of Operational Research*, *140*(1), 148–164. doi:10.1016/S0377-2217(01)00230-2

Beynon, M. (2005a). A method of aggregation in DS/AHP for group decision-making with the non-equivalent importance of individuals in the group. *Computers & Operations Research*, *32*(7), 1881–1896. doi:10.1016/j.cor.2003.12.004

Beynon, M. (2005b). Understanding local ignorance and non-specificity within the DS/AHP method of multi-criteria decision making. *European Journal of Operational Research, 163*(2), 403–417. doi:10.1016/j.ejor.2003.11.010

Beynon, M., Cosker, D., & Marshall, D. (2001). An expert system for multi-criteria decision making using Dempster Shafer theory. *Expert Systems with Applications, 20*(4), 357–367. doi:10.1016/S0957-4174(01)00020-3

Beynon, M., Curry, B., & Morgan, P. (2000). The Dempster–Shafer theory of evidence. An alternative approach to multicriteria decision modeling. *International Journal of Management Sciences, 28*, 37–50.

Bovee, M., Srivastava, R. P., & Mak, B. (2003). A conceptual framework and belieffunction approach to assessing overall information quality. *International Journal of Intelligent Systems, 18*(1), 51–74. doi:10.1002/int.10074

Bowersox, D. J., David, J. C., & Theodore, P. S. (1999). *21'' Century Logistics: Making Supply Chain Integration a Reality*. Chicago, IL: Council of Logistics Management.

Camerinelli, E. (2009). *Measuring the Value of the Supply Chain*. Ashgate Publishing Group.

Camerinelli, E., & Cantu, A. (2006). Measuring the value of the supply chain: A framework. *Supply Chain Practice, 8*, 40–59.

Chan, F. T. S. (2003). Performance measurement in a supply chain. *International Journal of Advanced Manufacturing Technology, 21*(7), 534–548. doi:10.1007/s001700300063

Chan, F. T. S., & Qi, H. J. (2003). An innovative performance measurement method for supply chain management. *Supply Chain Management: An International Journal, 8*(3), 209–223. doi:10.1108/13598540310484618

Chan, F. T. S., Tang, N. K. H., Lau, H. C. W., & Ip, R. W. L. (2002). A simulation approach in supply chain management. *Integrated Manufacturing Systems, 13*(2), 117–122. doi:10.1108/09576060210415455

D'Avanzo, R., Von Lewinski, H., & Van Wassenhove, L. N. (2003). The link between supply chain and financial performance. *Supply Chain Management Review, 7*, 40–47.

Dehning, B., & Stratopoulos, T. (2002). DuPont analysis of an IT-enabled competitive advantage. *International Journal of Accounting Information, 3*(3), 165–176. doi:10.1016/S1467-0895(02)00032-5

Drucker, P. (1995). The information executives truly need. *Harvard Business Review, 73*, 54–62.

Elgazzar, S., Tipi, N. S., Hubbard, N. J., & Leach, D. Z. (2011a). Linking supply chain operations' performance to the company's financial strategy: a case study of an Egyptian natural bottled water company. In *Proceedings from: The 16th International Symposium on Logistics*.

Elgazzar, S., Tipi, N. S., Hubbard, N. J., & Leach, D. Z. 2011b. A SWapplication system for measuring supply chain operations' performance using SCOR FAHP technique. In *Proceedings from: The 2011 International Conference on Business and Economics Research*. Retrieved November 2011 from: http://www.ipedr.com/vol16.htm

Ellram, L. M., Tate, W. L., & Billington, C. (2004). Understanding and managing the services supply chain. *Journal of Supply Chain Management*, *40*(4), 17–32. doi:10.1111/j.1745-493X.2004.tb00176.x

Forslund, H. (2007). The impact of performance management on customers' expected logistics performance. *International Journal of Operations & Production Management*, *27*(8), 901–918. doi:10.1108/01443570710763822

Gardner, D. (2004). *Supply Chain Vector Methods for Linking the Execution of Global Business Models with Financial Performance*. J. Ross Publishing.

Grant, R. M. (2005). *Contemporary Strategy Analysis*. Oxford, UK: Blackwell Publishing.

Gunasekaran, A., & Kobu, B. (2007). Performance measures and metrics in logistics and supply chain management: A review of recent literature (1995–2004) for research and applications. *International Journal of Production Research*, *45*(12), 2819–2840. doi:10.1080/00207540600806513

Gunasekaran, A., Patel, C., & McGaughey, R. E. (2004). A framework for supply chain performance measurement. *International Journal of Production Economics*, *87*(3), 333–347. doi:10.1016/j.ijpe.2003.08.003

Gunasekaran, A., Patel, C., & Tirtiroglu, E. (2001). Performance measures and metrics in a supply chain environment. *International Journal of Operations & Production Management*, *21*(1/2), 71–87. doi:10.1108/01443570110358468

Harrison, A., & New, C. (2002). The role of coherent supply chain strategy and performance management in achieving competitive advantage: An international survey. *The Journal of the Operational Research Society*, *53*(3), 263–271. doi:10.1057/palgrave.jors.2601193

Deloitte. (2003). Mastering Complexity in Global Manufacturing: Powering Profits and Growth through Value Chain Synchronization. Retrieved from https://www.entrepreneur.com/tradejournals/article/113306163.html

Huan, S. H., Sheoran, S. K., & Wang, G. (2004). A review and analysis of supply chain operations reference (SCOR) model. *Supply Chain Management*, *9*(1), 23–29. doi:10.1108/13598540410517557

Huang, S. H., Sheoran, S. K., & Keskar, H. (2005). Computer-assisted supply chain configuration based on supply chain operations reference (SCOR) model. *Computers & Industrial Engineering*, *48*(2), 377–394. doi:10.1016/j.cie.2005.01.001

Hutchison, P.D., Farris, M.T., & Gary, M.F. (2009). Supply chain cash-to-cash: a strategy for the 21st century. *Strategic Finance*, 41–48.

Kremers, L. (2010). The link between supply chain and finance. *Supply Chain Asia*, 22–25.

Kunadhamraks, P., & Hanaoka, S. (2008). Evaluating the logistics performance of intermodal transportation in Thailand. *Asia Pacific Journal of Marketing and Logistics*, *20*(3), 323–342. doi:10.1108/13555850810890084

Lambert, D. M., & Cooper, M. C. (2000). Issues in supply chain management. *Industrial Marketing Management*, *29*(1), 65–83. doi:10.1016/S0019-8501(99)00113-3

Lambert, D. M., García-Dastugue, S. J., & Croxton, K. L. (2005). An evaluation of processoriented supply chain management frameworks. *Journal of Business Logistics*, *26*(1), 25–51. doi:10.1002/j.2158-1592.2005.tb00193.x

Lockamy, A. III, & McCormack, K. (2004). Linking SCOR planning practices to supply chain performance. *International Journal of Operations & Production Management*, *24*(12), 1192–1218. doi:10.1108/01443570410569010

Marquez, A. C. (2010). *Dynamic Modelling for Supply Chain Management: Dealing with Front-End, Back-End and Integration Issues*. London: Springer. doi:10.1007/978-1-84882-681-6

Melnyk, S. A., Theodore, P. S., & David, J. C. (2000). Supply chain management at Michigan state university: The journey and the lessons learned. *Production and Inventory Management Journal*, *41*, 13–18.

Mentzer, J. T. (Ed.). (2001). *Supply Chain Management*. Thousand Oaks, CA: Sage Publications.

Mentzer, J. T., DeWitt, W., Keebler, J. S., Min, S., Nix, N. W., Smith, C. D., & Zacharia, Z. G. (2001). Defining supply chain management. *Journal of Business Logistics*, *22*(2), 1–25. doi:10.1002/j.2158-1592.2001.tb00001.x

Mikhailov, L. (2004). A fuzzy approach to deriving priorities from interval pair-wise comparison judgments. *European Journal of Operational Research*, *159*(3), 687–704. doi:10.1016/S0377-2217(03)00432-6

Morgan, C. (2004). Structure, speed and salience: Performance measurement in the supply chain. *Business Process Management Journal*, *10*(5), 522–536. doi:10.1108/14637150410559207

Morgan, C. (2007). Supply network performance measurement: Future challenges. *The International Journal of Logistics*, *18*(2), 255–273. doi:10.1108/09574090710816968

Naslund, D., & Williamson, S. (2010). What is management in supply chain management? – A critical review of definitions, frameworks and terminology. *Journal of Management Policy and Practice*, *11*, 11–28.

Neely, A. (2003). *Business Performance Measurement: Theory and Practice*. Cambridge: Cambridge University Press.

Neely, A. (2005). The evolution of performance measurement research developments in the last decade and a research agenda for the next. *International Journal of Operations & Production Management*, *25*(12), 1264–1277. doi:10.1108/01443570510633648

Neely, A., Gregory, M., & Platts, K. (2005). Performance measurement system design: A literature review and research agenda. *International Journal of Operations & Production Management*, *25*(12), 1228–1263. doi:10.1108/01443570510633639

Nissim, D., & Penman, S. H. (2001). Ratio analysis and equity valuation: From research to practice. *Review of Accounting Studies*, *6*(1), 109–154. doi:10.1023/A:1011338221623

Presutti, W. D. Jr, & Mawhinney, J. R. (2007). The supply chain-finance link. *Supply Chain Management Review*, *11*, 32–38.

Ramaa, A., Rangaswamy, T. M., & Subramanya, K. N. 2009. A review of literature on performance measurement of supply chain network. In *Proceedings from: Second International Conference on Emerging Trends in Engineering and Technology*. doi:10.1109/ICETET.2009.18

Shafer, G. (2008). Belief functions. In *Classic Works of the Dempster–Shafer Theory of Belief Functions*. Retrieved February 2010, from http://www.glennshafer.com/assets/downloads/rur_chapter7.pdf>

Shepherd, C., & Gunter, H. (2006). Measuring supply chain performance: Current research and future directions. *International Journal of Productivity and Performance Management*, *55*(3/4), 242–258. doi:10.1108/17410400610653219

Smarandache, F. (2003). Neutrosophic logic – A generalization of the intuitionistic fuzzy logic. In *Proceedings from: The Third Conference of the European Society for Fuzzy Logic and Technology*. University of Applied Sciences. Retrieved April 2010 from: http://www.arxiv.org/abs/math/0303009

Soliman, M. T. (2007). *The Use of DuPont Analysis by Market Participants*. Social Science Research Network. Retrieved June 2010 from: http://www.papers.ssrn.com/sol3/papers.cfm?abstract_id=1101981

Srivastava, R. K., Tasadduq, A. S., & Liam, F. (1999). Marketing, business processes, and shareholder value: An organisationally embedded view of marketing activities and the discipline of marketing. *Journal of Marketing*, *63*, 168–179. doi:10.2307/1252110

Supply-Chain Council. (2008). *Supply-chain Operations Reference-Model: SCOR Model – Version 9.* Retrieved May 2011 from: http://www.supply-chain.org

Tan, C., & Chen, X. (2011). Induced intuitionistic fuzzy choquet integral operator for multicriteria. *International Journal of Intelligent Systems*, *26*(7), 659–686. doi:10.1002/int.20489

Taticchi, P., Tonelli, F., & Cagnazzo, L. (2010). Performance measurement and management: A literature review and a research agenda. *Measuring Business Excellence*, *14*(1), 4–18. doi:10.1108/13683041011027418

Tejas, F., & Srikanth, K. (2007). Supply chain metrics that measure up. *Supply and Demand Chain Executive*, *8*, 50–53.

ADDITIONAL READING

Adams, S. M., Sarkis, J., & Liles, D. (1995). The development of strategic performance metrics. *Engineering Management Journal*, *17*(1), 24–32. doi:10.1080/10429247.1995.11414823

Alnestig, P., & Segerstedt, A. (1997). *Produktkalkyler*. Stockholm: Industrilitteratur.

Atkinson, A. A., Waterhouse, J. H., & Wells, R. B. (1997). A stakeholder approach to strategic performance measurement. *Sloan Management Review*, (Spring): 25–37.

Ayers, J. B. (2001). *Handbook of Supply chain management*. Florida, USA: St. Lucie Press.

Ballou, R. H. (2004). *Business logistics/Supply chain management: planning, organizing, and controlling the Supply chain* (5th ed.). Upper Saddle River, N.J.: Pearson Prentice Hall.

Bancroft, J. A., Seip, H., & Sprengel, A. (1998). *Implementing SAP R/3*. USA: Manning Publications Co.

Beamon, B. M. (1998). Supply chain design and analysis: Models and methods. *International Journal of Production Economics*, *55*(3), 281–294. doi:10.1016/S0925-5273(98)00079-6

Beamon, B. M. (1999). Measuring Supply chain performance. *International Journal of Operations & Production Management*, *19*(3), 275–292. doi:10.1108/01443579910249714

Bowersox, D. J. & Closs, D. J. (1996). Logistical Management – The Integrated Supply Chain Process, McGraw-Hill Companies Inc, New York.

Bowersox, D. J., Closs, D. J., & Stank, T. P. (2000). Ten mega-trends that will revolutionise Supply chain logistics. *Journal of Business Logistics*, *21*(2), 1–16.

Byrne, P. J., & Heavey, C. (2006). The impact of information sharing and forecasting in capacitated industrial Supply chains: A case study. *International Journal of Production Economics*, *103*(1), 420–437. doi:10.1016/j.ijpe.2005.10.007

Camp, R. C. (1989). *Benchmarking: The Search for Industry Best Practices That Lead to Superior Performance*. New York: Quality Resources.

Carter, J. R., Ferrin, B. G., & Carter, C. R. (1995). The effect of less-than truckload rates on the purchase order lot size decision. *Transportation Journal*, *34*(3), 35–44.

Chandra, C., & Kumar, S. (2000). Supply chain management in theory and practice: A passing fad or a fundamental change. *Industrial Management & Data Systems*, *100*(3), 100–113. doi:10.1108/02635570010286168

Chen, J. (1997). Achieving maximum Supply chain efficiency. *IIE Solutions*, *29*, 30–35.

Chibba, A. (2007). *Measuring supply chain performance measures – prioritizing performance measures. Lic.* Luleå University of Technology. Luleå.

Christopher, M. (1998). *Logistics and Supply Chain Management: Strategies for Reducing Costs and Improving Services* (2nd ed.). London: Pitman.

Christopher, M., & Gattorna, J. (2005). Supply chain cost management and value-based pricing. *Industrial Marketing Management*, *34*(2), 115–121. doi:10.1016/j.indmarman.2004.07.016

Christopher, M., & Towill, D. (2000). Supply Chain migration from lean and functional to agile and customised, *Supply Chain Management. International Journal (Toronto, Ont.)*, *5*(4), 206–213.

Collin, J. (2003). *Selecting the right Supply Chain for a Customer in project business.* Diss. Tekniska högskolan Helsingfors, Helsingfors

Cooper, M. C., & Ellram, L. M. (1993). Characteristics of Supply Chain Management and the Implications for Purchasing and Logistics Strategy. *The International Journal of Logistics Management.*, *4*(2), 13–24. doi:10.1108/09574099310804957

Cooper, M. C., Lambert, D. M., & Pagh, J. D. (1997). Supply chain management: More than a new name for logistics. *International Journal of Logistics Management*, *8*(1), 1–14. doi:10.1108/09574099710805556

De Meyer, A., Nakane, J., Miller, J. G., & Ferdows, K. (1989). Flexibility: The next competitive battle the manufacturing futures survey. *Strategic Management Journal*, *10*(2), 135–144. doi:10.1002/smj.4250100204

Dornier, P. P. (1998). *Global operations and logistics: text and cases*. New York: Wiley.

Ellram, L. & Cooper, M. (1990) Supply chain management, partnerships and the shipper – third party relationship. *The International Journal of Logistics Management,* 1(2), PP. 1-10.

Ericsson 2007 Tillgänglig: Retrieved 1 December 2007 from: WWW.ERICSSON.COM

Fernie, J. (1995). International comparisons of Supply chain management in grocery retailing. *Service Industries Journal*, *15*(4), 134–147. doi:10.1080/02642069500000053

Fisher, M. L. (1997). What is the right Supply chain for your product? *Harvard Business Review*, *75*(2), 105–116.

Geanuracos, J., & Meiklejohn, I. (1993). *Performance Measurement*. London: The New Agenda, Business Intelligence.

Ghalayini, A. M., & Noble, S. (1996). The changing basis of performance measurement. *International Journal of Operations & Production Management*, *16*(8), 63–80. doi:10.1108/01443579610125787

Goonatilake, L. (1990). Inventory management in the manufacturing sector in developing countries. *Engineering Costs and Production Economics*, *19*(1), 19–24. doi:10.1016/0167-188X(90)90020-I

Gunasekaran, A., Patel, C., & McGaughey, R. E. (2004). A framework for supply chain performance measurement. *International Journal of Operations & Production Management*, *87*, 333–347.

Gunasekaran, A., Patel, C., & Tirtioglu, E. (2001). Performance measurement and metrics in a Supply chain environment. *International Journal of Operations & Production Management*, *21*(1/2), 71–87. doi:10.1108/01443570110358468

Harrison, A., & New, C. (2002). The role of coherent Supply chain strategy and performance management in achieving competitive advantage: An international survey. *The Journal of the Operational Research Society*, *53*(3), 263–271. doi:10.1057/palgrave.jors.2601193

Haug, P. (1985). A multiple-period, mixed-integer-programming model for multinational facility location. *Journal of Management*, *11*(3), 83–96. doi:10.1177/014920638501100307

Hernández, N. H., Keogh, J., & Martinez, F. F. (2006). *SAP R/3 Handbook*. USA: McGraw-Hill.

Higginson, J. K., & Alam, A. (1997). Supply chain management techniques in medium-to-small manufacturing firms. *International Journal of Logistics Management*, *8*(2), 19–32. doi:10.1108/09574099710805646

Holmberg, S. (2000). *Supply chain integration through performance measurement.* Diss. Lund University. Lund

Holmes, G. (1995). *Supply chain management – Europe`s new competitive battleground.* Research report The Economist Intelligence Unit, London.

Hoole, R. (2005). Five ways to simplify your Supply chain. *Supply Chain Management: An International Journal, 10*(1), 3–6. doi:10.1108/13598540510578306

Hoover, W. E., Eloranta, E., Holmström, J., & Huttuen, K. (2001). *Managing the Demand-Supply Chain-Value Innovations for Customer Satisfaction*. New York: Wiley.

Horvath, L. (2001). Collaboration: Key to value creation in Supply chain management. *Supply Chain Management: An International Journal, 6*(5), 205–207. doi:10.1108/EUM0000000006039

Hosang, J., & Bongju, J. (2005). Decentralised production-distribution planning system using collaborative agents in a Supply chain network. *International Journal of Advanced Manufacturing Technology, 25*(1/2), 167–173.

Huan, S. H., Sheoran, S. K., & Wang, G. (2004). A review and analysis of Supply chain operations reference (SCOR) model. *Supply Chain Management: An International Journal, 9*(1), 23–29. doi:10.1108/13598540410517557

Johnson, H. T., & Bröms, A. (2000). *Profit beyond Measure*. New York: Free.

Johnston, P. (1995). Supply chain management: Past, the present and the future. *Manufacturing Engineer*, 213–217.

Kaplan, R. S., & Norton, D. P. (1996). *The Balanced Scorecard*. Boston: Harvard Business School Press.

Keebler, J. S. (1999). Keeping score: measuring the business value of logistics in the Supply chain, Oak Brook, Council of Logistics Management, cop.

Krantz, S. (1996). "What is it?" Purchasing Today, October, p. 4.

Kumar, S., & Kropp, J. (2006). Studying the operational efficiencies of a multiproduct Supply chain using excel spreadsheet model. *Technovation, 26*(10), 1186–1200. doi:10.1016/j.technovation.2005.08.001

LaLonde, B. J., & Maters, J. M. (1994). Emerging Logistics Strategies: Blueprints for the next century. *International Journal of Physical Distribution & Logistics Management, 24*(7), 35–47. doi:10.1108/09600039410070975

Lambert, D. M. (1998). *Fundamentals of logistics management*. Chicago, London: Irwin/McGraw-Hill.

Lambert, D. M., Cooper, M. C & Pagh, J. D. (1998). Supply Chain Management: implementation issues and research opportunities. *The international Journal of Logistics Management, 9(*2), pp. 1-19.

Lambert, D. M., & Pohlen, R. L. (2001). Supply chain metrics. *The International Journal of Logistics Management, 12*(1), 1–19. doi:10.1108/09574090110806190

Love, P. E. D., Irani, Z., & Edwards, D. J. (2004). A seamless Supply chain management model for construction. *Supply Chain Management: An International Journal, 9*(1), 43–56. doi:10.1108/13598540410517575

Mattson, S.-A. (2002). *Logistik i försörjningskedjor*. Lund: Studentlitteratur.

Mentzer, J., Flint, D., & Hult, T. (2001). Logistics Service Quality as a Segment-Customized Process. *Journal of Marketing, 65*(4), 82–104. doi:10.1509/jmkg.65.4.82.18390

Mentzer, J. T., & Konrad, B. P. (1991). An efficiency/effectiveness approach to logistics performance analysis. *Journal of Business Logistics*, *12*(1), 33–62.

Neely, A. (1999). The performance measurement revolution: Why now and what next? *International Journal of Operations & Production Management*, *19*(2), 205–228. doi:10.1108/01443579910247437

Neely, A., Gregory, M., & Platts, K. (1995). Performance measurement system design: A literature review and research agenda. *International Journal of Operations & Production Management*, *15*(4), 80–116. doi:10.1108/01443579510083622

Pagh, J., & Cooper, M. (1998). Supply Chain Postponement and Speculation Strategies: How to Choose the Right Strategy. *Journal of Business Logistics*, *19*(2), 13–33.

Parasuraman, A., Betty, L., & Zeithaml, V. (1991). Understanding Customer Expectations of Service. *Sloan Management Review*, (Spring): 39–48.

Parker, C. (2000). Performance measurement. *Work Study*, *49*(2), 63–66. doi:10.1108/00438020010311197

Paulsson, U., Nilsson, C.-H., & Tryggestad, K. (2003). *Flödesekonomi: Supply chain management*. Lund: Studentlitteratur.

Persson, U. (1997). *A conceptual and empirical examination of the management concept Supply chain management. Lic.* Luleå University of Technology. Luleå.

Quinn, F. Jr. (1998). Building a world-class Supply chain. *Logistics Management Distribution Report*, *37*(6), 37–41.

Sachan, A., Sahay, B. S., & Sharma, D. (2005). Developing Indian grain Supply chain cost model: System dynamics approach. *International Journal of Productivity and Performance Management*, *54*(3), 187–205. doi:10.1108/17410400510584901

SAP 2007: [On-line] Tillgänglig: Retrieved 1 December 2007 from: WWW.SAP.COM

Schary, P. B. Skjøtt-Larsen, T. (2001). *Managing the Global Supply Chain* Copenhagen: Handelshøjskolens forlag.

Schmitz, J., & Platts, K. W. (2003). Roles of supplier performance measurement: Indication from a study in the automotive industry. *Management Decision*, *41*(8), 711–718. doi:10.1108/00251740310496224

Schonberger, R. J. (1996). *World Class Manufacturing: The Next Decade: Building Power, Strenghth, and Value*. New York: The Free Press.

Segerstedt, A. (1999). Escape from the unnessessary-some guidelines for production management. *Production Planning and Control*, *10*(2), 194–199. doi:10.1080/095372899233343

Shapiro, J. F. (2001). *Modeling the Supply Chain*, Duxbury, Pacific Grove.

Simchi-Levy, D., Kaminsky, P. & Simchi-Levy, E. (2000), *Designing and Managing the Supply Chain*, McGraw-Hill, USA

Sink, D. S., & Tuttle, T. C. (1989). *Planning and Measurement in Your Organization of the Future*. Nocross, GA: Industrial Engineering and Management Press.

Solvang, W. D. (2001). *Architecture for Supply Chain Analysis and methodology for quantitative measurement of Supply chain flexibility. Ph.D.* Trondheim: Norwegian University of Science and Technology.

Splendolini, M. J. (1992). *The Benchmarking Book,* Amacom, New York. Su, J. C. P, Chang, Y.-L. & Ferguson, M. (2006), Evaluation of postponement structures to accommodate mass customization. *Journal of Operations Management, 23,* 305–318.

Svensson, G. (2003). Holistic and cross-disciplinary deficiencies in the theory generation of Supply chain management. *Supply Chain Management: An International Journal, 8*(4), 303–316. doi:10.1108/13598540310490062

Tummala, V. M. R., Phillips, C. L. M., & Johnson, M. (2006). Assessing supply chain management success factors: A case study. *Supply Chain Management: An International Journal, 11*(2), 179–192. doi:10.1108/13598540610652573

Van Hoek, R. I. (1998). Measuring the unmesurable – measuring and improving the performance in the Supply chain. *Supply, Chain Management. International Journal (Toronto, Ont.), 3,* 187–192.

Van Hoek, R. I., Chatman, R., & Wilding, R. (2002). Managers in the Supply chain management, the critical dimension. *Supply Chain Management: An International Journal, 7*(3), 119–125. doi:10.1108/13598540210436568

Vitale, M. R., & Mavrinac, S. C. (1995). How effective is your performance measurement system? *Management Accounting, 77*(2), 43–47.

Walker, B., Bovet, D., & Martha, J. (2000). Unlocking the Supply chain to build competitive advantage. *International Journal of Logistics Management, 11*(2), 1–11. doi:10.1108/09574090010806119

Chapter 4
System Dynamics Simulation of a Supply Chain Intelligence Model

Debasri Dey
IGNOU, Kolkata, India

D. Sinha
IIFT, India

ABSTRACT

Supply chains today are, primarily, measured by Key Performance Indicators (KPIs) such as order-fulfillment, availability to the consumer (percent in-stock) and cost reduction, as well as financial KPIs such as return on investment (ROI), return on brand equity and inventory. These KPIs measure the performance of supply chain as a whole. A supply chain is a network of nodes. The performances of individual nodes are measured with KPIs such as production rate, shipment rate, inventory and the like. These metrics may indicate the performance but may not indicate the cause of such performance. For example, a node whose production rate is below the desired level may be because of poor supply of inputs of production by its supplier node.Thus mere identification of KPIs and their evaluation will not enable to identify the root cause of a problem in a supply chain. Therefore, we need a business intelligence framework that will satisfy the objectives, namely, identification of outcome of each node of the supply chain and its cause. The existing Supply Chain Intelligence (SCI) frameworks aims at identifying metrics that reflect the performance of individual nodes and the total supply chain, but fail to identify the cause of such outcomes. It implies that the linkages or association between the KPIs of individual nodes are required to be identified and defined. In this paper, contingency and systems approach has been used to identify the dimensions of the firm, its internal environment, the complement and the external environment. A system dynamics based approach has been used to identify the causality and resulting behavior of the supply chain. The paper proposes a SCI framework and a System dynamics Model that help in identifying the reasons for supply chin performance and lead to the actions required to be taken for improvement in performance of the supply chain.

DOI: 10.4018/978-1-4666-9795-9.ch004

INTRODUCTION

Supply chain is a network of nodes that is responsible for providing the time and place utilities to the end customer. The measurement of performance of a supply chain is important from the perspectives of the customer satisfaction, organisation's profitability and stakeholder relationship. According to Chan (2003), performance measurement describes the feedback or information on activities with respect to meeting customer expectations and strategic objectives. It reflects the need for improvement in areas with unsatisfactory performance. Thus efficiency and quality can be improved.

Various metrics of measurement and frameworks are in use for the purpose of measuring performance. Key Performance Indicators (KPIs) such as order-fulfillment, availability to the consumer (percent in-stock) and cost reduction, as well as financial KPIs such as return on investment (ROI), return on brand equity and inventory measure the performance of supply chain as a whole. The KPIs such as production rate, shipment rate, inventory levels, inventory accuracy and the like measure the performance. The magnitude of the measures may indicate the deviations from desired performance levels but may not identify the cause of such performance. For example, a node whose production rate is below the desired level may be because of poor supply of inputs of production by its supplier node. Several appropriate performance metrics and measures of SCM have been identified and discussed by Gunasekaran et al. (2001), and Gunasekaran, Patel, Ronald, and McGaughey (2004). These include order path, reduction of lead-time, quality and the way the information is presented flexibility and other such metrics.

The model suggested by Edward Frazelle (2001) consists of four types of indicators: quality, time, financial, and productivity. To complete the analysis, all indicator types need to be considered, and they need to work together.

Quality: These indicators are often the simplest to implement and measure. Typically, they tell how well a supply chain or its constituent is performing a specific activity—a common logistics indicator in this classification is accuracy— including order accuracy, inventory accuracy, picking accuracy, etc.

Time: These indicators focus on the time the supply chain or its nodes takes to complete specific activities. They show where saving time during specific activities can improve the overall supply chain performance.

Financial: These indicators help managers identify the supply chain cost drivers and help move toward a more efficiently managed supply chain.

Productivity: These indicators examine how well resources are used. For example, filling vehicles to their capacity, instead of sending out vehicles half-full, could reduce costs and improve efficiency. Keebler (1999) suggests the qualities to look for in indicators, i.e., it should be quantitative, visible, uses economies of effort, facilitates trust, measures what is required and other similar properties.

However, focusing on only one type of indicator may actually have a negative impact on product availability. For instance, a decision to send vehicles on a distribution run only when they are filled to capacity could cause stock-outs at the next level down unless inventory policies are adjusted to compensate for reducing ad hoc shipments. It is very important to view these indicators holistically—to make sure they are harmonized and not working against each other—and to identify the tradeoffs required to strategically improve overall supply chain performance (USAID, 2010).

The SCOR-model specifies five performance metrics in two categories: customer- facing metrics that include reliability, responsiveness, and flexibility, and the internal-facing metrics that include cost and assets. Each of the performance attributes contains a set of metrics. As an example, the customer facing attribute named Flexibility is measured by the three metrics: Upside Supply Chain Flexibility, Upside

Supply Chain Adaptability, and Downside Supply Chain Adaptability. Upside refers to an increase in delivered quantities in the supply chain while downside refers to a decrease in deliveries. Flexibility is defined as the number of days required achieving an unplanned sustainable 20% increase in delivered quantities whereas adaptability is defined as the maximum sustainable percentage increase in delivered quantities that can be achieved in 30 days (SCOR, 2008). Each metric can be broken down to the lower levels (levels 2and3) and aggregated from the lower levels up to the top level (level1). The customer facing metric Perfect Order can work as an example.

Case studies are applied either to extend the SCOR-model to the service sector or to emphasize a specific decision area (Cavalieri et al. 2007; Naesens et al. 2007, 2009; Soffer & Wand 2007; Xu 2007). For example, Cavalieri et al. (2007) followed three Italian companies during their implementation of the SCOR-model in after-sales services. The SCOR-model was extended to both external processes (that includes customer care, marketing and pricing, network certification, service and order fulfillment) and internal processes (such as maintenance and warehousing). Naesens et al. (2007) used the SCOR-model to measure the strategic fit of inventory pooling in a horizontal collaborated supply chain. Additionally, Soffer and Wand (2007) focused on a single decision area of the SCOR-model, delivery decisions. By analyzing delivery activities for the make- to-order manufacturing strategy, they found that customer order, payment schedule, product, shipping documentation, product receiving and verification were key performance metrics (Erkan&Bac, 2011).

SCOR is relatively weak on addressing the issues in problem domain and does not take into account the limitations of methodology user, i.e., the modeler have a different mental construct formed by their beliefs, value sets etc. The modeling approach should account for the bias of the modeler in the process of model development. These two dimensions represent the social dimension of Leavitt's (1972) socio-technical model. Thus it can be said that SCOR is strong on the technical dimensions and relatively weak on the social dimensions. Leavitt's organizational change model emphasizes the importance of the social dimension for effective organizational change, thus SCOR needs to be strengthened on this dimension (Kasi, 2005).

The balanced scorecard (BSC) can provide a useful guidance for the practical managers in evaluation and measuring of SCM in a balanced way and proposes a balanced performance measurement system to map and analyze supply chains. While suggesting balanced scorecard, different SCM performance metrics have been reviewed and distributed into four perspectives. This helps managers to evaluate SCM performance in a much-balanced way from all angles of business.

A strategy is a set of assumptions about cause-and-effect. If cause-and-effect relationships are not adequately reflected in the BSC, it will not translate and communicate firm's vision and strategy. These cause-and-effect relationships can involve several or all four of the perspectives in the BSC framework. For example, flexibility of service systems to meet particular customer needs (internal business operations perspective) will be more likely to meet customer expectations (customer perspective). Higher level of customer expectations will lead companies to supply more innovative products and services (learning and growth perspective). This in turn will increase the market share and profitability (financial perspective) (Bhagwat & Sharma, 2007).

The above approaches and frameworks lack in the ability to provide for systemic approach that represents the holistic perspective.

Simulation models are dynamic representations of supply chains executed step-wise within a computer program. Transactions, events and time can be readily included in the simulation model to assess how the performance changes over time. A simulation model can easily capture the dynamic nature of supply chains. A simulation can also better handle the complexity involved in supply chains. Broadly speaking, two kinds of simulation models exist: discrete-event simulation models and continuous simulation models. Simulation approaches have been very successful in modeling and improving supply chains in industry even though such approaches have been out of favor with the general academic community for some time.

At every stage of Supply Chain, the Key Performance Indicators (KPI), which acts as performance measures, is evaluated separately to evaluate the performance of Supply Chain. This approach ensures that the internal control of the firm is well-regulated and corrective measures taken for any deviations. However, it is known that controlling the internal environment of a firm does not always ensure the corporate mission. This is because a firm's internal environment is surrounded by its complement, which comprises the business partners and customer, and above it, the firm is influenced by the external environment. The internal environment could be influenced as well as controlled, but the complement can only be influenced but cannot be controlled. The external environment is beyond influence and control (Coyle, 1977).Some approaches such as that prescribed by Bolstroff and Rosenbaum (2003) suggest preparing business context summary to understand the problem situation. A systems dynamics approach will enable a holistic and integrated way to incorporate the impact of external influence, the outcome of internal decisions and actions. It will also enable each modeler to have a different mental construct formed by their beliefs, value sets etc. The KPIs of the supply chain can be easily linked with the nodal performances identifying the root cause of a problem.

RESEARCH METHODOLOGY

In this paper a literature review has been carried out to identify the existing frameworks, the gaps in these frameworks and identification of a plausible approach to meet this gap. The relevance of such requirement has been explained with help of a case study. The proposed framework has been developed and simulated using the principles of Systems Dynamics.

LITERATURE REVIEW

Supply Chain

There are three distinct aspects regarding a supply chain of a firm. Firstly, supply chain is formed by many entities, each entity being a separate complex system. Secondly, the activities which take place implement in these entities should be taken into account, as these causes the flow of goods and customer service in the supply chain. Thirdly, supply chain must be treated as a whole system, the relationships through the entities should be considered, such as interaction relationship, dependent relationship and the supply and demand balance of finished-product and service (Ran, 2009) and customer. Supply chain activities transform natural resources, raw material and components into a finished product that is delivered to the end customer. A Simple supply chain shown in the figure 1.

Figure 1. Simple supply chain

Supply Chain Integration

Supply chain integration has two basic points: First, flow integration, the main aspect is the convergence of various functions to optimize the efficiency of integration; second, integrated sections, the main sections are the functions of the efficiency of the integration itself, usually happed as a whole system (Yang 2007). Supply Chain Management stakeholders includes suppliers, manufacturers, channel intermediaries and customers integration. Stevenson (1998) proposed four stages of integration: the benchmark organizations, the functions integration, internal integration, and external integration. In the study, the main consideration was in the optimization of resources, capacity, based on the lowest cost and fastest speed in the production of the best products, quick response to user needs in order to improve the responsiveness and efficiency of enterprises. External integration is the key to integration, the internal and external supply chain suppliers and users together to form an integrated supply chain network. Supply chain integration through information technology and management tools to achieve the supply chain of manufacturing processes and logistics seamlessly (Anderson & Katz, 1998).

The dynamics of the supply chain may be understood from the contingency, systems and the hybrid approach.

Contingency Approach

The study by (Stonebraker & Afifi, 2004) proposes that as supply chain technology evolves, it creates greater differentiation (decentralization and specialization), which, for organizational success, necessitates greater levels of integration (formality and collaboration). Thus, this study applies the classically defined relationship between internal differentiation and integration to external supply chain activities.

System Approach

Systems theory serves to identify the context in which organizations operate, thus enabling managers to understand the environment and how the parts, or subsystems, of the organization are interrelated (Sweeney, 1999).

Hybrid Approach

Hybrid systems use the advantages of its subsystems to reach a superior result to one system alone.

Whatever is the approach, at every stage of supply chain, the Key Performance Indicators (KPI), which acts as performance measures, are evaluated separately to evaluate the performance of Supply Chain. This approach ensures that the internal control of the firm is well-regulated and corrective measures taken for any deviations. However, the impact of the firm's complement and environment can be measured using derived supply chain KPIs. These derived KPIs are functions of KPIs of the connected nodes and not the individual nodes treated as silos of activities.

System Dynamics

System Dynamics, developed by (Forrester, 1961, 1968) identifies cause-effect relationships and structures them in a feedback control framework to understand the dynamic behavior of the systems. The approach professes causality doctrine associated with determinism. System Dynamics is a methodology that has ability to capture and model dynamic complexity of complex systems. Dynamic complexity refers to state where cause and effect are subtle and where effects over time interventions are not obvious (Senge, 1990). According to (Coyle,1977) a System Dynamics study aims at the following objectives: Explaining the systems behavior in terms of structure and policies, and suggesting changes in structure, policies or both, which will lead to an improvement in behavior .In practical system dynamics work, and as a conceptual framework, it is useful to look at systems in the light of how much certain knowledge about their workings it is possible to acquire, at any rate in principle, and how far it is possible to exert actual and direct control over what goes on. He proposed three sectors in the structure of a firm. These are the internal environment or the controller (can be controlled and influenced), the complement (cannot be controlled but may be influenced) and the environment (cannot be controlled or influenced). A firm but cannot control its complement.

HONDA MOTOR CO. LTD.: A CASE STUDY

Crisis Problems: Honda faced serious problem when they failed to provide dynamic manufacturing and assembling facilities due to delay in supplies of spare parts, interiors, etc. from other countries as well as from local venders.

Non-Crisis Problem: Natural Disaster Affects Demand in the ASEAN Market, such as the shortage of oil affected to soar gasoline prices. It has caused the price of almost the businesses -which need to transportations and vehicles- go up. Thus, the result was the automobile market to shrink. Similarly floods in Thailand disrupted production and supply chains. Honda Motor and many other companies have been forced to suspend production in central Thailand because of the floods. Most of the factory closures are due to disruptions to local supply chains.

Opportunity Problems: Honda trying to get significant organizational goal in appropriate actions like they sensed that they can get maximum profit and sales by opening their dealerships at strategic locations to maximize their customers.

Intelligence Measures: Honda is evaluating a series of "counter-measures" should supply chain issues start to impact on UK delivery following earthquake and Tsunami. The automaker has set up two emergency committees in Japan and the English town of Swindon to monitor components supply issues following these events that have devastated the country. The manufacturer has 113 suppliers in the north Japan and has spoken to around half of them to asses any delivery issues. Of those 50 or so, eight have reported serious problems.

Figure 2. A supply-chain-intelligence framework based on system dynamics principles

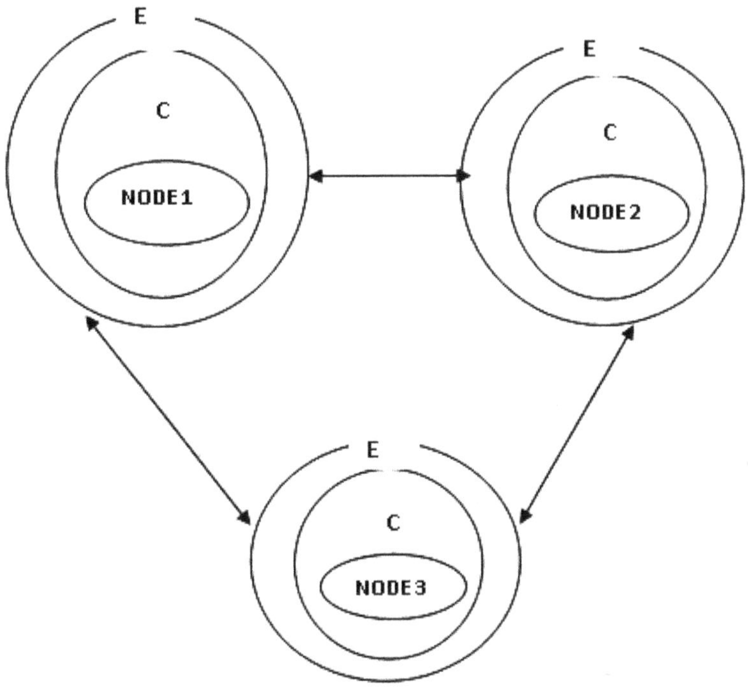

PROPOSED WORK

The interaction of the nodes governs the supply chain dynamics. The proposed frame work can be represented as shown in Figure 2.

The framework suggests that performance of one node is dependent on the other connected nodes. In the figure above a three node supply chain has been depicted. These three nodes are designated as NODE1, NODE2 and NODE3. Each of these nodes has its own internal environment, complement (C) and external environment (E). Any change in the environment, internal or external will not only impact the respective node's performance but also the node it connects. For example, stoppage or reduction in production in Thailand's car manufacturing unit due to flood is expected to impact Honda factory at Tapukara plant in India responsible for supply parts to Thailand. This in turn is expected to hinder supply of finished goods to the global market. Hence measuring performance on the KPIs of individual nodes may not make sense as they have to be associated with key variables and KPIs of other connected nodes. Data captured at each node should not be restricted only to the transactions in the supply chain but be extended to its complement and environment. The intelligence should be derived from the multi-dimensional causality framework embedded in the system.

THE SCI FRAMEWORK

Step 1: Define the objectives of the supply chain.
Step 2: Define the Key Performance Indicators (KPI) of Supply Chain.

Step 3: Identify the Nodes and its Key Performance Indicators.
Step 4: Establish the linkages of individual KPI with the KPI or key variables of other nodes.
Step 5: Evaluate the KPI of Supply Chain.
Step 6: Infer the results of evaluation and identify the node that needs correction.
Step 7: Evaluate the factors affecting the KPI of the node requiring correction.
Step 8: Simulate corrective action.
Step 9: Take corrective action.

Illustration

In a simple supply chain of three nodes as illustrated in figure 1, the supplier, the firm and the customer may be denoted as Node1, Node2 and Node3 respectively. The nature and outcome of the linkage between Node 1 and Node 2 is likely to impact the customer (Node3). The KPIs of Node1 may include Maximum Shipment Rate (MSHN1) and Actual Shipment Rate (ASR). Maximum Shipment Rate (MSHN1) of Node1 indicates the maximum achievable level of output of Node1 while ASR indicates the actual level. ASR is dependent on

Customer Order Rate (CORN2) from Node2. ASR is the minimum of MSHN1 and CORN2.

$$ASR = \min (MSHN1, CORN2) \quad (1)$$

If this indicator is considered, the performance of the firm against its maximum capacity (MSHN1) can be measured by plotting the MSHN1 and ASR over the period of time. If ASR continually remains below the MSHN1, one may infer Node1 as poor performer. On the other hand ASR may be equivalent to the order received (CORN2) from Node2.

Hence Node1 needs correction only if MSHN1 is less than CORN2, else Node2 needs correction. Thus the desired supply chain KPI (may be defined as "SC-KPI-1") is the difference between MSHN1 and CORN2.

$$SC\text{-}KPI\text{-}1 = MSHN1 - CORN2 \quad (2)$$

That is, a derived KPI is prescribed in lieu of KPIs of individual nodes.

The other such KPI of Node1 could be its "Inventory" (may be denoted as IN1); it is associated with its Shipment Rate (ASR). IN1 should, ideally, not exceed ASR, i.e., Shipment Rate of Node1 or not more than the desired level as decided by the supply chain manager. Shipment Rate of Node1 should be ideally equal to the CORN2 of Node2. Hence the Inventory (IN1) of Node1 is associated with CORN2, the KPI of Node2.The integrated supply chain indicator in this case may be defined as the inventory coverage (may be denoted as SC-KPI-2) which is the ratio of IN1 to CORN2.

$$SC\text{-}KPI\text{-}2 = IN1 / CORN2$$

It represents the number of weeks of inventory Node1 has as its stock. IfSC-KPI-2 is less than 1 (one), it will imply that Node1 needs correction to enhance its production and shipment rates. If it is greater than 1 (one) but less than desired level of inventory coverage, the Node1 can be considered as a good

Table 1. SCI framework illustration

SL.NO.	NODE1 KPI	ASSOCIATED - NODE2 KPI	SUPPLY CHAIN KPI (SC-KPI)	INFERENCE
1.	Maximum Shipment Rate (MSHN1)	Customer Order Rate (CORN2)	SC-KPI-1 (MSHN1 minus CORN2)	
			IF SC-KPI-1 is +VE (i.e., MSHN1 > CORN2)	NODE2 NEEDS CORRECTION
			ELSE	NODE1 NEEDS CORRECTION
2.	Inventory (IN1)	CORN2	SC-KPI-2 = (IN1 / CORN2)	
			If SC-KPI-2 > 1 ELSE	NODE1 NEEDS CORRECTION IF SC-KPI-2 EXCEEDS DESIRED VALUE AS PER INVENTORY POLICY. The firm needs to reduce its inventory level to the desired level.
				NODE1 NEEDSCORRECTION to increase its inventory level to meet the order rate from Node2.

performer. On the other hand if the value exceeds desired level of inventory coverage the Node1 may relook into its production and shipment activities, so as to reduce its inventory level below the desired level of inventory coverage.

The summary of two integrated KPIs discussed above is illustrated in Table 1.

SYSTEM DYNAMICS MODEL

Figure 3 shows the System Dynamics model capturing two nodes.

The state of level variable Backlog indicates the combined performance of Node1 and Node2. If the level increases then it can be concluded that Node1 is unable to meet order received from Node2. The order received is dependent on gap between the desired and actual capacity of Node 2. The actual capacity is a result of production rate of Node2 and its breakdown resulting in no production. The desired production or shipment level of Node1 is dependent on the average order and the Backlog correction.

Model Simulation

The above mentioned model has been simulated using Vensim software. Figure 4 shows the outcome of simulation run for a period of 30 days. The unit of measurements has been based on per day rates and simulated using Euler's integration.

Case I: Test 1 graph shows the graphical overview of production in plant-1 (Node1) without breakdown in node-2 (or plant-2).

Figure 3. SD model of two-node supply chain

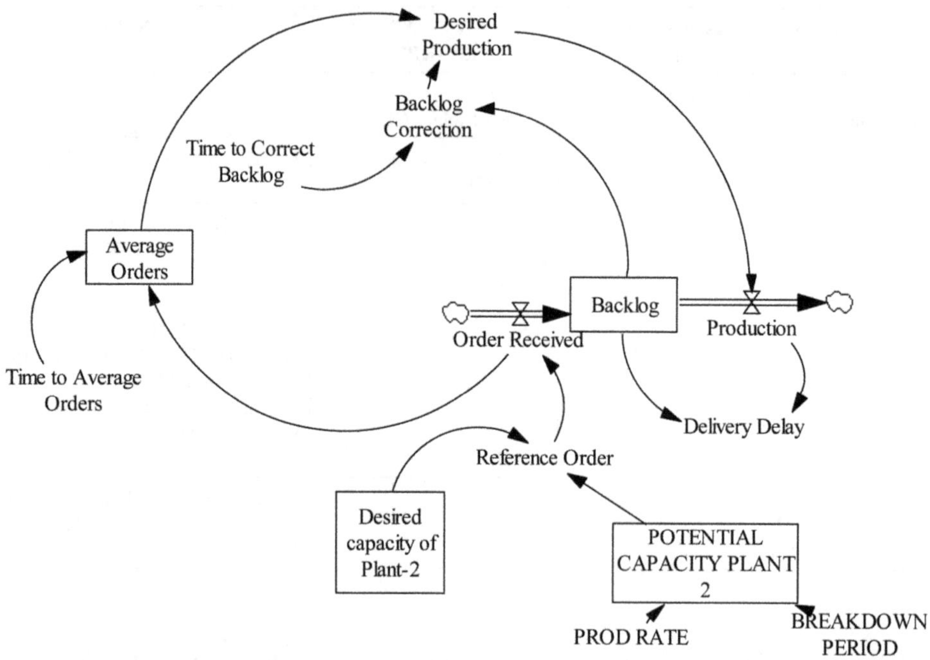

Figure 4. Red line (Test 1- Upper line): Model simulated for production from Node-1without breakdown in Node-2. Blue line (Test 2 – lower line): Model simulated for production from Node-1with a breakdown for a period of 15 days in Node-2.

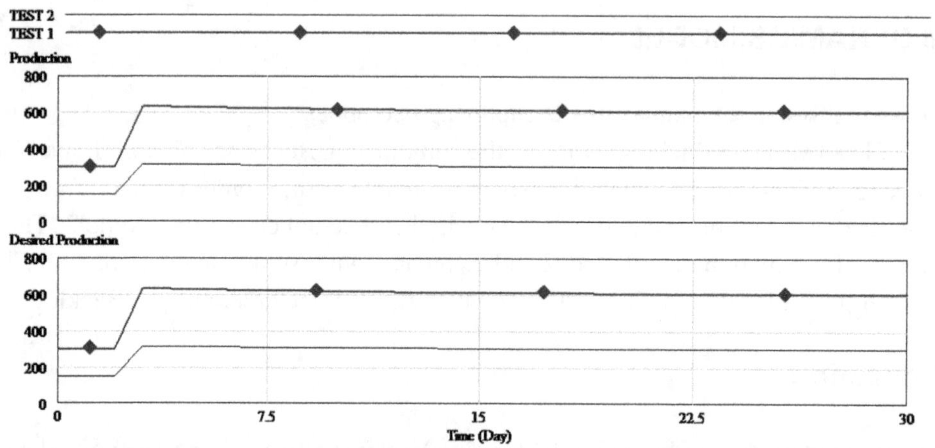

Case II: Test 2 graphs show the graphical overview of production in plant-1 with breakdown in node-2 or plant-2 for a period of 15 days. The production rate lowered at Node-1and continued over the period even though the break down was only for 15 days.

Thus it can be inferred that the poor performance of Node1 is due breakdown at Node2 producing finished goods for customer. Node1 is dependent on Node2 and the standalone KPIs of Node1 i.e., say actual shipment rate does not reflect the true position. The breakdown of Node2 may not due its internal causes but owing to external factors say floods. This can be validated with case of Honda motors where floods in car manufacturing plant in Thailand caused the production to drop in the component manufacturing plant in India.

CONCLUSION

This paper recommends contingency and systems approach to identify the dimensions of the firm, and the environmental variables.This paper proposes integrated KPIs (Key Performance Indicators) in lieu of standalone node based KPIs or KPIs of supply chain such as order-fill rate that not capture the root cause of a problem. It concludes that firm needs to account for its internal environment, its complement as well as its environment. The paper suggests that the intelligence should be derived from the multi-dimensional causality embedded in the system. It proposes a System Dynamics based model that captures the dynamics arising out of the firm's environments (both internal and external), defines the nodes' KPIs, its linkages with other nodes and establishes the supply-chain KPIs. The proposed SCI framework enables a decision maker to identify the root cause of a problem and simulate its proposed action to see its effect.

The SCI model simulated in this paper is only illustrative, but can be made exhaustive as further scope of research.

REFERENCES

Anderson, M. G., & Katz, P. B. (1998). Strategic sourcing. *The International Journal of Logistics Management*, *9*(1), 1–13. doi:10.1108/09574099810805708

Bhagwat, R., & Sharma, M. K. (2007, August). Performance measurement of supply chain management: A balanced scorecard approach. *Computers & Industrial Engineering*, *53*(1), 43–62. doi:10.1016/j.cie.2007.04.001

Bolstroff, P. & Rosenbaum, R. (2003). *Supply Chain Excellence*. American Management Association.

Cavalieri, S., Gaiardelli, P., & Ierace, S. (2007). Aligning strategic profiles with operational metrics in after-sales service. *International Journal of Productivity and Performance Management*, *56*(5–6), 436–455.

Chakravarty, S., Gulen, H., & Mayhew, S. (2005). Informed Trading in Stock and Option Markets. *The Journal of Finance*, *59*(3), 1235–1258. DOI: 10.1111/j.1540-6261.2004.00661.x

Chan, F. T. S. (2003, May). Performance Measurement in a Supply Chain. *International Journal of Advanced Manufacturing Technology*, *21*(7), 534–548. doi:10.1007/s001700300063

Chopra, S., & Meindl, P. (2001). *Supply chain management: Strategy, planning and operation*. Pearson Education Asia.

Christiaanse, E., & Kumar, K. (2000). ICT – enabled coordination of dynamic supply webs. *International Journal of Physical Distribution & Logistics*, *30*(3/4), 268–285. doi:10.1108/09600030010326019

Coyle, R. G. (1977). *Management System Dynamics*. London: John Wiley and Sons.

Erkan, T. E., & Bac, U. (2011). Supply chain performance measurement: A case study about applicability of SCOR model in a manufacturing industry firm. *International Journal of Business and Management Studies, 3*(1).

Forrester, J. W. (1961). *Industrial, Dynamics*. Cambridge, MA: MIT Press.

Forrester, J. W. (1968). *Principals of Systems*. Cambridge, MA: Wright Allen Press.

Frazelle. (2001). *Supply Chain Strategy*. Academic Press.

Gunasekaran, A., Patel, C., Ronald, E., & McGaughey, R. (2004). A framework for supply chain performance measurement. *International Journal of Production Economics, 87*(3), 333–348. doi:10.1016/j.ijpe.2003.08.003

Gunasekaran, A., Patel, C., & Tirtiroglu, E. (2001). Performance measures and metrics in a supply chain environment. *International Journal of Production and Operations Management, 21*(1/2), 71–87. doi:10.1108/01443570110358468

Haydock, P. M. (2003). Supply Chain Intelligence. *ASCET, 5*, 15–21.

Kasi, V. (2005). Systemic Assessment of SCOR for Modeling Supply Chains. *Proceedings of the 38th Hawaii International Conference on System Sciences*. doi:10.1109/HICSS.2005.574

Keebler, J. S., Manradt, K. B., Durtsche, D. A., & Ledyard, D. M. (n.d.). *Keeping score: Measuring the business value of logistics in the supply chain*. Accession No-00770539.

Leavitt, H. (1972). *The Volatile Organization: Everything Triggers Everything Else*. Chicago: University of Chicago Press.

Lummus, R. R., & Vokurka, R. J. (1999). Defining supply chain management: A historical perspective and practical guidelines. *Industrial Management & Data Systems, 99*(1), 11–17. doi:10.1108/02635579910243851

Muriati, Norleyza, Salha, & Zuraidah. (2009). A Framework for Analyzing e-Supply Chains. *European Journal of Scientific Research, 25*(4).

Naesens, K., Gelders, L., & Pintelon, L. (2007). A swift response tool for measuring the strategic fit for resource pooling: A case study. *Management Decision, 45*(3), 434–449. doi:10.1108/00251740710745061

Naesens, K., Gelders, L., & Pintelon, L. (2009). A swift response framework for measuring the strategic fit for a horizontal collaborative initiative. *International Journal of Production Economics, 121*(2), 550–561. doi:10.1016/j.ijpe.2007.04.004

Raisinghani, M., & Meade, L. L. (2005). Strategic decisions in supply-chain intelligence using knowledge management: An analytic-network-process framework. *Supply Chain Management: An International Journal, 10*(2), 114–121. doi:10.1108/13598540510589188

Ran, T. (2009). *Internal Logistics as a Part of Supply Chain, Case: Nokia-China, Dongguang Branch*. (Bachelor's Thesis). International Business Program, Lahti University of Applied Sciences, Faculty of Business Studies.

Reference. (n.d.). Retrieved from https://supply-chain.org/f/SCOR%2090%20Overview%20Booklet.pdf

Senge, P. M. (1990). *The fifth discipline: The art & Practice of the Learning organization*. Double Day Currency.

Soffer, P., & Wand, Y. (2007). Goal-driven multi-process analysis. *Journal of the Association for Information Systems*, *8*(3), 175–204.

Stevenson, G. (1998). "Integrating the Supply Chain", International Journal of Physical Distribution and Materials Management "Supply chain performance measurement: A literature review". *International Journal of Production Research*, *48*(17), 2010.

Sweeney, D.W., & Stulen, R.H. (1999). Quantum Electronics. *IEEE Journal of Sandia Nat. Labs., 35*(5), 694 – 699.

Stonebraker & Afifi. (2004). Toward a contingency theory of supply chains. *Management Decision*, *42*(5), 1131–1144.

USAID. (2010). *Agency Sustainability Plan*. U.S. Agency for International Development Earth Day Statement.

Van der Velde, L. N. J., & Meijer, B. R. (2003). *A System Approach to supply chain design with a multinational for colorant and coatings*. Retrieved September 20, 2006, from http://www.ifm.eng.cam.ac.uk/mcn/pdf_files/part6_5.pdf

Xu, H. (2010). Agile manufacturing in complex supply networks. In L. Wang & S. C. Lenny Koh (Eds.), *Enterprise networks and logistics for agile manufacturing* (pp. 39–66). London: Springer. doi:10.1007/978-1-84996-244-5_3

Z, Yang,., Hu, Z., Liu, J., & Zhao, P. (2007). The Relationships between Manufacturer Justice and Distributor Trust: A Case Study of Channel Relations in the Chinese Cell Phone Industry [In Chinese]. *Journal of Marine Science*, *3*(1), 91–105.

Chapter 5
Organizational Performance through Dairy Supply Chain Management Practices:
A Winning Approach

Rajeev Kumar
Banaras Hindu University (BHU), India

ABSTRACT

Organizational implementing supply chain management (SCM) has obtained improved performance. Cost savings, increased revenues and the reduction of defects in products are some of the main advantages of introducing dairy supply chain management. These are also mentioned as long-term goals of the supply chain. Business profitability is closely associated with market and business shares. Based on the long-term goals of the SCM, the organizational performance measures are identified as financial and market performance and customer satisfaction. In the context of SCM, the financial and market performance factor is operational zed in terms of market share, return of total assets, annuals sales growth (Tan et al, 1999).The research paper attempts to find out the relationship between the dairy supply chain management (DSCM) practices and organizational performance. Various practices of DSCM through extensive literature review is taken into account i.e. Information and Communication Technology Practices, Supplier Relationship Practices, Supply Chain Manufacturing Practices, Inventory management system, Warehousing Management System, Transportation Management System, Customer Relationship Management for establishing the relationship with organizational performance. The aim of this study is to recommend these findings to companies which are still at the infancy stage when it comes to dairy supply chain management and integration with customers and suppliers.

1. INTRODUCTION

1.1. Dairy Supply Chain Management and Organizational Performance

The supply chain is not a business function, it is a network of companies and supply chain management is the implementation of cross-functional relationship with key customers and suppliers in the network. It is a new business model necessary for an organization's success and every function needs to be involved.

DOI: 10.4018/978-1-4666-9795-9.ch005

In today's environment, there is an added pressure to be more socially and environmentally responsible and there are risks which need to be mitigated and managed. Then, there is the complexity created by ever increasing customer requirements and expectations, globalization, the pressure on cost and the availability and access to resources. On the top of management is expected to improve profitability, increase revenue growth and capture and protect larger market share. In order to succeed, management must recognize that the ultimate success of an organization depends on the ability to integrate the company's network of business relationships in a mutually beneficial way. The management of the network of relationships is supply chain management. Successful supply chain management requires cross-functional integration within the firm and across the network of firms that comprise the supply chain. It focuses on the improvements in performance that result from better management of key relationships. By understanding the supply chain management processes and how they should be implemented, management will better understand the value of more integrated supply chains and how the integration will lead to increased shareholder value and a sustainable competitive advantage. Organizational performance comprises the actual output or results of an organization as measured against its anticipated outputs or goals and objectives. According to Richard et al. (2009) organizational performance encompasses there specific areas of firm outcomes: (a) financial performance (profits, return on assets, return on investment, etc.); (b) product market performance (sales, market share, etc.) and (c) shareholder return (total shareholder return, economic value added, etc). The term organizational performance is broader. Specialists in many fields are concerned with organizational performance including strategic planners, operations, and finance, legal and organizational development. In recent years, many organizations have attempted to manage organizational performance using the balanced scorecard methodology where performance is tracked and measured in multiple dimensions such as:

- Financial performance (e.g. shareholder return).
- Customer service.
- Social responsibility (e.g. corporate citizenship, community outreach).
- Employee stewardship.

This paper mainly encompasses those parameters of organizational performance which are related with DSCM practices. The study included various performance parameters for improving organizational performance such as Procurement cost, Manufacturing cost, Inventory carrying cost, Distribution cost, Transportation cost, Order fulfillment cycle time, Inventory turnover (no. of times), One time delivery (%), Frequency of stock out, Product rejection rate, Backorder, Cash-to cash cycle time, Sales growth, profit margin, Return on investment (ROI), lead time, delivery speed, delivery performance etc.

1.2. Indian Dairy Scenario

Dairying in India has come a long way. From being written off as a basket case a few decades back, India has emerged today as the largest milk producer in the world with an annual production of 130 million tones (2012). More than 70 percent of our milk producers are small marginal farmers, the animal are low yielding, non-descript ones, and, land holding are small. The future of India dairy industry will can be only build on quality and quality alone. The Indian dairy industry must have the latest modern technology for milk processing and product manufacturer. One must define the standards necessary to achieve and maintain world class quality. Indian dairying must address itself to issues of productivity, efficiency and response to consumer demands. Side by side, it must upgrade and diversify the product mix into value added products to meet the challenges. So long the industry has been supply driven by demand but in future it should be driven by demand as disposable income rises. The white revolution created opportunities for the dairy farmers to improve employment and quality of life. The future of Indian dairy industry would also to a large extent depends on our ability to source new markets.one have the advantage of producing low cost milk and therefore the challenge is to lower processing and marketing costs without compromising on quality. Indian dairy industry need to improve product quality to compete with the best in the world.

1.3. Supply Chain Management and Indian Dairy Industry

The dairy industry in India has been on a steady path of progression since Indian independence. It has grown from producing 17 million tonnes of milk in 1951 to producing 127.3 million tonnes in 2012. Today, India is one of the largest milk producing countries in the world. This solid progress is primarily attributable to structural changes in the Indian dairy industry brought about by the advent of dairy cooperatives. The Indian dairy industry reported a market size of USD 48.5 billion in FY2011. With a Compound Annual Growth Rate (CAGR) of 16 percent, it is anticipated to reach USD 118 billion in 2017. On the back of a rise in disposable income, coupled with strong demand for dairy products, the Indian dairy industry is all set to experience high growth rates in the next five years. The consumption pattern of dairy products in India is quite unique as compared to some of the western countries. Consumption is primarily skewed towards traditional products; however, westernized products are gradually gaining momentum in the urban areas. Interestingly, buffalo milk accounts for the largest share of the total milk produced in the country. Since the pricing of milk is based on the fat content, buffalo milk offers higher profit margins as compared to cow milk as it contains higher fat. The Indian dairy sector is characterized by high fragmentation. It is dominated by the unorganized sector comprising of 70 million rural households. The per capita availability of milk in India stands at 289.4 grams per day. Backed by strong domestic demand, the per capita availability of milk is anticipated to reach 336 grams per day in FY 2017. Currently, the Indian dairy market is growing at an annual rate of 7 per cent. Despite the increase in production, a demand supply gap has become imminent in the dairy industry due to the changing consumption habits, dynamic demographic patterns, and the rapid urbanization of rural India. This means that there is an urgent need for the growth rate of the dairy sector to match the rapidly growing Indian economy. Despite being the one of the largest milk producing countries in the world, India accounts for a negligible share in the worldwide dairy trade. The ever increasing rise in domestic demand for dairy products and a large demand-supply gap could lead India to be a net importer of dairy products in the near future.

2. CRITICAL LITERATURE REVIEW ON DAIRY SUPPLY CHAIN MANAGEMENT (DSCM) PRACTICES

Some of the prominent research studies are highlighted in this research paper. The various research papers are further classified as per their focus area (table 1).

Table 1.

Authors	Study Title	Key Terms	Methodology Applied	Key findings
Gyan Prakash (2011)	Poor's Supply Chain: Indian Public Distribution System Revisited	Basic Dairy Supply Chain Management	The methodology adopted is a mix of literature review, document analysis such as government gazettes, interview with policy makers in government, officials of agency responsible for public distribution system.	Indian agro industry is fraught with many difficulties such as inefficiency, deterioration of perishable food items, unsatisfactory quality of commodities, mismatch of demand and supply, long waiting times, poor service delivery.
Karthikeyan and Teshome (2010)	Dairy supply chain management: A case of Ada'a Dairy Cooperative in Ethiopia.	Basic Dairy Supply Chain Management	Multiple Linear Regression	Multiple regression analysis indicates that out of seventeen variables taken, distance/proximity, transport services, quality of milk, distribution channel system, information system, milk collection centers, volume of milk production, management commitment, warehousing were found most important factors in enhancing the dairy supply chain effectively.
Subbaiah et al (2009)	Supply Chain Management in a Dairy Industry – A Case Study	Basic Dairy Supply Chain Management	Case Study	This paper revealed that how supply chain management is very important for the dairy industry which involves or comprises five main activities viz Purchase of materials from suppliers, transportation of materials from suppliers to facilities, production of goods at facilities, transportation of goods from facilitates to ware houses and transportation of goods from ware houses to customers.
Vilella et al (2008)	Strategic management in the Argentina dairy supply chain	Basic Dairy Supply Chain Management	This study on dairy industry based on primary and secondary information sources, personal interviews with experts in the sector (producers, industrialist, chamber representatives etc.).	The result suggest that product development, research and development, communication decision, training decision, institutional environmental coordination and adequacy decision are the key for the competitive advantage of the sector from the point of view of the coordination and management of the Argentina dairy chain.
Yigrem et al (2008)	Dairy production, processing and marketing systems of Shashemene–Dilla area, South Ethiopia.	Basic Dairy Supply Chain Management	Rapid market appraisal technique was employed to characterized dairy marketing system in the study area	This study show that the development of dairy production and marketing could be achieved with the contribution and integration of different stakeholders in a sustainable way, and also suggested following factors i.e. Extension services, processing, Marketing, Pricing, Veterinary services to be major constraints in the development of dairy supply chain
Hazen and Byrd (2012)	Toward creating competitive advantage with logistics information technology.	Information and Communication Technology (ICT)	Regression	The study defines the important role and relationship between logistics information technology (LIT) adoption and performance measures in terms of efficiency, effectiveness and resiliency. The study found that the implementation of LIT innovation generally produces positive performance outcomes for the adopting organization.
Asabere et al (2012)	A Review of the Roles and Importance of Information and Communication Technologies (ICTs) in Supply Chain Management (SCM) of Organizations and Companies	Information and Communication Technology (ICT)	Literature Review Paper	This research paper revealed the importance of information and communication technology (ICTs) in SCM and given various factors such as purchasing, e-procurement, operation, customer relationship, vendor management, transport and logistics are all can be improved through the effective use of information and communication technology (ICTs) in SCM.

continued on following page

Table 1. Continued

Authors	Study Title	Key Terms	Methodology Applied	Key findings
Fatehpuria (2013)	Indian cold supply chain: A case study	Warehousing Management (Cold Storage)	Basically two approaches were adopted to deliver into the issues and challenges faced by the industry. In the first phase of the research the cold supply chain was studied in Indian context with focus attention on one of the largest chilly producer in Guntur, A.P., and second part constituted studying the cold supply chain of basking Robbins which is a global ice-cream parlor to which cold supply chain is very much essential. The study was done in shilling by interviewing the mangers of basking Robbins in Shillong outlet.	The cold chain process is a special kind of logistics that handles the goods for both transportation and storage. The importance and necessity of the cold chain in agro sector such as fruits and vegetable, dairy industry etc. is very important or crucial and which shortage in the country damaging the perishable food items to the large extent.
Aung et al (2012)	Quality Monitoring and Dynamic Pricing in Cold Chain Management	Warehousing Management (Cold Storage)	Conceptual Paper	The key findings exposed that the cold chain monitoring system play a significant role in focusing on assessment of quality and dynamic pricing information about the perishable food items.
Qrunfleh et al (2012)	Examining alignment between supplier management practices and information systems strategy	Supplier Relationship Practices	Factor analysis, structural equation modeling (SEM) analysis using SMART Partial Least Square (PLS) software	The study finds that lean (agile) supplier management practices are positively associated with supply chain integration (flexibility). Further, alignment of lean supplier practices and IS for Efficiency enhances supply chain integration, as assessed by a positive moderating effect of IS for Efficiency on the relationship between lean supplier practices and supply chain integration.
Boniface (2011)	Producer relationships segmentation in Malaysia's milk supply chains	Supplier Relationship Practices	The study examines variation in relationship quality (trust, satisfaction, and commitment), loyalty and price satisfaction dimensions, using data from a survey of 133 dairy producers.	The study highlight different ways which milk buyers can develop and promote more appropriate and efficient marketing strategies with milk producers. The study has shown some implications for both policy and management in the milk industry in Malaysia. The main managerial implication is that buyers who want to promote sustainable and uninterrupted milk supplies should focus on different strategies for the different groups.
Singh et al (2012)	The use of carbon dioxide in the processing and packaging of milk and dairy products	Supply Chain Manufacturing Practices	Literature Review Paper	The key finding revealed the important role of modified atmosphere packing (MAP) which is being used with high carbon dioxide (CO_2) concentration as well as CO_2/N_2 gas mixes. Researcher also find that Modified atmosphere packaging or 'gas flushing' has become a popular technique which is used to extend the shelf life (both quality and safety) of a number of dairy products.

continued on following page

Table 1. Continued

Authors	Study Title	Key Terms	Methodology Applied	Key findings
National Bureau of Agricultural Commodity and Food Standards (2005)	Good manufacturing practices for milk collection center	Supply Chain Manufacturing Practices	Agricultural Standards Committee Report (2005)	This report revealed the various important guidelines for the good manufacturing practices for milk collection center which are milk collection center management, management of machines and equipment and tools for raw milk receiving, raw milk cooling system management, storage tank management, cleaning-in-place (CIP), maintenance of machines and equipment, utility management which involves electricity, water supply etc., management of transportation and transport vehicles which involves transportation, raw milk transport vehicle etc., management of raw milk purchasing system, personnel management, safety measures and environment management and management of raw milk quality control.
Smith et al (1995)	Good manufacturing practices for dairy processing plant	Supply Chain Manufacturing Practices	Dairy Practice Council Report (1995)	This report revealed that how various good manufacturing practices i.e. Personal Practices, Sanitary operation and production, Buildings and facilities, Warehousing areas practices, Equipment and utensil design practices play a significant role in improving and enhancing the quality and freshness of dairy products.
Saarijarvi et al (2013)	Extending customer relationship management : from empowering firms to empowering customers	Customer Relationship Management (CRM)	The paper reviews CRM literature published 2003-2011. Based on the literature review, it introduces a conceptual framework of the changing role of customer data in the CRM framework.	Literature has not adequately addressed the role of the emerging service orientation, value co-creation and the opportunities provided by new technology and communication channels. Drawing on a thorough CRM literature review, researchers argue that a fundamental change in CRM thinking is needed to shift the focus of CRM from empowering firms to empowering customers.
Wang and Feng (2012)	Customer relationship management capabilities Measurement, antecedents and consequences	Customer Relationship Management (CRM)	Structural Equation Modeling (SEM)	A three-factor (customer interaction management capability, customer relationship upgrading capability and customer win-back capability) measurement model of CRM capabilities is developed and tested. Furthermore, results support the hypothesized influences of customer orientation, customer-centric organizational system and CRM technology on CRM capabilities, as well as the influence of CRM capabilities on organizational performance.
Matapurkar & Sinha (2011)	Cold chain: finally warming up to India	Transportation Management	Article Paper- Infrastructure Today	The study also focuses on how the refrigerated transportation of milk and milk products is important in maintaining the quality and freshness of the agricultural products which involves refrigerated trucks, containers, ship, trains for the transportation of perishable products.
Garcia & Lunadei (2010)	Monitoring Cold Chain Logistics by means of RFID	Transportation Management	Researcher studied the literature based on role of RFID in cold chain, its advantage and synthesized the findings to develop a preliminary conceptual model.	Logistics service provider need to adopt more efficient logistics technologies i.e. RFID, Quality oriented tracking and tracing Systems (QTT), Safety monitoring and Assurance system (SMAS), FIFO (First In, First Out), FEFO (First Expire, First Out), to provide better supply chain services for their customers.

3. RESEARCH METHODOLOGY

Now a day's competition between Indian dairy cooperatives is increasing rapidly. In order to retain and sustain in such high competitive business environment dairy cooperatives are now trying to improve their organizational performance and achieve competitive advantage effectively and efficiently. In such situation supply chain management can play an important role in improving organizational performance and achieving competitive advantage. Through supply chain management practices dairy cooperatives can minimize their system wide costs and also provide maximum value to their customers. In such scenario the problem is to how strategize and manage the supply chain practices so that the Indian dairy cooperatives may improve their organizational performance and achieve competitive advantage in highly competitive Indian Dairy market?

3.1. Objectives of Study

The objectives of the study are:-

1. To assess the impact of transportation and distribution management on organizational performance.
2. To judge the impact of warehousing management system on organizational performance.
3. To appraise the impact of ICT (Information and Communication Technology) tools and techniques on organizational performance.
4. To judge the impact of supplier relationship management on organizational performance.
5. To judge the impact inventory management system on organizational performance.
6. To judge the impact of supplier chain manufacturing practices on organizational performance.

3.2. Research Hypotheses

The study has been carried out with following hypotheses:

H1: Information & communication technology (ICT) tools and techniques would serve as enabler in supply chain management to improve organizational performance.

H2: Supplier relationship practices would serve as enabler in supply chain management to improve organizational performance.

H3: Supply Chain Manufacturing System would serve as enabler in supply chain management to improve organizational performance.

H4: Warehousing management system would serve as enabler in supply chain management to improve operational organizational performance.

H5: Transportation management would serve as enabler in supply chain management to improve organizational performance.

H6: Customer relationship management practices would serve as enabler in supply chain management to improve organizational performance.

4. RESEARCH DESIGN

To explore and understand the supply chain management practices in Indian dairy industry, the study used descriptive cum analytical research. It requires a clear specification of the who (dairy cooperatives operated in India), what (DSCM practices), when, why (higher cost and less profit) and way (survey) of the research. It is used to describe the characteristics of Indian dairy industry, such as players in dairy market and their shares in market. It is useful to estimate the percentage of dairy cooperatives in a specified population i.e. in Indian dairy industry using SCM practices and their awareness about SCM practices. Whereas analytical research helps in determining the degree to which adopted SCM practices and performance of Dairy companies are associated. The study employed descriptive cum analytical research in the form of a survey undertaken to quantify the importance of the different SCM practices for improving organizational performance of Indian dairy cooperative. In the present study, structured questions were asked to respondents of Dairy companies about their agreement and adoption level of SCM practices in their organization and also the impact of these SCM practices on their organizational performance. Indian dairy cooperatives have not been able to minimize their supply chain cost efficiently. To minimize the supply chain cost and improve their performance, it is necessary to know the answers of why, how, when etc. that is why, the study, in order to know these entire why, how and when about the implementation of SCM practices, used the survey method.

4.1. Sampling of Indian Dairy Industry

Supply chain in Indian dairy industry start from milk production which is done by large scale, medium scale and small and marginal scale farmers, and then the milk is collected by various collection centers which is done by various milk cooperatives societies from there it moves to the dairy plants where storing, chilling, processing, packaging and transportation is done under the supervision of plant manager and after that processed milk and milk product are distributed and marketed to various retails outlets, supermarket etc. and from there it reach to their end consumers. The research will be carried out with one identified group of respondents, which are basically the supply chain members of dairy industry.

- Milk processing unit (Chilling, Processing, Packaging and Transportation).

4.1.1. Population

The proposed research is planned to cover one groups of respondent which are:

- First group of respondent (Milk Processing Unit): The entire milk processing unit operating in UP.

4.1.2. Sample Frame

The sample frame of the research is identified as:

- First group of respondent (Milk Processing Unit): All registered milk processing units situated in UP.

4.1.3. Sample Units

The sample units of the research will be as follows:

- First group of respondent (Milk Processing Unit): Individual registered milk processing unit.

4.1.4. Sampling Techniques and Size of Sample

- First group of respondent (Milk Processing Unit): Judgmental sampling, based on the market position of the organization, will be opted to select the various milk processing units. (Size of sample = 10).

5. DATA COLLECTION

- **Primary Data:** Structured questionnaire, observation and personal interviews of participant of Indian dairy industry.
- **Secondary Data:** Reports and documents will be taken from different magazines, journals, newspaper, central government agencies as well as of state government agencies.

6. DATA ANALYSIS

For the data analysis split-half technique is being used in order to check the reliability of scale which is used in the research questions. Cronbach's alpha value come out to be 0.870 and 0.890, respectively for both part of questionnaire while the Guttman-split coefficient value was .810. Correlation value come out to be .840 which seems to be far more than value 0.6 revealing that there is high correlation between two splits, which further indicate that the scale has high reliability. Mean value is used for the analysis of the agreement and adoption continuum questions in order to know the view of the respondent which is followed by paired T-test in order to check the difference in the opinion of the responses for both the agreement and adoption continuum questions.

7. FINDINGS

The result revealed that there is significant difference in opinion about agreement and adoption level of various DSCM practices in dairy industry which is leading to rejection of null hypothesis and acceptance of alternate hypothesis. Results of paired sample t-test are clearly shown in Table 2.

The research study show that there was less difference between agreement level and adoption level in the case of Information & Communication Technology Practices, Transportation Management System, Customer Relationship Management Practices and while in case of Supplier Relationship Practices, Supply Chain Manufacturing Practices, and Warehousing Management System, a significant difference was found. There were various reasons behind the differences between agreement and adoption level of different SCM practices such as lack of information and awareness about various ICT practices and

Table 2. Result of paired sample T- test

Sl.	Factor	T	P
1.	Information and Communication Technology Practices	5.112	.038
2.	Supplier Relationship Practices	7.324	.035
3.	Inventory Management System	6.103	.041
3.	Supply Chain Manufacturing Practices	5.928	.007
4.	Warehousing Management System	8.865	.006
5.	Transportation Management System	7.041	.024
6.	Customer Relationship Management	6.879	.038

tools, ICT enabled techniques such as ERP practices and understanding the utilities of such techniques among dairy companies. Companies did not adopt strategic sourcing and supplier relationship practices up to significant level because they were less aware about the strategic sourcing and strategic partnership practices and also worried that if they select few suppliers and one of those suppliers suddenly unable to deliver a product or service that is key to their core business, it could bring their operation to a close down. Dairy companies were adopted only few manufacturing practices as per their production objectives and customers. There was a significant difference in agreement and adoption of tracing and tracking system for vehicular movement because maximum Dairy companies outsourced their logistic activities through 3PLs and these 3PLs were monitoring and managing the entire logistic activities of companies. That is why Dairy companies did not adopt such system to a significant level. In case of customer relationship management practices, maximum small companies were in dilemma of its adoption. The result of trend analysis is clearly shown in Figure 1.

Figure 1. Trend analysis of agreement and adoption level of various SCM practices in dairy companies operating in India

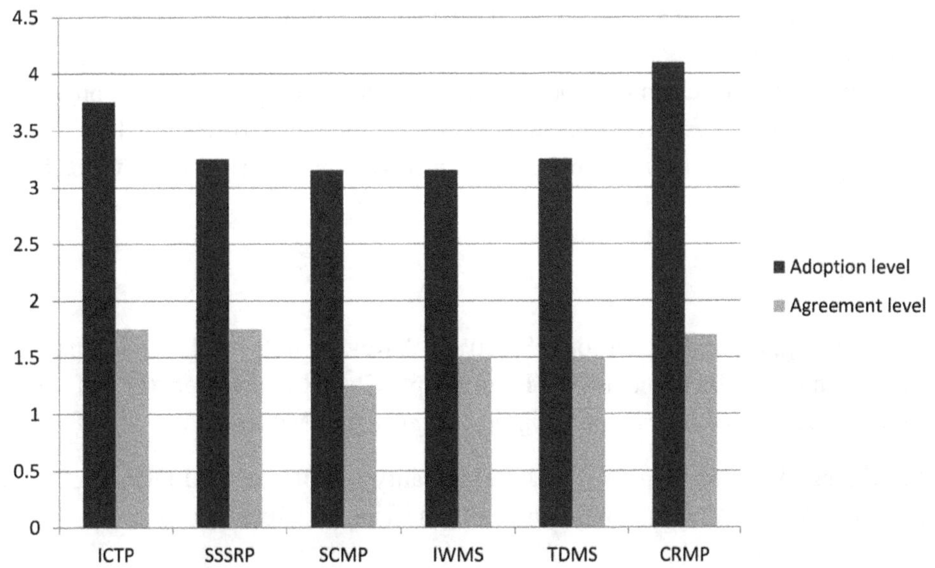

8. CONCLUSION

Various constructs of supply chain management were identified for the research study, which affect the organizational performance. The following section briefly focuses and explores the relationship between various DSCM constructs and organizational performance. Main SCM constructs are information & communication technology (ICT) practices, strategic sourcing & supplier relationship practices, manufacturing practices, inventory & warehousing management practices, transportation & distribution practices and customer relationship management practice. Dairy industry have to follow the following activities in order to achieve their objective like there should have a separate dairy supply chain management department which will be fully responsible for managing all the activities which take place in supply chain with reference to dairy. Secondly, the implementation of information technology (IT) for supply chain management is becoming more and more important in the context of an increasingly globalized and competitive economy. IT, by providing timely, accurate and reliable information, has greatly improved supply chain performance. ERP (enterprise resource planning) system should be adopted by the dairy industry in order to align business objective with latest technology solutions and for optimum utilization of organization's resources and assets. Thirdly, In order to improve organizational performance, firms within the supply chain have started establishing extended relationships with buyers and suppliers who are part of the supply chain (Pitta et al, 2004; Tan et al, 2002). It is essentially carried out with the sole purpose of offering lower cost products to customers with greater design and delivery flexibility (Tan & Wisner, 2003). Dairy industry should also focus on building more reliable supplier relationship because they are the only one who is responsible for distributing the milk and milk product to the end customer. Fourthly, Dairy industry should emphasis on safe and clean manufacturing practices especially from dairy product point of view because quality is the only and the only one parameter which can take India to the international level which can come only through good manufacturing practices. Fifth, Dairy companies should outsource their logistic activities through 3PLs which includes activities apart from transportation, warehousing and custom clearance a whole range of other activities such as freight bill payment, auditing, contract manufacturing and assembly operations, packaging and labeling etc. sixth, Dairy companies should use a combination of milk runs, cross-docking and TL & LTL carriers along with package carriers in some cases. High demand products to high demand retail outlets may be shipped directly, whereas low-demand products or shipments to low demand retail outlets are consolidated to and from the distribution center (DC) and last but not the least Dairy companies should provide platform to receive complains & feedback from customers. Dairy companies need to develop their own websites for receiving the customer complaints and feedback and providing all information required by customers.

REFERENCES

Asabere, N. Y., Oppong, D., & Kusi-sarpong, S. (2012). A Review of the Roles and Importance of Information and Communication Technologies (ICTs) in Supply Chain Management (SCM) of Organizations and Companies. *International Journal of Computer Science and Network*, *1*(4), 70–78.

Aung, M. M., Chang, Y. S., & Kim, W. R. (2012). Quality Monitoring and Dynamic Pricing in Cold Chain Management. World Academy of Science, Engineering and Technology.

Bhalla, S. K. (2001). *Quality - A Necessity to face global challenges in dairy industry.* National seminar on technology management for facing global challenge in dairy industry and alumni- convention, Anand, India.

Boniface, B. (2012). Producer relationships segmentation in Malaysia's milk supply chains. *British Food Journal, 114*(10), 1501–1516. doi:10.1108/00070701211263046

Brahmaprakash & Dinesh, K. (1997). Infrastructural requirements for the development for the development of Agro-processing industries in rural India. *Agril. Econ. Res. Rev., 10*, 325–329.

Burki & Khan. (2008). Milk Districts and Efficiency of Smallholder Dairy Producers in Pakistan. Academic Press.

CII–McKinsey & Co. (1997). *Modernizing the Indian Food Chain, Food & Agriculture Integrated Development Action Plan (FAIDA).* New Delhi: CII and McKinsey & Co.

Fatehpuria, S. (2013). Indian cold supply chain: A case study. *ASM's International E-Journal of Ongoing Research in Management and IT*, 1–10.

Gyan Prakash. (2011). *Poor's Supply Chain: Indian Public Distribution System Revisited.* Submission of Paper for the Biennial Supply Chain Management Conference, organized by IIM Bangalore.

Hazen, B. T., & Byrd, T. A. (2012). Toward creating competitive advantage with logistics information technology. *International Journal of Physical Distribution & Logistics Management, 42*(1), 8–35. doi:10.1108/09600031211202454

Kathikeyan, M. & Teshome, T., (2010). *Dairy supply chain management: A case of Ada'a Dairy Cooperative in Ethiopia.* LAP Lambert Academic Publishing GmbH & Co. KG.

Kulkarni, M. B. (1979). Collection of Milk: Some Lacunae and Remedies. *Dairy Guide, 1*, 31–34.

Kulkarni, S. (1997). A case for the review of present food laws relating to dairy products in India. *Indian Journal of Dairy and Biosciences, 8*, 45–47.

Matapurkar, A., & Sinha, A. (2011). Cold chain: finally warming up to India. *Infrastructure Today*, 22-25.

NBACFS. (2005). Good manufacturing practices for milk collection center. *Royal Gazette, 122*, 1-20.

Qrunfleh, S., Tarafdar, M., & Ragu-Nathan, T. S. (2012). Examining alignment between supplier management practices and information systems strategy. *Benchmarking: An International Journal, 19*(4/5), 604–617. doi:10.1108/14635771211258034

Rangasamy, N., & Dhaka, J. P. (2007). Marketing procurement cost for cooperative and private dairy plants in Tamil Nadu: A comparison. *Indian Journal of Agricultural Economics Research Review, 62*(4), 679–69.

Roy, B. C. (1997). "Growth and prospects of fruit and vegetables processing industry in India," *The Bihar J. Agril. Mkting, 5*, 356–365.

Saarijarvi, H., Karjaluoto, H., & Kuusela, H. (2013). Extending customer relationship management : From empowering firms to empowering customers. *Journal of Systems and Information Technology, 15*(2), 140–158. doi:10.1108/13287261311328877

Singh, P., Wani, A. L. I. A., Karim, A. A., & Langowski, H. (2012). The use of carbon dioxide in the processing and packaging of milk and dairy products. *International Journal of Dairy Technology, 65*(2), 161–177. doi:10.1111/j.1471-0307.2011.00744.x

Singh, P. J., & Power, D. (2009). The nature and effectiveness of collaboration between firms, their customers and suppliers : A supply chain perspective. *Supply Chain Management: An International Journal, 14*(3), 189–200. doi:10.1108/13598540910954539

Smith, G.B. (1995). *Good manufacturing practices for dairy processing plant*. The Dairy Practice Council.

Subbaiah, K. V., Rao, K. N., & Nookesh, K. (2009). *Supply Chain Management in a Dairy Industry – A Case Study*. Academic Press.

Tan, K. C., Kannan, V. R., Handfield, R. B., & Ghosh, S. (1999). Supply chain management: An empirical study of its impact on performance. *International Journal of Operations & Production Management, 19*(10), 1034–1052. doi:10.1108/01443579910287064

Technopak. (2010). *Public Private Partnership in India Dairy Industry*. Retrieved from: http://www.technopak.com/resouces/food/PPP%20indian%20dairy%20industrytechnopak_CII_background%20papermay08,2010.pdf

Vilella, A., & Neves, F. (2008). Strategic management in the Argentina dairy supply chain. *8th International Conference on Management in AgriFood Chains and Networks Ede*, The Netherlands.

Walters, D., & Lancaster, G. (2000). Implementing value strategy through the value chain. *Management Decision, 38*(3), 160–178. doi:10.1108/EUM0000000005344

Wang, Y., & Feng, H. (2012). Customer relationship management capabilities Measurement, antecedents and consequences. *Management Decision, 50*(1), 115–129. doi:10.1108/00251741211194903

Yigrem, . (2008). *Dairy production, processing and marketing systems of Shashemene–Dilla area, South Ethiopia*. Nairobi, Kenya: International Livestock Research Institute.

Zsidisin, G., Jun, M., & Adams, L. (2000). The relationship between information technology & service quality in the dual-direction supply chain. *International Journal of Service Industry Management, 11*(4), 312–328. doi:10.1108/09564230010355359

Chapter 6
Pure Curd Supply Chain:
An Indian Case on Modern Food Safety Management

Rajneesh Mahajan
Apeejay School of Management, India & Delhi Technological University, India

Suresh Garg
Delhi Technological University, India

P. B. Sharma
Delhi Technological University, India

ABSTRACT

The modern food safety management comprises of ISO 22000:2005. It has modified the hazard analysis critical control point (HACCP) by embedding food safety management. It has created a standardized management system. The objective of current chapter is to endow a systematic approach for the ground level implementation of ISO 22000 in Indian pure curd supply chain management. The chapter is prepared utilizing combination of qualitative research and case study method. A case of Milsh Dairy Ltd. (MDL, organisation's name is disguised) was discussed to shed light on ISO 22000 features, comparative analysis between HACCP and ISO 22000. The research is limited to professional pure curd manufacturing sector. Authors have adopted the research methodology which can be applied to other sectors also.

INTRODUCTION

India is the largest milk producer in the world. More than 15 million milk producers sell their milk to 44,246 dairy cooperatives nationally. The collected milk is processed in 177 district co-operative unions and marketed by 22 state marketing federations cum dairies, ensuring a better milk quality (Animal and Dairy Tech Asia, 2013). The milk and milk products are customarily a crucial source of human nutrition. Milk products comprises of fresh milk, milk based products such as butter, butter milk of lassi, pure curd or yogurt, cheese, cottage cheese etc. The pure curd or yogurt is a crucial source of nourishment in India, and rest of world. Tamime et al. (2004), Valli et al. (2005) talked about consumption of yoghurt

DOI: 10.4018/978-1-4666-9795-9.ch006

and it's by products showing continuously increasing trend. Roughly four out of every five fast food advertisements (79%) associated their product with fruit. Dairy products (53.3%) and sugared drinks (52.6%) included associations with fruit in more than half of their advertising. Common examples include fruit-flavoured yogurt products (Castonguay et al. 2013). Regular intake of pure curd produced by lactobacillus delbruceckii ssp. bulgaricus and streptococcus thermophilus looks effective both in prevention and treatment of various illnesses in human being like gastrointestinal, hypercholesterolemia, etc. (Saeed et al,2013).

The objective of current chapter is to endow a systematic approach for the ground level implementation of ISO 22000 in Indian pure curd supply chain management. The chapter is prepared utilizing combination of qualitative research and case study method. A case of Milsh Dairy Ltd. (MDL, organisation's name is disguised) was discussed to shed light on ISO 22000 features, comparative analysis between HACCP and ISO 22000

The chapter is started with introduction of topic. Then, elaborates HACCP practice, explains ISO 22000 and shed light on comparative analysis of HACCP and ISO22000. It is followed by a case of MDL and suggestion for ISO 22000 implementation process.

HAZARD CRITICAL CONTROL POINTS (HACCP)

The food sector faces numerous challenges concurrently, to supply better food quality and safe food at affordable prices in sufficient quantity, to cater greater demand from growth of population. There are three issues that stand out from a health angle like first, ensuring food safety from chemical or biological food contamination. The second is organic food preferred by consumers (Dangour et al., 2009). The third, impact of food offered general health of consumers (McCarthya et al., 2013). HACCP is a food safety management system (FSMS) that is recognized in the international food safety community as a worldwide guideline for controlling food borne safety hazards (Kafetzopoulos et al., 2013). The industrialised countries reported a progressing trouble of food borne illness (Flint et al., 2005; Anonymous, 2007). Ultimately all the efforts can support in achieving quality and safety.

The major landmark strategy in the food sector is introduction of HACCP (History of HACCP, 1959). The Pillsbury Company joined the programme as contractor to produce food for astronauts in 1959. HACCP was first presented 1971. It was based on three principles. HACCP-based ingredients and product specifications were completed in 1975. In 1985, interest in HACCP was restored, when a subcommittee of the Food Protection Committee of NASA issued a report on microbiological criteria. They recognised the advantages of HACCP over traditional quality procedures in food safety management, progressive food companies. The present status of HACCP i.e. Seven Principles are shown in figure 1.

In the figure 2, the six benefits pertaining to the HACCP are depicted.

The HACCP is a globally recognised as a FSMS. To introduce the HACCP to Indian processed food sector (PFS), monetary support in the form of grant is being offered by the Government of India (GOI). The Ministry of Food Processing of Industries (MOFPI, 2012) is offering financial incentives for the benefit of the PFS. To inspire the food processers for adoption of food safety and quality assurance mechanisms such as HACCP, good manufacturing process (GMP) and good hygienic practices (GHP). MOFPI makes efforts to organize food operators to face the global competition as per guidelines

Figure 1. HACCP Principles
Note: CCPs implies critical control points

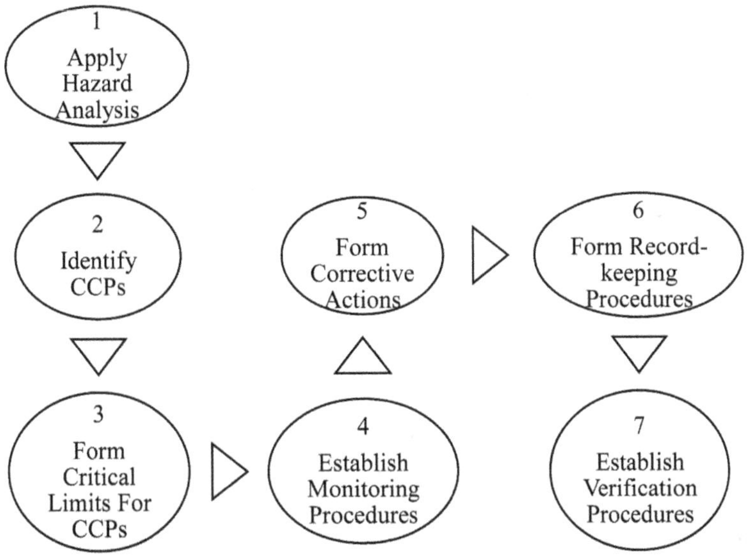

Figure 2. Benefits of HACCP

of WTO. Therefore, Indian processed food sector can adhere to stringent hygiene and quality norms. It will improve product recognition by foreign buyers and maintain sector technologically abreast to best practices. The pattern of financial assistance, up to 50% of the total cost of the project towards implementation of quality management system including HACCP, GMP and GHP in general areas and up to 75% in difficult areas. The absolute ceiling will remain $20,000 and $30,000 respectively ($@Rs. 50) (Ministry of Food Processing Industries, 2012).

FOOD SAFETY CONTROL SYSTEM

The prerequisite for deployment and implementation of HACCP philosophy is to introduce sanitation standard operating procedures (SSOP) and standard operating procedures (SOP) in processed food sector. Therefore, Food Safety and Standards Act (2006) have made provisions in governing food sanitation requires that food HACCP, GMP and GHP ought to be implemented in Indian PFS. Indian viewpoint, the global harmonisation of food regulations poses many issues and challenges in improving the overall food and nutrition security of the population and the food trade within as well as outside the country (Prakash, 2013).With a growing population of over 1.2 billion, Indian food law is intently focused on strengthening food security, and increasingly on promoting nutrition and food safety. The resultant food safety control system is three fold in India. It comprises of HACCP, GMP and GHP (figure 3).

FASSI laid-down nine SOPs for processed food manufacturing and food processing sector for implementation of HACCP in India. These SOPs should be in black and white and describes daily procedures. It shall be conducted before and during procedures to prevent product contamination or adulteration. In addition, the SOPs also include the frequency at which each procedure will be done. SOPs in HACCP, GMP and GHP should be agreed and signed by an official with overall authority at plant. Even for miner alteration, initiation and modification, the SOPs must be signed. The microbiological testing including the total aerobic count, coliform and E.coli, is often being done to ensure that each particular SOP achieves an tolerable level of performance in controlling microbial hazard. Table 1 is depicting SOP for food manufacturers, processed food processors and food services in India.

On the above mentioned lines other SOPs can be developed separately for storage, delivery, testing, measuring, customer complaints, product recall and documentation.

Figure 3. Food safety control system

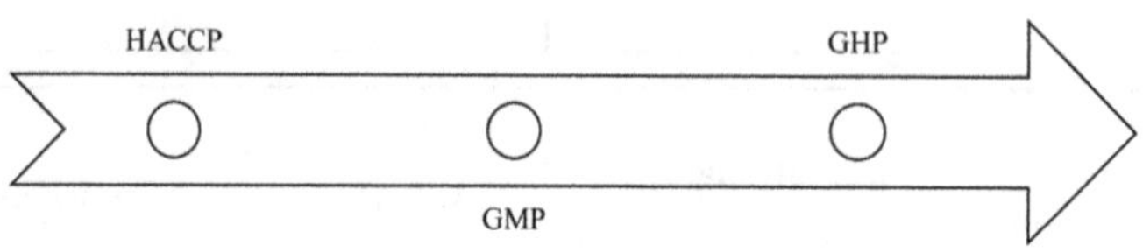

Table 1. SOPs for food manufacturers, processors and services

Standard Operating Procedures (SOPs) for food manufacturers			
Food Manufacturers and Processors		Food Services	
S.No.	SOP for Sanitation Management	S.No.	SOP for Sanitation Management
1	Factory, Building etc.	1	Factory, Building etc.
2	Cleaning of tools and containers	2	Cleaning of tools and containers
3	Sanitation management of employees	3	Sanitation management of employees
4	Administration of chemicals and detergents	4	Administration of chemicals and detergents
5	Waste product treatment process	5	Waste product treatment process
6	Man in charge for sanitation management	6	Man in charge for sanitation management

GOODS MANUFACTURING PROCESS (GMP)

GMP is defined as a bundle of feasible methods and practices that ensure quality manufacturing system. It can provide consistent manufacturing by controlling product quality through qualitative and conformity assessing criteria. It has a clear goal as desired by International, national legal authorities specification, marketing channels and consumers of the product. The planned outlay is desired to ensure that a food handling facility is suitably constructed to offer safe and quality food. This may be stating the obvious but the lack of effective GHP and GMP has often been overlooked (Jackson, 2006). The advance manufacturing techniques are used by production plants for performing tasks related to quality assurance systems such as GMP, GHP and HACCP (Konecka-Matyjek et al., 2005).

Key conditions of GMP are as follows. The current GMPs consist of seven subparts. The requirements are purposely general to allow individual variation by manufacturers to implement the requirements in a manner that best suit their needs (FAO, 2004).

GOOD HYGIENE PRACTICE (GHP)

Good hygiene practice consists of procedures and processes that revisit the processing environment to its original condition, maintain building and equipment in efficient operation, control of cross-contamination during manufacturing. GMP and GHP as laid down in the relevant code of practice as seen in figure 4 (IFST, 1991; Grigg and McAlinden, 2001).

INTERNATIONAL STANDARD ORGANIZATION

International Standard Organization (ISO 22000:2005) combines main constituents to make certain food safety along with the processed food supply chain management (PFSCM). ISO 22000 ensures the effective production of safe products fit for human consumption through three major prerequisites, a).Pre Requisites Programs b).Operational Pre Requisite Programs (O-PRPs) and c).Detailed and implementable HACCP Plan.

Figure 4. Eight principles of GHP

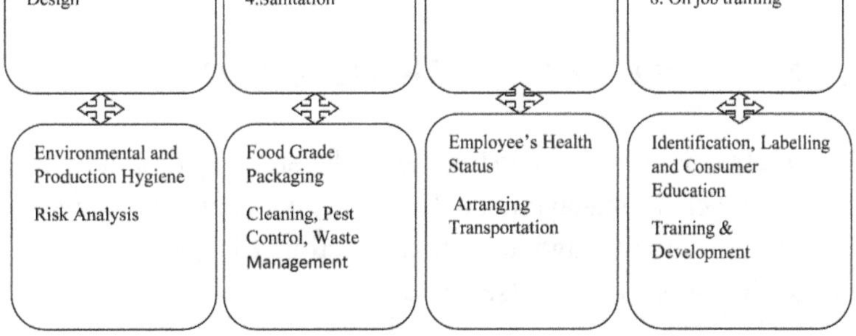

Benefits of ISO22000

- Most advantageous allocation of resources within PFSCM firm.
- Valuable communication with all the business associates such as suppliers, consumers etc.
- Spotlight on the prerequisite programmes, hygiene measures, planning of preventive actions to minimise failures.
- Better Paper work for all procedures.
- Establishment of trust in system and procedure.

Advance and new ISO22000, Food safety management systems is needed for PFSCM firms' aims at the implementation of the principles of HACCP from the food chain organisations to offer safe food to the consumers.

ISO22000 specifies the requirements of a FSMS, surrounded by the range of food firms involved in the PFSCM from farmers to catering businesses. ISO22000 creates a uniform and homogeneous platform of requirements, acceptable to authorities globally. ISO22000 was proposed in 2005. Categories of food organisations

1. The directly involved with PFSCM firms such as food producers, processors, food additives manufacturers, raw and auxiliary raw materials for the food industries, food services, food distributors, distributors, carrying and forwarding agents, logistics service providers and warehousing companies.
2. The indirectly involved with PFSCM such as suppliers of raw materials, equipment, cleaning and packaging materials and other materials that come directly or indirectly into contact with food (Arvanitoyannis & Tzouros 2006, Varzakas et al. 2008).

ISO 22000 FSMS has been designed to overcome the problems that can be encountered when implementing following procedures:

- Lack of strong pre-requisite programmes.
- Complex and unmanageable systems with many CCPs, resulting from a misunderstanding of the role of prerequisite hygiene programs (PRPs) and an inability to conduct proper hazard analysis.
- Weak monitoring and corrective actions due to poor training and verification procedures.
- Excessive documentation and lack of focus due to over-complex systems.
- Poor validation and verification due to lack of expertise.
- Over complication of HACCP implementation.

After acquiring an understanding of food safety principles, firm can have the commitment and resources to implement food safety measures. This may lead to form an excellent Food Safety Management System.

COMPARATIVE ANALYSIS OF HACCP AND ISO22000

It is true that neither ISO22000 nor HACCP can function independently to achieve the herculean objective of food safety in food supply chain management. Thus, it is marriage two can able to offer consumers 100% hazard free food experience worldwide. In this direction it is necessary to understand the differences between two HACCP and ISO 22000 (see table 2).

Table 2. Comparative analysis of HACCP and ISO22000

Hazard Analysis Critical Control Points HACCP	ISO 22000 Food Safety System for PFSCM
HACCP is based on seven principles	Improved food safety procedures
Implemented is in isolatation	Implementation on bases of process management
Well documented standard operating procedures (SOPs) desired	Food safety plan required to be integrated with SOPs
Implementable integrated food safety procedures	Offer food safety system designed, operated and continuously updated
HACCP is qualitative in nature	ISO 22000 has quantitative targets based system. It offers a hazard free food
HACCP ensures following safety control procedures: a) Prerequisite Programs (PRPs) b) HACCP plan	ISO 22000 standard compulsory deploy following procedures at control level, because of this its created a three-level safety control hierarchy, namely:- a) PRPs b)Operational Prerequisite Programs (O-PRPs) c) HACCP plan

A CASE OF PURE CURD MANUFACTURING

Milsh Dairy Ltd. (MDL, organisation's name is disguised), is India's dairy products manufacturing and marketing firm turnover of US$ 2.5 billion annually. Its daily milk procurement is around 12 million litres per day from more than 12000 village milk cooperatives and 3.18 million milk producers. The objective of MDL is to provide remunerative returns to the milk producers and serve the interest of Indian consumers by providing excellent quality and safe products which are good value for money. Its success has been emulated in India.

MDL product range comprises milk, milk powder, pure curd, ghee, butter, cheese; ice-cream etc. MDL has received numerous awards for its consistent adherence to quality, customer focus and dependability.

Pure curd is prepared from purest, freshest pasteurized milk and it is free from added sugar. It does not contain any preservatives. Pure curd is prepared from certified and selected bacterial culture which gives excellent curd. It is hygienically processed and packed in a reusable plastic tub which is free from possible adulteration. It is high on nutrition and taste. MDLs have built an excellent and strong deal network in India. It operates through 48 sales offices and has a dealer network of more than 3000 dealers and million retailers across India.

The pure curd supply chain is temperature sensitive in nature. Pure curd supply chain required to optimal temperature below plus six degree celsius. Therefore, immediately after its production settling down as curd, its temperature must be maintained below plus six degree celsius. It has to be maintained right from warehousing of firm, during logistics, at distributors, retailers and even at consumer home before its final consumption. If desired temperature is not maintained then pure curd may become sour, smelly and bad bacteria formation in the product can convert it to most harmful for human consumption. To cater to Delhi and national capital region (NCR), MDL has plant located at Manesar, Gurgaon in India. The pure curd plant is the most advanced one installed in this production facility of MDL. The most of machines in dairy belong to Tetra pak. It is one of the best manufacturer of dairy equipment. This also provides a guarantee that MDL is deploying best of equipment to produce pure curd.

A Typical Supply Chain of MDL in Delhi, India

The pure curd is produced through implementing the HACCP. There is tremendous scope of implementing the ISO 22000 in the plant. That will further reduce the possible of any sort of food hazards and this approach leads to the comprehensive food safety management.

In the section 4, the pure curd or yogurt manufacturing process is explained. It has 11 steps depicted in figure 6. A proposed plan for implementation of ISO 22000 is explained. Table 4, table 5 elaborates the HACCP for Pure curd Preparation Plan and Operational prerequisite programs (O-PRPs) for Pure curd production.

PURE CURD MANUFACTURING PROCESS

Pure curd or Yogurt is classified as dairy item and prepared by fermentation of fresh milk in combination with the addition of a desirable culture or starter or Khatta or Jaman and coagulation. Milk used comes from various animals such as cows and buffaloes in India. The milk fat and total solid composition is

Table 4. Operational prerequisite programs (O-PRPs) for pure curd preparation

O-PRPs	Measures	Hazard Control	Critical Parameters	Limits	Strategy
Step1. Fresh Milk Receipt at MDL plant	Periodic verification of suppliers conformance	Aflatoxin (C) Heavy metals (C) Dioxins (C) Microbial growth (B)	Toxin presence Pb presence Dioxins presence Staphylococcus count Bacillus cereus count Listeria count Salmonella count	>0,05 mg/kg >.0,02 mg/kg >3 pg/g of milk fat >10 cfu/g	Discussion with Suppliers Supplier rejection
Step2. Milk Powder Suppliers delivered at MDL Plant	Periodic verification of suppliers Conformance	Aflatoxin (C) Heavy metals (C) Dioxins (C) Microbial growth (B) Antibiotics (C)	As per O-PRP 1 Antibiotic presence	As per O-PRP 1	Discussion with suppliers Supplier rejection
Step3. Raw milk and powder storage at MDL Plant	Process control	Microbial growth (B)	Temperature/time period	>4°C for >4 hours	Control measure as per CCP 2
Step4. Logistics from Plant to warehouse and then distributors	Process control	Microbial growth (B)	Temperature/Time period	>6°C for> 8 hours	Finished product withdrawal
Step5. Retailers	Periodic verification of finished products status	Microbial growth (B)	-		Finished product recall Discussion with retailers

usually 10 percent higher than original milk, while pH varies between 4-4.4. The product is packaged in plastic polystyrene cups and during preservation, temperature below plus 8°C is maintained until its expiration date 15 days from the production date in India (Codex, 2003). Pure curd can be consumed by consumers of all ages, from young to elderly people. Dairy products are delivered to end consumers through retailers. They are categorized into large supermarket, local retailers, smaller stores, kiosks, canteens, food catering services, hotels etc. The storage and preservation processes of pure curd are the crucial task for food supply chain management. Figure 5, shows a typical pure curd or yoghurt production process.

Pure curd or yogurt is ready and stored under refrigeration below 6°C temperature to keep its quality intact (Arvanitoyannis & Mavropoulos, 2000, Council of European Communities - CEC, 1992). Sandrou and Arvanitoyannis, 2000, cold chain maintenance prevent bacteria growth. The finishing product is distributed with 24 hours of production. The pure curd is loaded in trucks, transported and distributed to the retailers under proper cold chain (Codex, 2003).

The comprehensive scheme is adopted by the ISO 22000 standard based on the application of process management principles. A number of management philosophies, like Total Quality Management and Six Sigma, are based on these principles (Llore´ns-Montes & Molina, 2006). Deming gave Plan-Do-Check-Act (PGDA) for an effective implementation of process management.

Figure 5. MDLs pure curd or yogurt supply chain

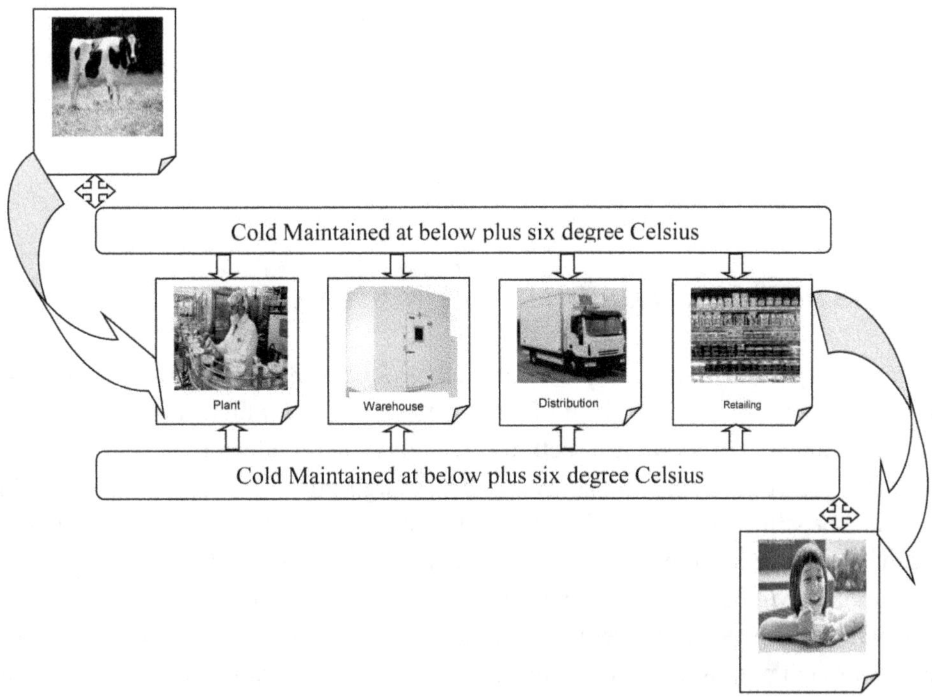

Figure 6. Manufacturing of pure curd or plain yogurt

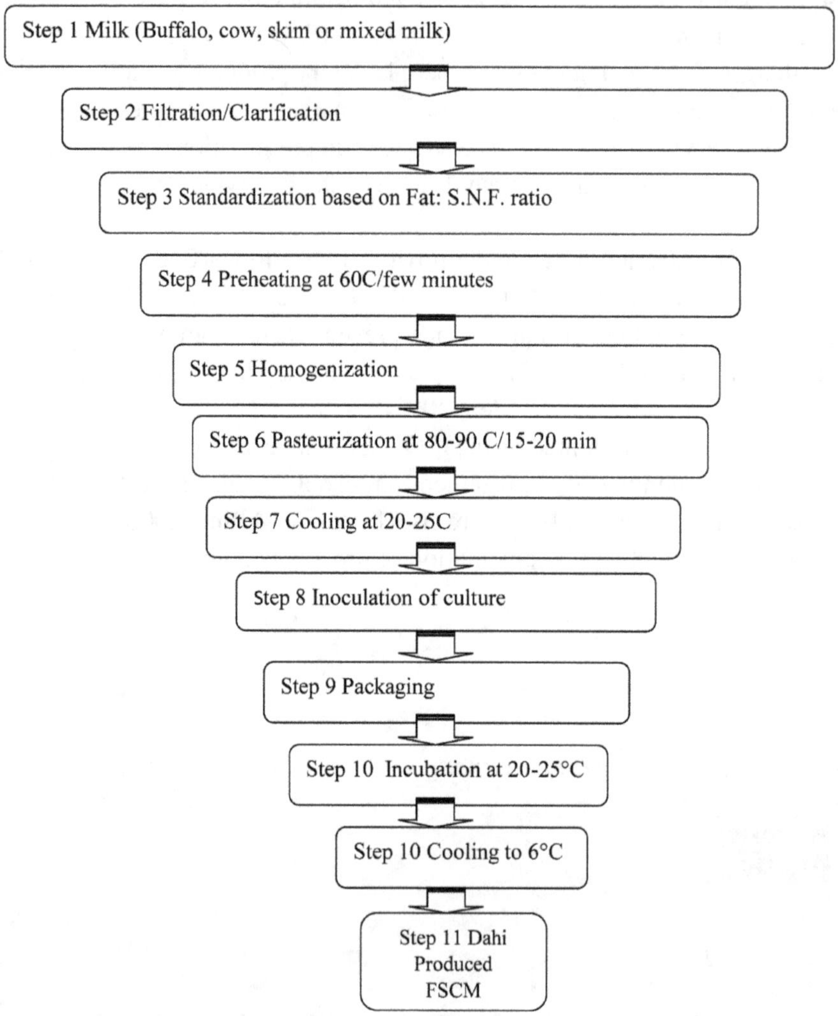

DISCUSSIONS AND CONCLUSION

ISO 22000 standard introduction is a step ahead to achieve food safety objective in a professional manner. ISO2007, over a million ISO 9001 and ISO 14001 certificates were awarded at the end of 2005-2006 worldwide. The most of food manufactures, marketing and food supply chain management firms have taken a step further to adopt food safety system through implementing ISO 22000 with HACCP. is standardized management systems and a certification achieving practice developed . In this paper, an attempt is made to achieve objective through briefly explaining role of HACCP, defining and explaining ISO22000 system, drawing a comparative analysis between HACCP and ISO 22000. A case of MDL is utilized to explain the step by step process of manufacturing of pure curd. A clear cut plan of food safety was introduced through proposing HACCP plan and ISO 22000, Operational prerequisite programs (O-PRPs) for Pure curd production at MDL.

Table 3. HACCP for pure curd preparation

Production Stages	Measures	Hazard Control	Critical Parameters	Limits	Strategy
Step1 Fresh Milk Receipt at MDL plant	Accept the Fresh Milk if its Temperature > 4 °C	Microbial growth (B) Antibiotics (C)	Total colony count Antibiotic presence	>105 cfu/g $>4* 10^5$ cfu/g	Rejection of raw milk
Step2 Milk Storage at MDL Plant	Allowed upto declared capacity	Microbial growth (B)	Total colony count	$>10^5$ cfu/g (fresh milk)	Rejection of fresh milk
Step3 Processing: Heat Treatment at MDL Plant	Process Control	Pathogens (B)	Temperature/Time period	<75°C for 20 seconds	Procedure stop – restart heat treatment
Step4 Adding Culture in the microbiology lab of MDL Plant	Acceptance control of starter culture ingredients	Microbial growth (B) Non-vital culture (B)	Coliform count Yeasts count Moulds count Lactic acid quantity	>1 cfu/g >10 cfu/g >1 cfu/g <0.85-95>%	Rejection of starter culture ingredients
Step5 Packaging	Process Control procedure	Foreign bodies (P)	Adhesives	Any Issue	Rejection of finished product
Step6 Finished Pure curd Storage and logistics	Acceptance control of outgoing Pure curd after settling 24 Hours	Microbial growth (B)	Check a sample	Any Issue	Rejection of finished product

Figure 7. Plan Do Check Act (PGDA)

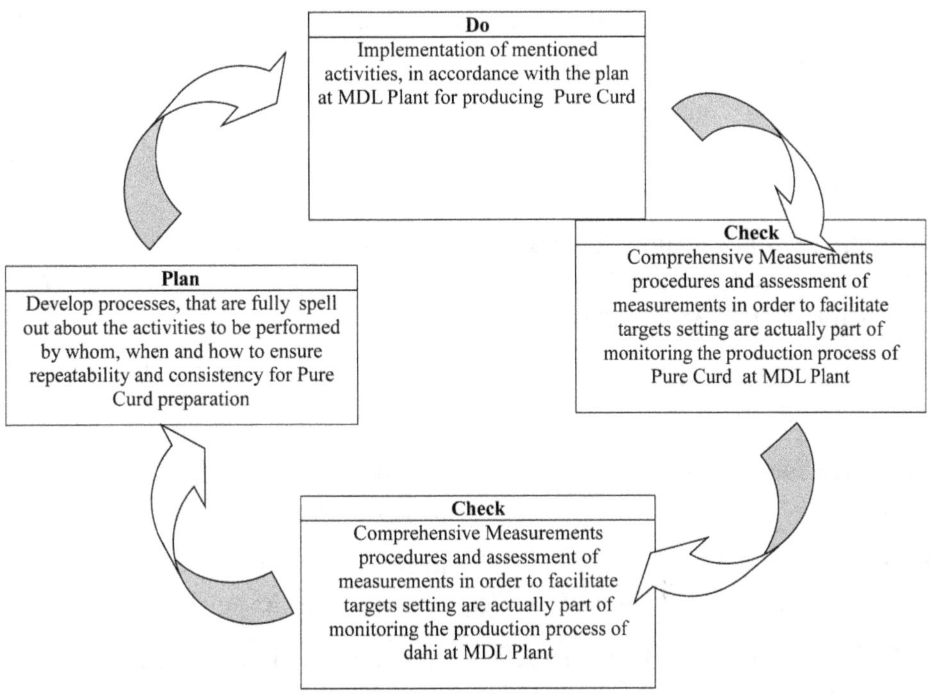

LIMITATION

The research paper has deployed a combination of qualitative and case study method to study the food safety system implementation in a professional manner under managerial scanner. It is a longitudinal research work and inclined to single product pure curd or yogurt only. It may be due to limited time available to carry the research on food safety field.

MANAGERIAL IMPLICATIONS

The learning derived from present case study can be applied to rest of processed food sector in ensuring a comprehensive food safety. This case studies learning can be further applicable other sectors such as general management, operations research, knowledge management etc.

FUTURE SCOPE OF RESEARCH

It is desired that an empirical study can fill the gap in this regard. There is tremendous scope of empirical research to be carried out on ISO 22000. An empirical research should be conducted on food safety system of India using questionnaire.

REFERENCES

Animal and Dairy Tech Asia. (2013). Retrieved on August, 15, 2013, from http://www2.kenes.com/agritech-asia-animal/Pages/Home.aspx

Anonymous, . (2007). The community summary report on trends and sources of zoonoses, zoonotic agents, antimicrobial resistance and food borne outbreaks in the European Union in 2006. *The EFSA Journal*, *130*(5), 1–352.

Arvanitoyannis, I. S., & Mavropoulos, A. A. (2000). Implementation of the hazard analysis critical control point (HACCP) system to Kaseri/Kefalotyri and Anevato cheese production lines. *Food Control*, *11*(1), 31–40. doi:10.1016/S0956-7135(99)00061-4

Arvanitoyannis, I. S., & Tzouros, N. H. (2006). *ISO 22000 the New Food Quality and Safety Standard*. Athens: Stamoulis S.A.

Castonguay, J., McKinley, C., & Kunkel, D. (2013). Health-Related Messages in Food Advertisements Targeting, Children. *Health Education*, *113*(5), 420–432. doi:10.1108/HE-09-2012-0047

CEC. (1992). Hygienic rules for the production and marketing of raw milk, heat treated milk and products based on milk, Council Directive 92/46/EEC. *Official Journal of the European Communities*, *268*, 1–32.

Chountalas, P., Tsarouchas, D., & Lagodimos, A. (2009). Standardized food safety management: The case of industrial yoghurt. *British Food Journal*, *111*(9), 897–914. doi:10.1108/00070700910992835

Codex. (2003). *Codex Standard for Fermented Milks*. Codex STAN 243-2003. Codex Alimentarius Commission.

Dangour, A. D., Dodhia, S. K., Hayter, A., Allen, E., Lock, K., & Uauy, R. (2009). Nutritional quality of organic foods: A systematic review. *The American Journal of Clinical Nutrition*, *90*(3), 680–685. doi:10.3945/ajcn.2009.28041 PMID:19640946

Flint, J. A., Van Duynhoven, Y. T., Angulo, F. J., DeLong, S. M., Braun, P., Kirk, M., & Walke, H. B. P. et al. (2005). Estimating the burden of acute gastroenteritis, foodborne disease, and pathogens commonly transmitted by food, an international review. *Clinical Infectious Diseases*, *41*(5), 698–704. doi:10.1086/432064 PMID:16080093

Food and Agriculture Organization (FAO) of the United Nations. (2004). Retrieved on September,18, 2012, from http://www.fao.org/docrep/006/Y5160E/Y5160E00.htm

Grigg, N. P., & McAlinden, C. (2001). A new role for ISO 9000 in the food industry? Indicative data from the UK and mainland Europe. *British Food Journal*, *103*(9), 644–656. doi:10.1108/00070700110407003

History of HACCP. (1959). Retrieved on August, 15, 2013, from http://foodsafety.unl.edu/haccp/start/gettingstarted.html

IFST. (1991). *Food and Drink-Good Manufacturing Practice: A Guide to Its Responsible Management, 3*. Institute of Food Science and Technology.

Jackson, L. (2006). Towards a South African model for food safety auditor registration. *British Food Journal*, *108*(1), 16–26. doi:10.1108/00070700610637607

Kafetzopoulos, D. P., Evangelos, L., Psomas, E. L., & Kafetzopoulos, P. D. (2013). Measuring the effectiveness of the HACCP Food Safety Management System. *Food Control*, *33*(2), 505–513.

Konecka-Matyjek, E., Turlejska, H., Pelzner, U., & Szponar, L. (2005). Actual situation in the area of implementing quality assurance systems GMP, GHP and HACCP in Polish food production and processing plants. *Food Control*, *16*(1), 1–9. doi:10.1016/j.foodcont.2003.10.007

Llore'ns-Montes, F. J., & Molina, L. M. (2006). Six Sigma and management theory: Processes, content and effectiveness. *Total Quality Management and Business Excellence*, *17*(4), 485–506. doi:10.1080/14783360500528270

McCarthya, M., Cluzel, E., Dressel, K., & Newton, R. (2013). Food and health research in Europe: Structures, gaps and futures. *Food Policy*, *39*, 64–71. doi:10.1016/j.foodpol.2012.12.005

Ministry of Food Processing Industries. (2012). Retrieved on July 10, 2012, from http://mofpi.nic.in

Prakash, J. (2013). The challenges for global harmonisation of food safety norms and regulations: Issues for India. *Journal of the Science of Food and Agriculture*. doi:10.1002/jsfa.6147 PMID:23553336

Saeed, M., Anjum, F. M., Khan, M. R., Khan, M. I., & Nadeem, M. (2013). Isolation, characterization and utilization of starter cultures for the development of wheyghurt drink. *British Food Journal*, *115*(8), 1169–1186. doi:10.1108/BFJ-10-2011-0274

Tamime, A. Y., & Robinson, R. K. (2004). Yoghurt Science and Technology. Woodhead Publishing.

Valli, C., & Traill, W. B. (2005). Culture and food: A model of yoghurt consumption in the EU. *Food Quality and Preference*, *16*(4), 291–304. doi:10.1016/j.foodqual.2004.05.006

Varzakas, T. H., & Arvanitoyannis, I. S. (2008). Application of ISO22000 and comparison to HACCP for processing of ready to eat vegetables: Part I. *International Journal of Food Science & Technology*, *43*(10), 1729–1741. doi:10.1111/j.1365-2621.2007.01675.x

ADDITIONAL READING

Mahajan, R., Garg, S. & Sharma, P.B. (2013), Food safety in India: a case of Deli Processed Food Products Ltd., International Journal of Productivity and Quality Management (Inderscience), Vol. X, No.xx, has been accepted for publication, September, 2013 (In press), (International Journal, ranked in ABDC Journal Ranking List-Business and Management)

Mahajan,R., Garg,S. & Sharma, P.B.(2013), An Illustration of Logistic Regression Technique: A Case of Processed Food Sector, International Journal of Business Excellence, has been accepted for publication, October, 2013(In press), (International Journal, ranked in ABDC Journal Ranking List-Business and Management)

Mahajan, R., Garg, S., & Sharma, P. B. (2013). A System Dynamic Framework for Identifying Demand Gaps in Indian Processed Food Sector. *International Journal of Intercultural Information Management*, *3*(4), 324–343. doi:10.1504/IJIIM.2013.059974

Mahajan,R., Theuvsen,L., Garg,S. & Sharma, P.B. (2013), Comparative Study of Indian and German Food Supply Chain: A case of Tomato Supply Chain, International Marketing and Management Research Series - Palgrave Macmillan, Published, April 2014.

KEY TERMS AND DEFINITIONS

Dairy Products: Dairy products are categorized as fresh milk, curd, butter, cheese and rest of dairy cultural products of India like butter milk (lassi), cottage cheese etc.

HACCP: Hazard critical control points (HACCP) is based on following seven principles: Apply Hazard Analysis, Identify CCPs, Form Critical Limits For CCPs, Establish Monitoring Procedures, Form Corrective Actions, Establish Verification Procedures and Form Record-keeping Procedures.

Pure Curd Supply Chain: It is defined as the connect among all the pillars of proud supply chain. It begins with starting from souring of raw material for curd production. It will be delivered at plant, processing of raw milk takes place, final product is packed and dispatched to warehouses amd from t.

Quality Standards: They are the benchmarked quality standards for food producers.

Total Food Safety Management: It is a comprehensive food safety system contains implementation of HACCP, GHP, GMP and ISO22000.

Chapter 7
A New Approach to Supply Chain Performance Measurement:
An Empirical Study of Manufacturing Organizations

Amit Kumar Marwah
SGSITS, India

Girish Thakar
SGSITS, India

R. C. Gupta
SGSITS, India

ABSTRACT

Existing research work has established that many of today's manufacturing organizations have failed to develop a comprehensive supply chain performance measures. In this chapter, the authors intend to empirically assess the effects of supplier buyer relations and human metrics on supply chain performance in the context of Indian manufacturing organizations. After rigorous literature review, total 18 variables have been identified which are later on reduced in number by factor analysis. As a pilot study, primary data is collected from 100 manufacturing organizations by means of a questionnaire and a scale is developed. On a sample size of 100, the proposed hypotheses are tested by applying two-tailed tests. t-test and factor analysis resulted in 5 factors, 2 related to supplier-buyer relations and 3 related to human metrics. The overall reliability of the scale comes out to be 0.697. The research work provides a new approach to the manufacturing organizations to understand the factors affecting supply chain performance. The present study is limited to Indian manufacturing organizations.

1. INTRODUCTION

Improvement in the supply chain is critical to a company's bottom line in the current era of global sourcing and global competition. Going a step further, the maturity of the supply chain governs a company's performance, affecting the top line as well as the bottom line. Indian industry is facing competition both from multinational companies and imports in the domestic markets. The new competition is in terms of

DOI: 10.4018/978-1-4666-9795-9.ch007

improved quality, products with higher performance, reduced cost, a wider range of products and better service; all delivered simultaneously (Dangayach & Deshmukh, 2003).

The excellent overview of performance measurement provided by Neely et al. (1995) has been widely cited in research into supply chain performance measurement systems (e.g., Beamon, 1999; Beamon & Chen, 2001; Gunasekaran et al., 2001, 2004). These, and other studies, have highlighted how the majority of the limitations cited by Neely et al. are still valid in the case of performance measurement systems for supply chains. Moreover, they have stressed the need for new measurement systems and metrics which address these deficiencies. It is an established fact that many companies have not succeeded in developing the performance measures and metrics needed to fully integrate their supply chain to maximize effectiveness and efficiency (Gunasekaran et al., 2004). Also, it has been argued that measuring supply chain performance can result in understanding of the supply chain and improve overall companies' performance (Chen & Paulraj, 2008).

Measurement has been recognized as a crucial element to improve business performance (Taticchi et al., 2010). Various performance metrics are in place for measuring effectiveness of SC. Different perspectives of Supply Chain Performance Measures (SCPM) are cost and non-cost perspective; strategic, tactical or operational focus (Gunasekaran et al., 2001); business process perspective and financial perspective (Beamon, 1999). The earlier focus of performance measurement was on financial perspective which is gradually changing to non-financial perspectives. Most of the models have gone through some empirical testing and some have only theoretical developments (Taticchi et al., 2010).

The manufacturing sector is growing rapidly in India and China and has shrunk in most advanced economies. The growth will require several changes, which include significant increase in productivity and quality at the plant levels, pursuit of worldwide competitive manufacturing strategies and operations and successful integration into the global supply chains (Deloitte, 2007). The purpose of this chapter is to propose a model of supply chain performance measurement for manufacturing organizations. A new approach is put forward, evaluating the effects of supplier-buyer relations as well as human metrics on supply chain performance.

2. BACKGROUND

In this chapter, the authors propose a conceptual model by establishing the relationships of supplier-buyer relations and human metrics with SC performance in a single study in the context of Indian manufacturing organizations.

2.1 Supplier Buyer Relations (SBR)

In supply chain management (SCM) strategies, supplier relationship activities play an important role (Wisner, 2003). Long-term relationships refer to intention that the arrangement is not going to be temporary (Chen & Paulraj, 2008). A successful strategic alliance and integrated relationship with suppliers and buyers is very much needed (table 1). It should be revolved around trust, loyalty, positive sum game (a win-win relationship), cross-functional teams, achieving common goals and collaboration (Chandra & Kumar, 2000). A firm's success is linked to the strength of its relationship with supply chain partners and it could reduce and increase revenue (Spekman, Kamauff & Myhr, 1998).

Table 1. Construct definitions and literature support

Construct	Definitions	Literature
Supplier-Buyer Relations	"The long-term relationship between the organization and its suppliers. It is designed to leverage the strategic & operational capabilities of individual participating organizations to help them achieve significant ongoing benefits" *(Li et al., 2006, p. 109)*	Li et al., 2005; Gunasekaran et al., 2001; Balsmeier and Voisin, 1996; Monczka et al., 1998; Noble, 1997; Stuart, 1997; Lamming, 1993; Sheridan, 1998; Tan et al., 2002, Humphreys et al., 2004, Autry et al., 2010

2.2 Human Metrics (HM)

There is a heavy influence of behavioral issues while establishing and implementing the key Performance Measures (PMs) and metrics. Organizations sharing values in terms of extreme trust, commitment and collaboration, organizational capability and top management support are essential for an effective SCM (Mello & Stank, 2005). It is suggested that human factor is significantly affecting the SCM effectiveness (Tony and Kelvin, 2007) and the role of SCM managers are critical in achieving strategic and operational objectives and changes in the supply chain(van Hoek, 1998).

It is found that firms lacking in the appropriate cultural elements such as shared assumptions and values tend to fail when implementing SCM initiatives (Mello and Stank, 2005). On top of that, the organizational commitment and governance for supply chain success are being studied (Fawcett, Ogden, Magnan & Cooper, 2006). The findings indicated that four types of managerial supports are needed to achieve the highest levels of supply chain success: top management support, broad-based functional support, channels support and infrastructural/governance support. Few more research works (Robinson & Malhotra, 2005; Wouters, 2009) clearly support the need for a performance measurement system taking the holistic picture, including the human side and organizational issues.

The present study involves behavioural determinants of SCM viz. continuity, communication, power and trust and related variables are chosen from the existing body of literature.

2.3 Supply Chain Performance (SCP)

Researchers suggest that an appropriate performance measurement system (PMS) is a critical requirement for the effective management of a supply chain (Liang, Yang, Cook & Zhu, 2006). There are studies about the performance measurement systems and metrics of supply chains (SCs) by critically reviewing the contemporary literature and suggesting possible implications for future research (Shepherd and Gunter, 2006).

SCM needs to be assessed for its performance in order to evolve an efficient and effective supply chain (Gunasekaran, Patel & Tirtiroglu, 2001). For effective management in a SC, measurement goals must consider the overall SC goals and the metrics to be used. These should represent a balanced approach and should be classified at strategic, tactical and operational levels. There should be financial as well as non-financial measures (Gunasekaran et al.,2001). Various research papers that deal with performance measurement in a SC context (Van Hoek, 1998) have appeared in the literature (table 2).

However, most of them are prescriptive and not based on historical facts, changing market and operations environments or well grounded empirical analysis. In addition, they lack a complete coverage of

Table 2. Construct definitions and literature support

Construct	Definitions	Literature
Supply Chain Performance	The overall efficiency and effectiveness of a supply chain	Beamon, 1998; 1999; Harland, 1996; Garwood, 1999; Gunasekaran et al., 2001; Holmberg, 2000; Tompkins and Ang, 1999; van Hoek, 1998; Bechtel and Jayaram, 1997; Kiefer & Novack, 1999; Narasimhan and Jayaram, 1998; Hewitt, 1999; Spekman et al., 1998

all the performance measures and metrics in new organizational environments. An overview of PMSs in SCMs environments highlights the justification for the selection of suitable metrics based on the current and emerging new enterprise environments.

3. PROPOSED CONCEPTUAL MODEL

There is a model to examine relationship between supply chain performance and degree of linkage among supplier, internal integration and customer (Lee, Kwon and Severance, 2007). But, there is no model incorporating human involvement and related issues. This initiates the authors to propose a conceptual model.

3.1 Model Development

Due to the lack of any previous research work done in finding the combined effect of supplier-buyer relations (SBR) and human metrics (HM) on supply chain performance (SCP), the authors propose that a model for manufacturing organizations can be developed using co-relational links between SCM performance and its determinants (Figure 1).

3.2 Scale Development

Various items or questions used in questionnaire are taken from previous studies (Table -3). There are 6 items for SBR and 12 items for HM, totalling to 18 items overall.

3.3 Development of Hypothesis

Based on the conceptual model, following hypotheses are proposed:

Hypothesis I

Ho: Supplier-buyer relations have no effect on supply chain performance.

Hypothesis II

Ho: Human Metrics has no effect on supply chain performance.

Figure 1.

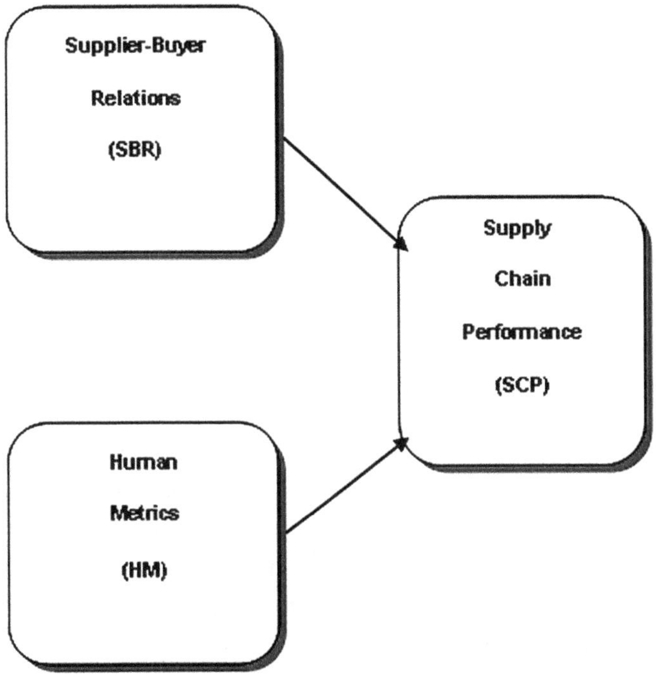

Table 3. Questionnaire Items

1	SBR1	We consider quality as our number one criterion in selecting suppliers
2	SBR2	We regularly solve problems jointly with our suppliers
3	SBR3	We have helped our suppliers to improve their product quality
4	SBR4	We have continuous improvement programs that include our key suppliers
5	SBR5	We include our key suppliers in our planning and goal- setting activities
6	SBR6	We actively involve our key suppliers in new product development processes
7	HM1	The degree of dealings between us and our supply chain partner is very high
8	HM2	The relationship between us and our supply chain partner is very stable
9	HM3	We are quite involved in the marketing and planning efforts of our supply chain partner
10	HM4	We and our supply chain partner seek advice for each other when doing marketing analysis
11	HM5	We and our supply chain know the strengths and weaknesses of each other very well
12	HM6	Our firm is powerful enough to ask our supply chain partner to readjust price strategy
13	HM7	Our firm is powerful enough to ask our supply chain partner to readjust their product
14	HM8	Our firm can provide training support to our supply chain partner
15	HM9	Our supply chain partner perceives that our firm is perfectly honest and truthful
16	HM10	Our supply chain partner perceives that our firm is perfectly having high integrity
17	HM11	We would like to inform our supply chain partner everything about new developments
18	HM12	We willingly share all information that might help your supplier make better decisions

4. METHODOLOGY

This study is part of a larger research project exploring SC related practices, their relevance to managers, and their impact on firm performance and eventually on firm competitiveness (Marwah, Thakar & Gupta, 2014). Methodology used in this study is based on the views of Beamon (1999), that is, PMS should develop a reliable metrics to provide feedback on various performance areas by eliminating the overlapping metrics and to include the most important metrics of supply chain management.

4.1 Sample Profile

The sample (sample size = 100) of this pilot study focuses on departments of purchasing, production, logistics and distribution in the Indian manufacturing companies. The survey covers various cities across India (Figure 2).

Questionnaire (both online and offline) is the main instrument of this study. Questionnaire items are designed after an extensive literature review. 5–point Likert scale is used. There are total 35 items or questions. First 18 questions are related to the quantitative aspects (6 for SBR and 12 for HM, as shown in Table-3). Next 17 are demographic questions. Questionnaire has been emailed to various respondents and out of 108 responses which are received, 100 responses are complete in all respects and therefore taken for this pilot study. SPSS is used for analysis of the data.

4.2 Sampling Parameters

With factor loadings higher than 0.55 (for sample size = 100), it is found that the sample is adequate for 95% confidence level and 5% level of significance (Hair et al., 2007, p 152)

Figure 2.

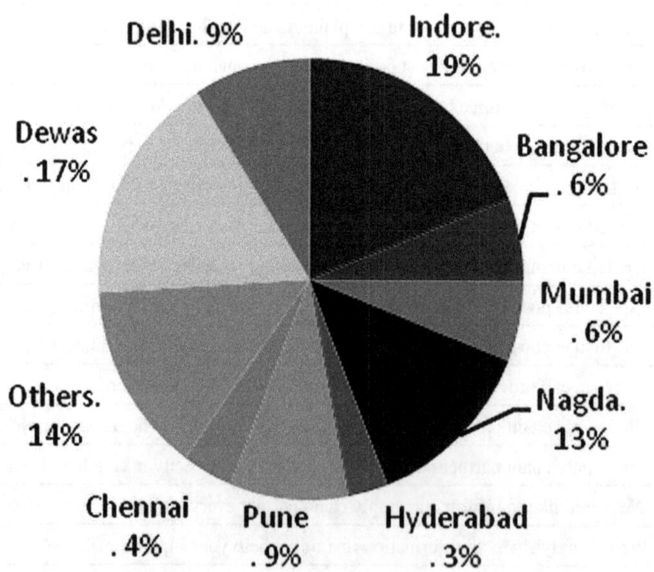

Table 4. Reliability statistics (initial)

Cronbach's Alpha	N of Items
.711	18

The internal consistency of the questionnaire has been checked and the value of Cronbach's Alpha is found to be 0.711 (Table 4). It is more than 0.6 for the exploratory research utilizing new scale. Hence the questionnaire can be used for analysis purposes (Hair et al., 2007)

5. RESULTS

The hypotheses proposed earlier are now properly phrased and tested. Factor analysis is also performed on each construct using SPSS to reduce unnecessary items from the questionnaire. Reliability analysis is done eventually to test how well the items in a set are positively correlated to one another.

5.1 Hypothesis Testing

On a sample size of 100, the proposed hypotheses are tested by applying two-tailed tests as both the hypotheses are directional in nature.

Hypothesis I

Ho (Null hypothesis): Supplier-buyer relations have no effect on supply chain performance.
H1 (First alternative hypothesis): Supplier-buyer relations have significantly positive effect on supply chain performance.
H2 (Second alternative hypothesis): Supplier-buyer relations have significantly negative effect on supply chain performance.

The hypothesis includes six variables as shown in Table -5. One sample t-test has been applied on these variables. It tests that whether the mean of a distribution differs significantly from our assumed test value. A test value of 3 is chosen as it is the mid value of the five point scale used in questionnaire. Table - 5 summarizes the results of this test.

The null hypothesis (Ho) is rejected and the first alternative hypothesis (H1) is accepted for all the variables. Thus, the results show strong influence of SBR on SCP.

Hypothesis II

Ho (Null hypothesis): Human Metrics has no effect on supply chain performance.
H1 (First alternative hypothesis): Human Metrics has significantly positive effect on supply chain performance.

Table 5. Results for Hypothesis I

One-Sample Test (Test value = 3)						
Variables	**t- Value***	**p- Value****	**MD**	**Remarks*****	**95% Confidence Interval of the Difference**	
					Lower	**Upper**
SBR1	21.51	0	1.34	SA	1.22	1.46
SBR2	17.93	0	1.14	SA	1.01	1.27
SBR3	14.17	0	1.14	SA	0.98	1.3
SBR4	10.62	0	0.94	A	0.76	1.12
SBR5	11.82	0	1.14	SA	0.95	1.33
SBR6	13.02	0	1.06	SA	0.9	1.22

* calculated value of t- statistics
** minimum level of significance at which null hypothesis is rejected
*** the responses are categorized as: Disagree (D) if mean difference (MD) is negative; Agree (A) if MD ranges from 0 to +1; and Strongly Agree (SA) if MD > +1

H2 (Second alternative hypothesis): Human Metrics has significantly negative effect on supply chain performance.

The null hypothesis (Ho) is again rejected and the first alternative hypothesis (H1) is accepted for all the variables. Thus, the results show strong influence of HM on SCP. Table - 6 summarizes the results of this test.

Table 6. Results for Hypothesis II

One-Sample Test (Test Value = 3)						
Variables	**t- Value***	**p- Value****	**MD**	**Remarks*****	**95% Confidence Interval of the Difference**	
					Lower	**Upper**
HM1	22.48	0	1.22	SA	1.11	1.33
HM2	8.22	0	0.78	A	0.59	0.97
HM3	7.71	0	0.84	A	0.62	1.06
HM4	9.50	0	0.84	A	0.66	1.02
HM5	9.99	0	0.82	A	0.66	0.98
HM6	13.38	0	1.04	SA	0.89	1.19
HM7	16.34	0	1.06	SA	0.93	1.19
HM8	10.78	0	1.08	SA	0.88	1.28
HM9	12.27	0	1.12	SA	0.94	1.3
HM10	21.66	0	1.36	SA	1.24	1.48
HM11	10.7	0	1.02	SA	0.83	1.21
HM12	18.86	0	1.16	SA	1.04	1.28

5.2 Factor Analysis

The previous section leaves us with all the 18 variables as determinants of SCP. The factor analysis is conducted on these variables separately for both the constructs viz. SBR and HM.

5.2.1 Construct I (SBR)

In a factor analysis of 6 variables, a correlation matrix is obtained. The KMO and Bartlett's test results are shown in Table 7(a), which shows a value of 0.574 which is more than 0.5, suggesting an acceptable value.

The inspection of diagonal values of anti-image correlation matrix obtained from SPSS gives assessment of the sampling adequacy of the variables. The diagonal values below 0.5 are not acceptable. So, SBR1 is removed as its value is 0.442 and again factor analysis is performed. The revised KMO and Bartlett's test results are shown in Table 7(b).

The inspection of diagonal values of anti-image correlation matrix now reveals all values above 0.5. Now, the data set is ready for factor extraction. The method of principal component analysis is used, the results of which are shown in Table 8.

Table 7. KMO and Bartlett's Test for SBR

(a) KMO and Bartlett's Test		
Kaiser-Meyer-Olkin Measure of Sampling Adequacy.		**.574**
Bartlett's Test of Sphericity	Approx. Chi-Square	67.564
	df	15
	Sig.	.000
(b)KMO and Bartlett's Test		
Kaiser-Meyer-Olkin Measure of Sampling Adequacy.		**.634**
Bartlett's Test of Sphericity	Approx. Chi-Square	49.798
	df	15
	Sig.	.000

Table 8. Total variance explained (SBR)

Component	Initial Eigenvalues			Extraction Sums of Squared Loadings			Rotation Sums of Squared Loadings		
	Total	% of Variance	Cumulative %	Total	% of Variance	Cumulative %	Total	% of Variance	Cumulative %
1	1.896	37.914	37.914	1.896	37.914	37.914	1.758	35.168	35.168
2	1.093	21.853	59.767	1.093	21.853	59.767	1.230	24.599	59.767
3	.806	16.116	75.882						
4	.680	13.598	89.480						
5	.526	10.520	100.00						
Method: Principal component Analysis									

Before factor rotation it is necessary to assess the explanatory power (model fit) of the model constructed by SPSS. There are 80% non-redundant residuals with absolute values greater than 0.05, which is a good value. The factors are now rotated using Varimax rotation method (Table 9).

Factor analysis has extracted 2 factors, thereby reducing initial 6 items. It can also be seen in scree plot (Figure 3). On the basis of variables associated with them, the factor labels are now given to the factors. The percentage of variance explained by each factor as obtained from Table - 8 is put in parenthesis next to the factor label.

Table 9. Rotated component matrix (SBR)

Variables	Component	
	1	2
SBR2	.800	
SBR4	.734	
SBR5	.591	
SBR3		.879
SBR6		.591

Extraction Method: Principal Component Analysis.
Rotation Method: Varimax with Kaiser Normalization.

Figure 3.

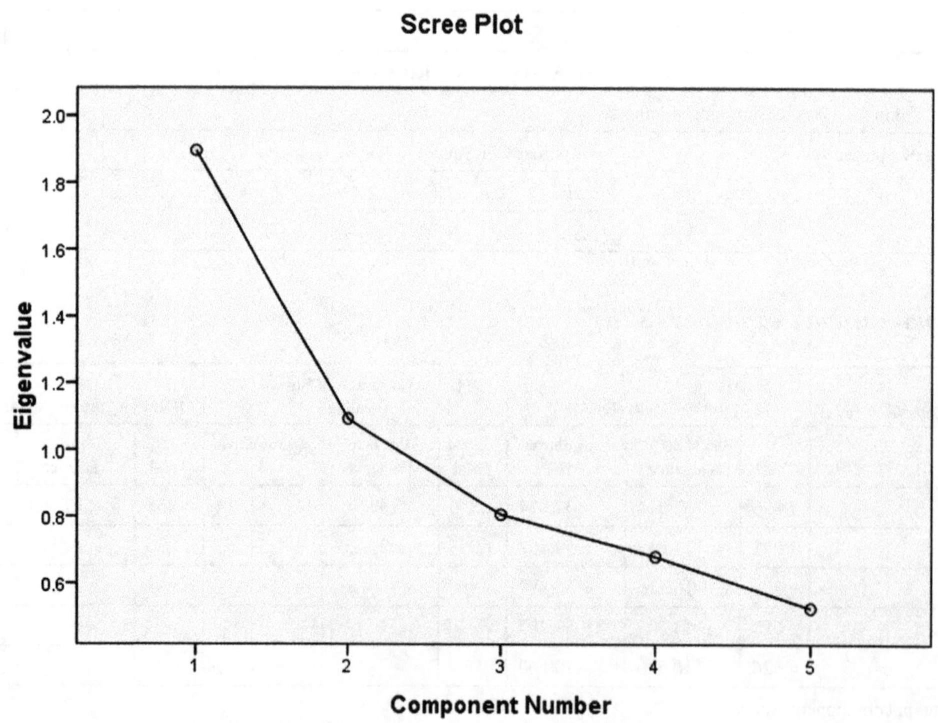

Factor 1: We have joint planning, problem solving and improvement programs with our suppliers (35.17%).

Associated Variables:

- SBR2 *(0.800)*: We regularly solve problems jointly with our suppliers.
- SBR4 *(0.734)*: We have continuous improvement programs that include our key suppliers.
- SBR5 *(0.591)*: We include our key suppliers in our planning and goal- setting activities.

Factor 2: We help our key suppliers in developing new and better quality products (24.6%).

Associated Variables:

- SBR3 *(0.879)*: We have helped our suppliers to improve their product quality.
- SBR6 *(0.519)*: We actively involve our key suppliers in new product development processes.

5.2.2 Construct II (HM)

In a factor analysis of 12 variables, a correlation matrix is obtained. The KMO and Bartlett's test results are shown in Table 10(a), which shows a value of 0.534 which is more than 0.5, suggesting acceptable value.

The inspection of diagonal values of anti-image correlation matrix obtained results in removal of HM8 (MSA= 0.416) and subsequently of HM11 (MSA=0.460). Again factor analysis is performed. The revised KMO and Bartlett's test results are shown in Table 10(b).

The inspection of diagonal values of anti-image correlation matrix now reveals all values above 0.5. Now, the data set is ready for factor extraction. The method of principal component analysis is used, the results of which are shown in Table 11. A scree plot is also shown in Figure 4.

There are 73% non-redundant residuals with absolute values greater than 0.05. The factors are now rotated using Varimax rotation method (Table -12).

Table 10. (a). KMO and Bartlett's Test for HM

(a) KMO and Bartlett's Test		
Kaiser-Meyer-Olkin Measure of Sampling Adequacy.		**.534**
Bartlett's Test of Sphericity	Approx. Chi-Square	223.888
	df	66
	Sig.	.000
(b) KMO and Bartlett's Test		
Kaiser-Meyer-Olkin Measure of Sampling Adequacy.		**.626**
Bartlett's Test of Sphericity	Approx. Chi-Square	143.079
	df	45
	Sig.	.000

Table 11. Total variance explained (HM)

Component	Initial Eigenvalues			Extraction Sums of Squared Loadings			Rotation Sums of Squared Loadings		
	Total	% of Variance	Cumulative %	Total	% of Variance	Cumulative %	Total	% of Variance	Cumulative %
1	2.216	22.158	22.158	2.216	22.158	22.158	1.918	19.182	19.182
2	1.916	19.159	41.317	1.916	19.159	41.317	1.915	19.146	38.328
3	1.231	12.307	53.625	1.231	12.307	53.625	1.530	15.296	53.625
4	.938	9.375	63.000						
5	.834	8.338	71.338						
6	.796	7.955	79.293						
7	.600	6.001	85.294						
8	.594	5.941	91.235						
9	.472	4.716	95.951						
10	.405	4.049	100.00						
Extraction Method: Principal Component Analysis.									

Figure 4.

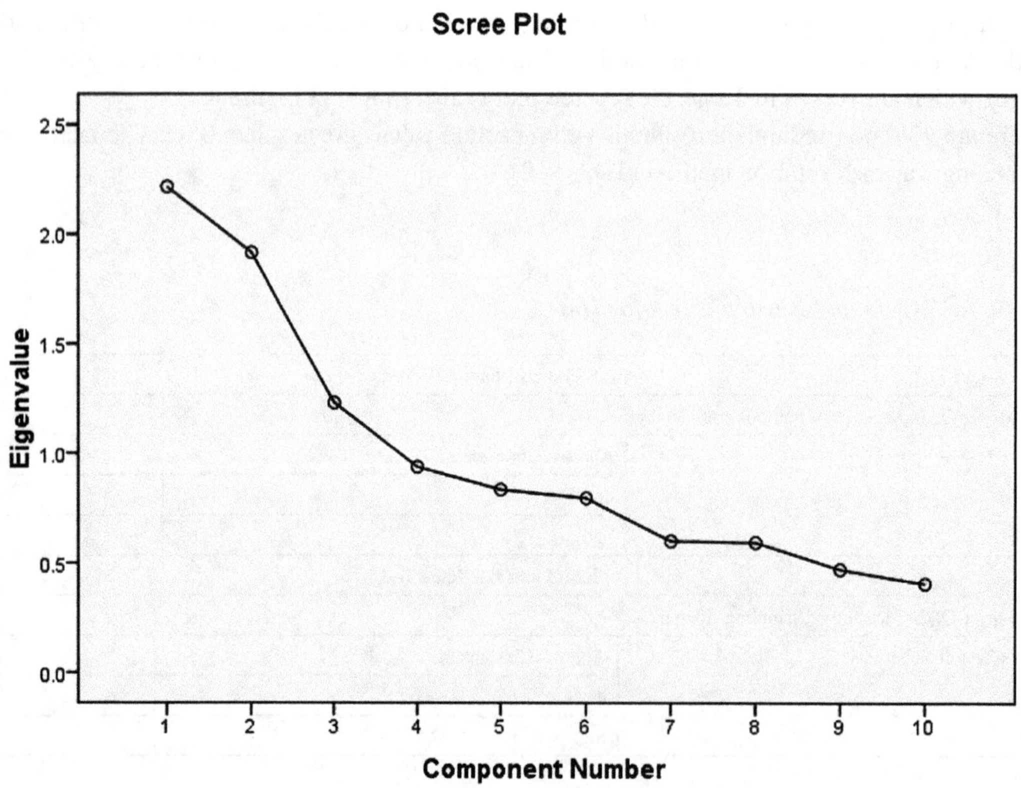

Table 12. Rotated component matrix (HM)

Variables	Component		
	1	2	3
HM1	.679		
HM5	.675		
HM6	.651		
HM2	.587		
HM4		.757	
HM7		.686	
HM3		.659	
HM9			.648
HM12			.615
HM10			.598
Extraction Method: Principal Component Analysis. Rotation Method: Varimax with Kaiser Normalization.			

Factor analysis has extracted 3 factors, thereby reducing initial 12 items. On the basis of variables associated with them, the factor labels are now given to the factors. The percentage of variance explained by each factor as obtained from Table 11 is put in parenthesis next to the factor label.

Factor 1: Trust and commitment (19.18%).

Associated Variables (see Table 13).

Factor 2: Joint strategy and planning (19.15%).

Associated Variables (Table 14).

Factor 3: Transparency and honesty in information sharing (15.30%).

Associated Variables (Table 15).

Table 13.

HM1 *(0.679)*	The degree of dealings between us and our supply chain partner is very high
HM5 *(0.675)*	We and our supply chain know the strengths and weaknesses of each other very well
HM6 *(0.651)*	Our firm is powerful enough to ask our supply chain partner to readjust price strategy
HM2 *(0.587)*	The relationship between us and our supply chain partner is very stable

Table 14.

HM4 *(0.757)*	We and our supply chain partner seek advice for each other when doing marketing analysis
HM7 *(0.686)*	Our firm is powerful enough to ask our supply chain partner to readjust their product
HM3 *(0.659)*	We are quite involved in the marketing and planning efforts of our supply chain partner

Table 15.

HM9 *(0.648)*	Our supply chain partner perceives that our firm is perfectly honest and truthful
HM12 *(0.615)*	We willingly share all information that might help your supplier make better decisions
HM10 *(0.598)*	Our supply chain partner perceives that our firm is perfectly having high integrity

Table 16. Reliability statistics (final)

Cronbach's Alpha	Cronbach's Alpha Based on Standardized Items	N of Items
.697	.710	15

Thus, the entire initial items for SBR (6 Nos.) and HM (12 Nos.) have been reduced to 5 factors and 15 items. The final questionnaire is tested for reliability. The internal consistency reliability will be higher if the Cronbach's alpha is closer to 1 (Sekaran, 2003). The final questionnaire shows the value of Cronbach's alpha as 0.697, which is acceptable for a new scale (Table 16).

6. FUTURE RESEARCH DIRECTIONS

This study is a part of a larger research project exploring SC related practices. The methodology of critical evaluation involves literature review of empirical research articles on performance measurement and SCM. The authors' intention is to fill up the gap about the lack of research in supply chain management which investigates the role of critical success factors in manufacturing organizations of India (Marwah et al., 2013, 2014). Furthermore, the study to be carried out resulting from the proposed model is expected to investigate the critical determinants of SCM performance of the Indian manufacturing organizations.

The present study is limited to Indian manufacturing organizations. It can be extended to cover other sectors and organizations too. Various other factors contributing to supply chain performance can be considered simultaneously using structural equation modeling (SEM) approach.

7. CONCLUSION

The increasingly global nature of competition requires that firms utilize all of their available resources in order to survive and succeed. Consequently, their supply chains need to be very efficient. The present work aims at narrowing down the different variables leading to SC performance. At this stage, pilot study results indicate the need of an exhaustive model to assess the SC performance. However, the results show that in case of 5 SBR variables except SBR4, on all other variables the respondents show strong agreement. And, in case of 12 HM variables except HM2, HM3, HM4 & HM5, on all other variables the respondents show strong agreement. Overall, the respondents show significant agreement towards all 18 initial variables under study. Factor analysis has yielded 5 major factors comprising of 15 variables. And, overall reliability of the scale is 0.697.

Thus, the chapter has empirically tested the effect of supplier-buyer relationships and human metrics on supply chain performance measurement for manufacturing organizations. A new approach to measure supply chain performance has been put forward which can be incorporated along with the traditional measures, which are mostly financial in nature.

REFERENCES

Autry, C. W., & Susan, G. (2010). Evaluating buyer-supplier relationship–performance spirals-A longitudinal study. *Journal of Operations Management*, *24*(2), 87–100. doi:10.1016/j.jom.2009.07.003

Beamon, B. M. (1999). Measuring supply chain Performance. *International Journal of Operations & Production Management*, *19*(3), 275–292. doi:10.1108/01443579910249714

Chandra, K., & Kumar, S. (2000). Supply chain management in theory and practice: A passing fad or a fundamental change? *Industrial Management & Data Systems*, *100*(3), 100–113. doi:10.1108/02635570010286168

Chen, J. (2008). Inter-organizational communication as a relational competency: Antecedents and performance outcomes in collaborative buyer–supplier relationships. *Journal of Operations Management*, *26*(1), 45–64. doi:10.1016/j.jom.2007.04.001

Dangayach, G., & Deshmukh, S. (2003). Evidence of manufacturing strategies in Indian industry: A survey. *International Journal of Production Economics*, *83*(3), 279–298. doi:10.1016/S0925-5273(02)00372-9

Deloitte. (2007). A Report on the Summit on Indian Manufacturing Competitiveness. *Deloitte Research: Competing in Global Manufacturing and Service Networks.* Retrieved from http://www.deloitte.com/research

Fawcett, S. E., Ogden, J. A., Magnan, G. M., & Cooper, M. B. (2006). Organisational commitment and governance for supply chain success. *International Journal of Physical Distribution & Logistics Management*, *36*(1), 22–35. doi:10.1108/09600030610642913

Gunasekaran, A. (2004). Supply Chain management: Theory and applications. *European Journal of Operational Research*, *159*(2), 265–268. doi:10.1016/j.ejor.2003.08.015

Gunasekaran, A., & Kobu, B. (2007). Performance measures and metrics in logistics and supply chain management:a review of recent literature (1995-2004) for research and applications. *International Journal of Production Research*, *45*(12), 2819–2840. doi:10.1080/00207540600806513

Gunasekaran, A., Patel, C., & McGaughey, R. A. (2004). Framework for supply chain performance measurement. *International Journal of Production Economics*, *87*(3), 333–347. doi:10.1016/j.ijpe.2003.08.003

Gunasekaran, A., Patel, C., & Tirtiroglu, E. (2001). Performance measures and metrics in a supply chain environment. *International Journal of Operations & Production Management*, *21*(1/2), 71–87. doi:10.1108/01443570110358468

Hair, J., Andeson, R., Black, B., Barry, B., & Tatham, R. (2007). *Multivariate Data Analysis* (5th ed.). New York, NY: Macmillan Publishing Company.

Humphrey, P. K., Li, W. L., & Chan, L. Y. (2004). The impact of supplier development on buyer–supplier performance. *International Journal of Management Sciences*, *32*(2), 131–143.

Lee, C. W., Kwon, I. G., & Severance, D. (2007). Relationship between supply chain performance and degree of linkage among supplier, internal integration, and customer. *Supply Chain Management: An International Journal*, *12*(6), 444–452. doi:10.1108/13598540710826371

Liang, L., Yang, F., Cook, W. D., & Zhu, J. (2006). DEA models for supply chain efficiency evaluation. *Annals of Operations Research*, *145*(1), 35–49. doi:10.1007/s10479-006-0026-7

Marwah, A. K., Thakar, G., & Gupta, R. C. (2013). An Empirical Study of Supplier-Buyer Relations as a Key Determinant of Supply Chain Performance of Indian Manufacturing Organizations. In *Proceedings of the International Conference on Managing Supply Chain for Global Competitiveness.* IIIE.

Marwah, A. K., Thakar, G., & Gupta, R. C. (2014). A Confirmatory Study of Supply Chain Performance and Competitiveness of Indian Manufacturing Organizations. *International Journal of Quality Research*, *8*(1), 23–38.

Mello, J. E., & Stank, T. P. (2005). Linking firm culture and orientation to supply chain success. *International Journal of Physical Distribution & Logistics Management*, *35*(8), 542–554. doi:10.1108/09600030510623320

Robinson, C. J., & Malhotra, M. K. (2005). Defining the concept of supply chain quality management and its relevance to academic and industrial practice. *International Journal of Production Economics*, *96*(3), 315–337. doi:10.1016/j.ijpe.2004.06.055

Sekaran, U. (2003). *Research methods for business: A skill-building approach.* John Wiley & Sons.

Spekman, R. E., Kamauff, J. W. Jr, & Myhr, N. (1998). An empirical investigation into supply chain management:A perspective on partnerships. *Supply Chain Management: An International Journal*, *3*(2), 53–67. doi:10.1108/13598549810215379

Taticchi, P., Tonelli, F., & Cagnazzo, L. (2010). Performance Measurement and Management: A literature review and a research agenda. *Measuring Business Excellence*, *14*(1), 4–18. doi:10.1108/13683041011027418

Tony, W., & Kevin, H. (2007). The human factors in managing China supply chain. In *Proceedings of the 92nd Annual International Supply Management Conference* (pp. 1-6). Nokia (China) Investment Co. Ltd. Shanghai Branch

Van Hoek, R.I. (1998). Measuring the unmeasurable - Measuring and improving performance in the supply chain. *Supply Chain Management: An International Journal*, *3*(40), 187-192.

Wisner, J. D., & Fawcett, S. E. (1991). Linking firm strategy to operating decisions through performance measurement. *Production and Inventory Management Journal*, *32*(4), 5–11.

Wouters, M. (2009). A developmental approach to performance measures- results from a longitudinal case study. *European Management Journal*, *27*(1), 64–78. doi:10.1016/j.emj.2008.06.006

ADDITIONAL READING

Bititci, U. S., Mendibil, M., Nudurupati, S., Turner, T., & Garengo, P. (2006). Dynamics of Performance measurement and organizational culture. *International Journal of Operations & Production Management*, *26*(12), 1325–1350. doi:10.1108/01443570610710579

Cigolini, R., Cozzi, M., & Perona, M. (2004). A New Framework for Supply Chain Management: Conceptual Model and Empirical Test. *International Journal of Operations & Production Management*, *24*(1), 7–14. doi:10.1108/01443570410510979

Ellram, L. M., & Cooper, M. C. (1990). Supply Chain Management, Partnerships, and the Shipper-Third-Party Relationship. *The International Journal of Logistics Management*, *1*(2), 1–10. doi:10.1108/95740939080001276

Kalwani, M., & Narayandas, N. (1995). Long-Term Manufacturing-Supplier Relationships: Do They Pay-Off For Supplier Firms? *Journal of Marketing*, *59*(1), 1–16. doi:10.2307/1252010

Kennerley, M., & Neely, A. (2003). Measuring performance in a changing business environment. *International Journal of Operations & Production Management*, *23*(2), 213–229. doi:10.1108/01443570310458465

Lambert, D. M., & Cooper, M. C. (2000). Issues in Supply Chain Management. *Industrial Marketing Management*, *29*(1), 65–83. doi:10.1016/S0019-8501(99)00113-3

Li, S., Rao, S., Ragunathan, T., & Ragunathan, B. (2005). Development and Validation of A Measurement Instrument for Studying Supply Chain Management Practices. *Journal of Operations Management*, *23*(6), 618–641. doi:10.1016/j.jom.2005.01.002

Neely, A., Gregory, M., & Platts, K. (1995). Performance measurement system design- a literature review and research agenda. *International Journal of Operations & Production Management*, *15*(4), 80–116. doi:10.1108/01443579510083622

Neely, A., & Powell, S. (2004). The challenges of performance measurement. *Management Decision*, *42*(8), 1017–1023. doi:10.1108/00251740410555515

Sahay, B. S., Gupta, J. N. D., & Mohan, R. (2006). Managing Supply Chains For Competitiveness: The Indian Scenario. *Supply Chain Management: An International Journal*, *11*(1), 15–24. doi:10.1108/13598540610642439

Sahay, B. S., & Mohan, R. (2003). Supply Chain Management Practices In Indian Industry. *International Journal of Physical Distribution & Logistics Management*, *33*(7), 582–606. doi:10.1108/09600030310499277

KEY TERMS AND DEFINITIONS

Bartlett's Test of Sphericity: It is a test statistic used to check the hypothesis that the variables are uncorrelated in the population (i.e. $r = 0$).

Correlation Matrix: It shows correlation between all possible pairs of variables included in the analysis.

Empirical Study: It is a way of gaining knowledge by means of direct and indirect observation or experience.

Factor Analysis: It is a tool to reduce the correlated variables to a manageable level.

Hypothesis: It is a formal statement that presents the expected relationship between an independent and dependent variable.

KMO Measure of Sampling Adequacy: It is an index used to examine the appropriateness of factor analysis (a score of 0.5 and more is desirable).

Manufacturing: It is the process of converting raw materials, components, or parts into finished goods that meet a customer's expectations or specifications.

Percentage of Variance: Percentage of the total variance attributed to each factor.

Reliability: It is the overall consistency of a measure.

Supplier-Buyer Relationship: A relationship characterized by cooperation, trust and commitment.

Chapter 8
Logistics for Global Automobile Manufacturing Industry

Ravi Potti
JDA Software, India

Swapnil Saurav
JDA Software, India

ABSTRACT

The automotive industry is truly a global industry. Manufacturing plant of the company is set up in one corner of the earth, sourcing the automobile parts from many companies located all over the globes, and selling the final product in another country. This leads to complex supply chain. Production strategies like just-in-time (JIT) and just-in-sequence (JIS) creates testing times but presents opportunities for the logistics management. But currently, the industry is going through a rough phase. Manufacturing costs are increasing, rise in fuel prices is having its own affect on the industry, competitive landscape is changing, and globalization presents new political and economic challenges. Currently, the company needs to manage its supply chains efficiently, reduce cost and improve the accuracy of processes delivery. Today, the automobile industry is facing significant other challenges due to complex logistics. Logistics is about co-coordinating various activities like materials handling, storage, inventory management, freight transport and the related information flow and processing in a way that meets customer requirements at minimum cost. This paper will talk about the current challenges, opportunities, growth drivers and future strategies of the companies' logistics management. It will also cover how companies are adapting to the changing economic environment. The study is based on the qualitative analysis of secondary data. The challenges and opportunities for the companies exist from the current operations management and risks of globalization. Authors have looked at the challenges and opportunities in detail in this chapter. Improving the organization's performance and securing competitive advantage over others greatly depends on the supply chain management as it is seen that the competition is no longer between the organizations, rather they are amongst supply chains. From the research, it has been found that companies believe that for sustainable transportation and logistics, green supply chain is a must for long term sustainability of the company. Companies finds export compliance to the emerging markets as another major challenge. Transport and warehouse systems are, more or less, stabilized and companies are looking to get most out of them. Though there is connectivity between automobile companies and its suppliers but information is not available real time and hence it's difficult to collaborate. Visibility into logistics has come out as the biggest need for the companies.

DOI: 10.4018/978-1-4666-9795-9.ch008

INTRODUCTION

In 1933, Harlow H. Curtice, president of General Motors had said- "General Motors has no bad years, only good years and better years." General Motors ruled automobile industry for over 50 years. In 1980, Japan for the first time surpassed the United States as the leading automobile-producing nation in the world. What this means is that, there is always something which customers want and even the leaders of the industry fail to foresee. Ever since then, the world order for leadership in the automobile sector has changed. One of the important factors that led to this is logistics. Logistics management is the part of supply chain management that plans, implements, and controls the efficient, effective, forward, and reverse flow and storage of goods, services, and related information between the point of origin and the point of consumption in order to meet customer's requirements.

Efficient and effective logistics management has become a survival factor for the automotive sector. As conditions of post-crisis impacts the automotive industry and due to strong competitive pressures of automakers in global market, the flexibility to manage of materials and information flow in automobile assembly plants are declared as the key specification to future growth (IBM, 2012). Automobile manufacturers must schedule the supply at few thousand sub-assemblies and components in vehicle, with billions of possible combinations to car outfit. Key trend in the automotive production is standardization of modules of construction to common platforms (Fathi 2011). Modules refer to groups of related components and systems serving for the same or connected tasks (e.g. the front/rear axle, complete front-section of a body or the steering system).

Automobile companies are looking for reducing total costs not just manufacturing cost or labor cost. There is a tremendous pressure on suppliers from emerging markets as total supply chain cost would increase when equipment parts are sourced from all over the world. Today, as Original Equipment Manufacturers (OEMs) design and build their vehicles globally, automotive industry supply chains have become increasingly complex (Fathi 2011). Automotive supply chains need to become more responsive and flexible to remain competitive. Automotive OEMs and suppliers require capability to manage shorter production lifecycles using processes such as Make-To-Order, Assemble-To-Order, and Configure-To-Order (PRTM Management Consultants, 2008). One of the important factors which changed the face of logistics management was Information Technology. But to the get maximum benefit out of the system, suppliers must cultivate a system which is easy to use, integrated with other systems, scalable for future use and secure to protect the data. Logistics has thus, gained a very importance place in today's automobile supply chain.

This paper attempts to bring out the challenges faced by the global automobile industry related to the logistics management and also look at the industry growth drivers and opportunities for logistics management. This paper is based on studied opinions, published inputs, expert views available in public domain and authors own analysis. The information and data in this study about the supply chain management is collected from 30 global automobile organizations. This study will show the aspects of supply chain management in the automobile industry with focus on logistics management. It will also highlight the challenges faced in global logistics management, capabilities and the priorities for the next 2-3 years. While logistics management in global scenario is extremely complex, but organizations involved across the globe share many priorities — balance the availability of the inventory; control the costs and customer service. The idea behind the research has been to explore how all the stakeholders are managing these priorities and how are they changing in current times.

BACKGROUND

This research, based on the automobile companies of every revenue size and geographical areas, shows that the companies have turned to the supply chain professionals in order to experience further growth by managing global logistics. The supply chain professionals take up the challenge as the essential role in the organization's success. The automobile executives are finding it difficult to cope up with the increasing complexity of the business due to globalization and are in pressure to reduce costs as the businesses face more global pressure.

These professionals need to understand the current scenario and the future growth prospects. The research findings will help these executives and professionals to understand the challenges and the growth opportunities that the sector has to offer. The data collected, for the study, is a combination of companies' views expressed in their annual report and other disclosures and practical observations being carried out in the logistics sector of automobile industry. The research goal is to be able to recognize the essential role that the supply chain management plays in the success of managing global logistics. The result will show the current capabilities, immediate challenges, and supply chain priorities for automobile making organizations.

The purpose of this study is to explore the critical role that logistics management plays in the automobile industry and to study the changing landscape of automobile supply chain management. The study addresses three key research questions:

- What are the current logistics capabilities of automobile companies and the logistics factors companies think they are managing well?
- What logistics challenges are of greatest concern to the automobile industry in next year or two and the logistics factors that companies think they need to manage immediately?
- What capabilities automobile manufacturing companies must develop and leverage to achieve logistics excellence and the logistics factors where the company has already started making or thinking about investing?

The study is aimed at providing information to various stakeholders to facilitate future planning and decision making.

METHODOLOGY OF RESEARCH

The research in this study included studying the current logistics challenges and capabilities of automobile industry players and their priorities for the next 2-3 years. . Research was done by reading the well-known automobile companies' strategies from their annual reports, earnings call, executive interviews, and other company presentations, referring to other articles from top automobile magazines and also by talking to industry experts. Then the results were organized into emergent themes and the major issues in Automobile Supply Chain Management were identified. Both qualitative and quantitative methods were used for the study. Study methodology is listed in Table 1.

Research included collecting the supply chain related information from publically available data for world's top Automobile companies as given in Table 2.

Table 1. Analytical steps

Step 1	Prepared a list of Automobile companies which would be part of the study.
Step 2	Read through their Annual reports, Earnings call and other company reports.
Step 3	Looked for supply chain related post/comments by leading magazines, top executives comments/interviews and solution providers' press release.
Step 4	Data collection- Data about Automobile companies supply chain capabilities, its challenges and future supply chain strategies.
Step 5	Data Analysis

Table 2. Companies used for study

S.No.	Company	Country	Revenue
1	Ashok Leyland	India	$2.3 B
2	Audi AG	Germany	$62.7 B
3	Bayerische Motoren Werke	Germany	$98.7 B
4	Chrysler Group	US	$65.784 B
5	Daihatsu Motor	Japan	$21.3 B
6	Eicher Motors	India	$1.2 B
7	Fiat SpA	Italy	$107.9 B
8	Force Motors	India	$269.3 M
9	Ford Motor	US	$134.3 B
10	GB Auto Group	Egypt	$1.4 B
11	General Motors	US	$152.3 B
12	Hindustan Motors	India	$106 M
13	Hino Motors	Japan	$18.6 B
14	Honda Motor	Japan	$119 B
15	Hyundai Motor	S. Korea	$75 B
16	Isuzu Motors	Japan	$20 B
17	Kia Motors	S. Korea	$42 B
18	Mahindra & Mahindra	India	$12.7 B
19	Maruti Suzuki	India	$8.3 B
20	Mazda Motor	Japan	$26.6 B
21	Mitsubishi Motors	Japan	$21.9 B
22	Nissan Motor	Japan	$116 B
23	Paccar Inc	US	$17 B
24	Peugeot SA	France	$71.3 B
25	Porsche Automobile	Germany	$4.9 B
26	SAIC Motor	China	$76.2 B
27	Tata Motors	India	$34.9 B
28	Tesla Motors	US	$413.3 M
29	Toyota Motors	Japan	$265.9 B
30	Volkswagen AG	Germany	$247.6 B

Table 3. Screen of the analysis

	Toyota	Ford	Tata Motors	General Motors	Volkswagen	%
Current Challenges						
Visibility	1	1	1		1	83.3%
Rising Fuel Costs	1	1	1	1	1	86.7%
Cut on emission		1		1		53.3%
Current Capabilities						
Transportation Management		1		1		53.3%
Warehouse Management	1		1	1		43.3%
Investment Strategies						
Supplier Sustainability	1		1		1	73.3%
Green Logistics/Supply Chain	1	1		1	1	83.3%

The information from each company then divided into 4 groups — current logistics challenges, current capabilities, logistics investment strategies and long term strategies adopted by the companies. Then the data based on points was grouped under the discussion headers. Screenshot of the analysis is shown below in Table 3.

MAIN FOCUS OF THE CHAPTER

Automobile landscape is constantly changing. Players win or lose based on how well they manage their critical supply chain links. Supply Chain Management requires a logistics model to meet the customers' demanding and specific requirements. Expectations from the logistics operations are: organizational flexibility and responsiveness, internal and external teamwork and demands the use of processes and technology. This paper will focus on these things and understand their impact. Most of the global sourcing is done through sea routes and sea transportation is becoming increasingly competitive. Companies expect sea transportation to offer decreased risk of quality and safety, increase in visibility leading to improved reliability for material inventories and decrease in inventory levels and all this at lower supply chain cost. If supply chain cost goes up then it will offset the lower manufacturing cost. The focus here is to understand how logistics can work to create effective and productive solutions to meet various challenges faced by the manufacturers, suppliers and logistics providers.

The analysis of the acquired information in the context of global supply chain management scenario and challenges has been used towards the key findings presented in this section. The findings of the study show the capabilities of supply chain management. The results from the study have been divided into:

1. Current Challenges and immediate concerns of the automobile companies in terms of managing supply chain specifically logistics,
2. Automobile companies current logistics capabilities,
3. Logistics areas where companies are looking to invest in near future, and
4. Logistics areas where companies have already or will be looking to invest in next 3 to 5 years.

Solutions and Recommendations

A. Supply Chain Challenges for Immediate Attention

Current challenges of automobile companies and their impact on logistics management and results are presented in the Table 4.

Most of the companies studied, have talked about various economic factors like globalization and rising energy costs as their primary challenges while managing the logistics. One of the major consequences of globalization is a loss of visibility to global operations, and therefore a loss of control. Many companies are actively investing in technologies to regain control through visibility and alignment across functions and trading partners. Increasing oil demand from the emerging markets and continuously rising extraction costs are driving the energy costs. Along with these costs, impact of globalization is having huge impact. Further to this, automotive companies are finding it hard to pass this increase in costs on to the buyers due to stiff competition and are taking hit on their operating margins. Export compliance poses more risks for the companies in newer market.

B. Current Supply Chain and Logistics Capabilities of the Automobile Manufacturers

Table 5 depicts top important current logistics management capabilities which the companies investing in the current situation.

We found that that more and more companies are outsourcing warehouse and transportation management functions to the 3rd party logistics (3PL) logistics providers.

Companies are putting considerable focus on working with dealerships and logistics providers to improve inventory management and delivery reliability. Dealers are being encouraged to make at least once or twice weekly stock orders, which also benefits logistics companies as it allows them to plan for larger loads. The companies are also sharing statistical reports with logistics providers and monitoring their performance to improve delivery services. This inventory management, data sharing initiatives are enabling companies to reduce the rushed orders which usually move by air and thereby reducing the operational costs.

Companies which already have these capabilities may not think of making any investment for next 2-3 years but these factors become immediate cause of concerns for others.

Table 4. Current challenges for logistics management

Rank	Current Challenges	Score
1	Rising fuel cost	87%
2	Visibility Control	83%
3	Globalization	67%
4	Cut on emission	53%
5	Export compliance risk	50%

Table 5. Important logistics management capabilities

Rank	Current Capabilities	Score
1	Inventory Management	63%
2	Transportation Management	53%
3	Warehouse Management	43%
4	Data sharing	43%
5	Delivery Reliability	37%

C. Current Logistics Investment Strategies of the Automobile Manufacturers

Table 6 presents investment strategies of the companies in short terms (0-2 years). The Investment in Green Supply Chain turned out to be the strategy getting maximum investment. Supply chain carbon optimization and alternative energy vehicles are some of the investment areas to move towards the green logistics. Supplier sustainability is another area of the strategies which most of the companies are employing. Some of the initiatives which companies are taking in order to achieve sustainability in their operations are entering into agreements with supply chain partners and making them adhere to highest ethical standards and designing procurement framework for supply chain partners.

As more and more companies are looking at emerging markets for their future growth, there are multiple challenges in terms of infrastructure and distribution which companies are looking at solving. Companies are looking to diversify production locations, opportunities to adopt local dealer city distribution centre (LDC DC) and micro LDC to counter the problem of infrastructure congestion.

Rising cost of conventional fuels is forcing automobile and 3PL companies to invest in improving their capacity utilization. They are looking at various optimization initiatives such as multi modal transport and network optimization applications.

Companies are also investing in number of initiatives such as collaborating with key customers in value chain and in initiatives to gain end to end supply chain visibility to achieve higher rate of delivery performance. Also companies are looking ways to gain cost efficiencies by moving away from low cost country sourcing model to the best cost country sourcing and practicing.

D. Long Term Supply Chain Growth Strategies of the Automobile Manufacturers

Based on the factors discussed under sections A, B and C, Top 5 long term strategies that automobile companies will adopt or in some cases have already started adopting has been derived (Table 7) Companies think these factors to be their competitive advantage over its peers.

There is a consolidation of global vehicle platforms across the regions, combined with a number of global vehicle launches, continue to drive inter-regional material flow. These changes require the need for fast response in logistics and network planning. Companies are looking at Logistics Engineering strategy to plan logistics in ways that lower daily transport and inventory costs, as well as how to use logistics data to better inform decisions on sourcing and production locations.

Table 6. Investment strategies automobile logistics management

Rank	Investment Strategies	Score
1	Green Logistics/supply chain	83%
2	Supplier Sustainability	73%
3	International growth	62%
4	Sourcing strategy	43%
5	Cost Management and Delivery performance	37%

Table 7. Elements of logistics management for future growth

Rank	Strategy for Long Term
1	Multi Modal Transport
2	Digital Logistics
3	Build-to-Distribute
4	Logistics Engineering
5	Emergence of 4PL

Growing customer demands coupled with lack of coordination among various supply chain partners are creating new challenges for the companies. Companies are looking to invest in various digital initiatives like visibility tools such as RFID technology, network and satellite communications to achieve best in industry service level and to increase visibility throughout the value chain.

Another major industry trend which will impact the industry is the emergence of 4th party logistics providers (4PL). As initiated by GM, by outsourcing all of its logistics to 4PL player Gefco. More and more companies in future will outsource its design and procurement functions for inbound and outbound logistics functions including the control of inbound material from suppliers or ports to assembly and powertrain plants to 4PL players in order to save costs. There is a scope of global companies embracing this idea in coming years to save costs beside this gives them space to concentrate on their core capabilities.

With the global car production growth taking place in new and emerging markets; there have been significant changes to OEM outbound distribution networks. Companies are taking 'build-for-distribution' approach – making distribution a priority in the production system before cars are built – which has allowed the carmaker, working closely with its service providers, to gain stability and timely delivery. Prioritizing customers in its ordering system and working on a closer interaction between production and distribution, companies are able to bring down the order-to-delivery time and reducing dispatch space, transport costs and process time.

With the growing concerns of automobile companies over transport logistics capacity issues, added costs from taxes and regulations, companies are looking at multi modal transportation as future strategy to solve the issue. While there is a concern over the environmental footprint for embracing this strategy; it also allows efficiency in terms of low/traffic and more labor turnover compared to conventional transports methods.

FUTURE RESEARCH DIRECTIONS

There are hundreds of suppliers and spare parts units working for Automotive industry. Globalization has brought a revolution in the way Automotive industry works. This Industry is constantly looking for innovative methods to operate by reducing operational cost, inventory and lead time and at the same time improve forecast accuracy and fill rate. Technological advancement and globalization has helped the emerging economies. Suppliers and sourcing partners, based in emerging economies are working together with automotive companies to better manage inventory, bring more visibility and control into the supply chain process. This is possible when there is a co-operation and coordination across the supply chain partners. These suppliers based in emerging economies have their own set of challenges like high raw material costs, energy costs and poorer infrastructure. Future research work aimed at integrating the entire supply chain should identify and capture key challenges involved in integrating and managing the integrated supply chain and suggest strategies to overcome these challenges.

CONCLUSION

Companies believe that green supply chain is must for long term sustainability of the company. Environment friendly strategies are most talked about by the companies. It was also found that export compliance to the emerging markets is another major challenges faced by the companies. Transport and warehouse systems are, more or less, stabilized and companies are looking to get most out of them. Though there is a connectivity between automobile companies and its suppliers but information is not available real time and hence it is difficult to collaborate. Visibility into logistics has come out as the biggest need for the companies.

Managing labor costs as a part of overall cost reduction initiatives is a big challenge for the automotive companies. Globalization has presented the companies with increased profitability but also a different set of problems along with risks which companies have never experienced before. Focus is thus to increase the capacity utilization to reduce cost and also the emissions. Network optimization in global environment is the need of the hour.

Due to all these challenges, companies are taking up few growth strategies like Green Logistics and Supplier sustainability very seriously. International growth to emerging markets is pushing the sales upward so companies are looking to have manufacturing setup closer to "home". Automobile companies are working more closely with the supplier from these regions. There is a focus on local sourcing as these not only decrease the cost but also reduces the logistics related risks (delivery on time).

The logistics and supply chain executives of the automobile companies are keen to look into the above discussed issues for further planning and are preparing to deal with the issues. Emergence of 4PL, which takes over complete logistics division of a company, will gain prominence in automobile industry due to the discussed challenges faced by the industry.

REFERENCES

Babu, J. (2012). Study on Supply Chain Practices with References to Automobile Industry. *International Journal of Marketing, Financial Services & Management Research, 1*(9).

Barry, J. (2004). Supply chain risk in an uncertain global supply chain environment. *International Journal of Physical Distribution & Logistics Management, 34*(9), 695–697. doi:10.1108/09600030410567469

Deveshwar, A., & Rathee, R. (2010, July). Challenges For Supply Chain Management In Today's Global Competitive Environment. *International Review of Business Research Papers, 6*(2), 194–203.

Duschek, S. (2004). Inter-Firm Resources and Sustained Competitive Advantage. *Management Review.*

Fathi, T. (2011). *Automotive Supply Chain Trends – Innovation and Solutions*. DHL.

Fleischmann, B., Ferber, S., & Henrich, P. (2006). Strategic Planning of BMW's Global Production Network. *Interfaces.*

Global Automotive Logistics. (2013). *Report.* Research and Markets.

Harrison, A., & van Hoek, R. (2008). *Logistics Management and Strategy: Competing Through The Supply Chain*. London: Prentice-Hall.

Holweg, M. (2008). The Evolution of Competition in the Automotive Industry. In G. Parry & A. Graves (Eds.), *Build To Order*. London: Springer. doi:10.1007/978-1-84800-225-8_2

IBM. (2012). *The Smarter Supply Chain of the future*. IBM.

Interview articles of logistics executives from various companies

Klug. (2006). *Logistics implications of global production networks in car manufacturing*. University of Applied Sciences Munich.

Little, A. D. (2007). *Innovation Excellence in Logistics*. European Logistics Association.

PRTM Management Consultants. (2008). *Global Supply Chain Trends 2008 – 2010*. Author.

Orton, J. D., & Weick, K. E. (1990). Loosely coupled systems: A reconceptualization. *Academy of Management Review*.

PWC. (2012). *Next-generation supply chains Efficient, fast and tailored*. PWC.

Supply Chain Value. (n.d.). UPS Whitepaper.

Tseng & Yue. (2005). *The role of transportation in logistics chain*. Eastern Asia Society for Transportation Studies.

World Economic Forum. (2012). *Outlook on the Logistics & Supply Chain Industry 2012*. Author.

Yiakoumi. (2014). Overcoming Automotive Logistics Challenges. *Supply Chain Nation*.

Zielinski. (n.d.). *Logistics Challenges in a Changing Auto Industry*. CEVA Logistics.

ADDITIONAL READING

Autry, C. W., & Bobbitt, M. L. (2008). Supply chain security orientation: Conceptual development and a proposed framework. *International Journal of Logistics Management*, *19*(1), 42–64. doi:10.1108/09574090810872596

David, B., & Florian, K. (2012). Logistics supplier integration in the automotive industry. *International Journal of Operations & Production Management*.

de Leeuw, S., Holweg, M., & Williams, G. (2011). The impact of decentralised control on firm-level inventory: Evidence from the automotive industry. *International Journal of Physical Distribution & Logistics Management*, *41*(5), 435–456. doi:10.1108/09600031111138817

Gereffi, G., Humphrey, J., & Sturgeon, T. (2005). The Governance of Global Value Chains. *Review of International Political Economy*, *12*(1), 78–104. doi:10.1080/09692290500049805

Omera, K., & Bernard, B. (2007). Risk and supply chain management: Creating a research agenda. *The International Journal of Logistics Management*, *18*(Issue: 2), 197–216. doi:10.1108/09574090710816931

Storey, J., Emberson, C., Godsell, J., & Harrison, A. (2006). Supply chain management: Theory, practice and future challenges. *International Journal of Operations Production Management Volume*, *26*(7), 754–774. doi:10.1108/01443570610672220

Sturgeon, T., Memedovic, O., Van Biesebroeck, J., & Gereffi, G. (2009). "Globalisation of the automotive industry: main features and trends", International Journal of Technological Learning, Innovation and Development.

Sturgeon, T., & Van Biesebroeck, J. (2011). "Global value chains in the automotive industry: an enhanced role for developing countries", International Journal of Technological Learning, Innovation and Development.

KEY TERMS AND DEFINITIONS

Actual Demand: Actual demand is composed of customer orders (and often allocations of items, ingredients, or raw materials to production or distribution). Actual demand nets against or "consumes" the forecast, depending upon the rules chosen over a time horizon.

Back Order: Product ordered but out of stock and promised to ship when the product becomes available.

Barrier to Entry: Factors that prevent companies from entering into a particular market, such as high initial investment in equipment.

Base Demand: The percentage of a company's demand that is derived from continuing contracts and/or existing customers. Because this demand is well known and recurring, it becomes the basis of management's plans. Synonym: Baseload Demand.

Bottleneck: A constraint, obstacle or planned control that limits throughput or the utilization of capacity.

Break-Even Point: The level of production or the volume of sales at which operations are neither profitable nor unprofitable. The break-even point is the intersection of the total revenue and total cost curves.

Bulk Storage: The process of housing or storing materials and packages in larger quantities, generally using the original packaging or shipping containers or boxes.

Capacity Planning: Assuring that needed resources (e.g., manufacturing capacity, distribution center capacity, transportation vehicles, etc.) will be available at the right time and place to meet logistics and supply chain needs.

CAPEX: A term used to describe the monetary requirements (CAPital EXPenditure) of an initial investment in new machines or equipment.

Consortium: A group of companies that work together to jointly produce a product, service, or project.

Dedicated Contract Carriage: A third-party service that dedicates equipment (vehicles) and drivers to a single customer for its exclusive use on a contractual basis.

Domestic Trunk Line Carrier: An air carrier classification for carriers that operate between major population centers. These carriers are now classified as major carriers.

Duty Free Zone (DFZ): An area where goods or cargo can be stored without paying import customs duties while awaiting manufacturing or future transport.

Fourth-Party Logistics (4PL): Differs from third party logistics in the following ways; 1)4PL organization is often a separate entity established as a joint venture or long-term contract between a primary client and one or more partners; 2) 4PL organization acts as a single interface between the client and multiple logistics service providers; 3) All aspects (ideally) of the client's supply chain are managed by the 4PL organization; and, 4) It is possible for a major third-party logistics provider to form a 4PL organization within its existing structure. The term was registered by Accenture as a trademark in 1996 and defined as "A supply chain integrator that assembles and manages the resources, capabilities, and technology of its own organization with those of complementary service providers to deliver a comprehensive supply chain solution.", but is no longer registered.

Integrated Carrier: A company that offers a blend of transportation services such as land, sea and air carriage, freight forwarding, and ground handling.

Integrated Logistics: A comprehensive, system-wide view of the entire supply chain as a single process, from raw materials supply through finished goods distribution. All functions that make up the supply chain are managed as a single entity, rather than managing individual functions separately.

Obsolete Inventory: Inventory for which there is no forecast demand expected. A condition of being out of date. A loss of value occasioned by new developments that place the older property at a competitive disadvantage.

Outbound Logistics: The process related to the movement and storage of products from the end of the production line to the end user.

Third-Party Logistics (3PL): Outsourcing all or much of a company's logistics operations to a specialized company. The term "3PL" was first used in the early 1970s to identify intermodal marketing companies (IMCs) in transportation contracts.

Chapter 9
Supply Chain Management:
Large vs. Small and Medium Enterprises (SMEs)

Y. Ramakrishna
Vignana Jyothi Institute of Management, India

ABSTRACT

Majority of research conducted in Supply Chain Management worldwide relates to areas of supply chain like, definitions, its frameworks and strategies, supply chain design, achievement of competitive advantage through its successful implementation etc. But, vast part of this research is limited to only large scale enterprises and very limited research is done in supply chain related to SMEs sector. Also, the strategies and frameworks of SCM that are developed for these large enterprises are found to be not suitable and implementable in small scale enterprises. This lack of research with special focus on SCM is more pertinent to Indian SMEs and its much more less in small enterprises belonging to manufacturing sector. It is found that most of the strategies of SCM applicable and implementable in large enterprises are either difficult to implement or SMEs do not view those strategies seriously due to lack of understanding, implementation feasibility and awareness. Thus, there are many differences in the way the SCM is viewed and understood in SMEs when compared to that of large enterprises. This paper focuses on these differences of SCM with reference to challenges and difficulties in implementations when compared with large enterprises with the help of literature review. It is found that there exists a huge gap in understanding, appreciating and utilizing the practices of SCM in SME sector. The research is limited only to the secondary data available in published national / international journals. It is identified that SCM can be a potential tool for competitive advantage, cost minimization and optimal decision-making, risk-sharing in SMEs also. The paper identified the gaps, challenges and opportunities for SMEs in the implementation of SCM.

DOI: 10.4018/978-1-4666-9795-9.ch009

INTRODUCTION

Small and Medium Enterprises

As there is no standard way of defining an SME, it is defined in different manner by different authors and authorities. The important ones among them are definitions based on geographical location, size of the enterprise in terms of no. of employees, no. of products manufactured, supplier or OEM (Original Equipment Manufacturer), structure, volume of sales, worth of assets, type of ownership, adoption of technology etc. (Rahman, 2001). In India, the Industries Development and Regulation Act, 1951 (IDR), defined the small scale enterprises, but no formal definition for medium scale enterprises. The MSMED (Micro, Small and Medium Enterprises Development Act, 2006) defined medium enterprise as an enterprise which has investment more than Rs. 50 million but not exceeding Rs. 100 million in plant and machinery.

Supply Chain Management

Supply Chain Management (SCM) is defined by several authors in different ways. About 173 unique definitions of SCM are available in the literature. An encompassing definition of SCM was provided (James R. Stock and Stefanie L. Boyer, 2009) through a review of these 173 definitions in a systematic manner by analyzing the terms involved in each of these definitions. SCM is management of upstream and downstream relationships with suppliers and customers to deliver superior customer value at lesser cost to the chain as a whole (Martin Christopher, 1992)

SCM constitutes a set of approaches utilized to efficiently integrate suppliers, manufacturers, warehouses, and retailers, so that merchandize is produced and distributed in the right quantities, to the right locations, and at the right times, in order to minimize system-wide costs while satisfying service level requirements (Simchi-Levi et al., 2000).

Research Methodology

For the purpose of the study, a thorough literature review was done using the published articles in various national / international journals related to the study. The paper is organized into three phases. The first phase identifies the gaps, difficulties, challenges and obstacles in the implementation of supply chain practices in SMEs compared to the large enterprises. The second phase highlights the role and contribution of SME sector to the overall economic growth of any country. The third phase summarizes the gaps and recommends strategies for successful implementation of SCM.

SCM IN LARGE VS. SMES: GAPS AND CHALLENGES

Many studies were carried out related to supply chain management worldwide. Majority of these studies were focused on definitions, frameworks, supply chain strategies and design, achievement of competitive advantage through effective management of SCM etc. A thorough observation of literature reveals that majority of researchers have focused on supply chain management issues related to large enterprises. The literature review also indicates that predominantly the scope of the research was restricted to large companies in developed counties. A very little literature is available with special emphasis and focus on SCM in medium size manufacturing companies, especially in developing countries like India. The needs and operating environment of SMEs is very different from large enterprises. There is a scarcity of literature on the usage of practices related to SCM and its impact on the overall performance of SMEs in emerging market economies (Lenny Koh et. al, 2007).

Inspite of its great benefits, the successful SCM implementation incurs costs, faces challenges and runs through hazards. Compared to large enterprises, the small and medium enterprises are unable to take leverage of SCM and get benefits out of it. Rather, they face obstacles in implementing those practices. Several studies have identified that there is a negative correlation of SME performance with SCM implementations (Arend & Wisner, 2005).

Findings of a research of 152 Indian SMEs reveal that absence of a dedicated supply chain management team; lack of clear alliance guidelines and lack of proper supply chain training are the major barriers for the successful implementation of SCM (Rouhollah Zahedirad & Shivraj, 2011)

A study of about 120 small and medium enterprises (Maguire & Magrys, 2001) found that most of the methodologies, techniques and tools related to information and communications technology were developed keeping in view the large firms rather than small firms. The study used metrics like Finished Product Tardy Delivery (FPTD) to measure the delivery performance, ERP applications, CRM applications, Knowledge management and found that size of medium sized enterprises influences these metrics. It is found that there are very important differences between SMEs and large enterprises. The top management in SMEs lack formal and professional training and they generally involve themselves only in short-term operational issues rather than creating and developing a long-term vision and mission (McCartan-Quinn & Carson, 2003). This lack of vision and clarity in long term strategies is hindering the growth of SMEs (Gunasekaran et al., 1996). It is found that inconsistent operational processes, Small Internal Team, lack of Project management capabilities, lack of training, lack of continuous management support, heavy dependencies on consultants are major obstacles for proper implementation of SCM in SMEs (Brent Snider, 2008). Many recent studies of supply chain management focus on the practices of large firms, while small firms are treated mostly from the view point of larger firms (Chopra & Meindl, 2001; Kukalis, 1989; Lambert & Cooper, 2000)

It is observed that majority of the schemes developed for SMEs are based on competencies of large organization management functions which are not suitable for SMEs management competencies (Westhead & Storey, 1996).

Whilst SMEs have a strong focus on the requirements of the markets they serve, for financial reasons, this focus tends to be directed at maximizing short-term sales opportunities rather than optimizing long-term supply chain performance (Mezgar et al., 2000; Wyer & Mason, 1998).

SMEs face several challenges in terms of SCM implementation. These are divided into three categories: Challenges related to country, related to industry and related to the firm specifically. Broadly the challenges identified in the study were efficiency and global reach, lack of innovation, lack of risk taking attitude and pro-activeness (Todd & Javalgi, 2007). A study based on SMEs belonging to different sectors revealed that they give less importance and consideration to planning and control methods than large enterprises. They show less concern for i) SCM methods applied, ii) methods that support SCM on product quality, iii) balancing the operations and capital cost, iii) integration of systems with other entities in supply chain, iv) EDI and v) e-based solutions (Valand & Heide, 2007). One of the reasons attributed to this is the significant difference in the implementation processes of it (Arend & Wisner, 2005). Similar gaps and differences were found in the study of SMEs in UK too (Quayle, 2003).

Implementation of new technology, research and e-commerce were considered to be of low importance by SMEs (Wagner et al. 2003).

It is also found that the SMEs do not compete with large firms and majority of the SMEs are not very optimistic about their growth in the near future. Small and Medium Enterprises are unable to adopt the latest online / e-business supply chain solutions. The barriers to this adoption are found to be negative perception about these solutions, discouraging economic perceptions and dependency of SMEs on their business partners in supply chain related decision-making (Norm Archer, Shan Wang and Claire Kang, 2008). The barriers to achieve competitiveness and become competitive are insufficient technology, lack of expert man-power, scarcity of financial resources, lack of environment which provides growth, inadequate support from customers as well as government (Singh, Garg & Deshmukh, 2008).

SMEs face innumerable challenges and obstacles like limited financial and human resources in the implementation of new technological systems like ERP (McAdam, 2002; Achanga et al., 2006). The Critical Success Factors (CSFs) influencing the implementation of ERP in SMEs is studied by using case study approach method of five enterprises (Brent Snider et al. 2008). Fragmented approaches, lack of integration, lack of management buy-in, difficulties in the measurement and availability of information and inadequate information systems can all act as barriers, preventing organizations from implementing SCM (Monckza & Morgan, 1997). A study of 250 SMEs in Merseyside, UK (Meehan & Muir, 2008) to identify barriers revealed that SMEs realize the benefit of SCM implementation but they were unable to do so due to barriers mentioned below.

1. Scarcity of skilled manpower.
2. Lack of initiatives that implement SCM.
3. Inadequate experience in the execution of development programs.
4. Disinterest in the participation throughout the supply chain.
5. Absence of trust among the different entities of supply chain.
6. Inadequate knowledge in managing trading online.
7. Wide spread distances between customers and suppliers.

Like many other implementations, the ERP implementation was also successful in large enterprises rather than small enterprises (Snider, 2008). The reason attributed to this gap in ERP was lack of research focus on SMEs in this aspect and more research focus on large enterprises. It is found that SMEs did not give much importance to development of strategy for achieving competitiveness (Singh, et. al, 2008) as they focused more on short-term goals and have no formal framework towards long-term goals.

The report of the symposium (Dec, 2001) on SCM in SMEs has found that the following issues affect the implementation of SCM in SMEs in Asia and India. They are,

- Low Awareness of SCM practices.
- Supply Chain practices are implemented slowly in SMEs.
- The scope of supply chain in SMEs is limited local level.
- There is no information sharing among SME vendors and customers.
- Information Technology is not yet used for achieving competitive advantage.
- SMEs is look for governmental support for the development of SCM.

In the same report (Dec, 2001), some of the barriers for implementation and development of SCM in SMEs in Nepal are identified to be,

- Natural resources, Geographical restrictions and country's Infrastructure.
- Difficulty in the achievement of Economies of Scale.
- Policies of Government, Stability of Governments, Its Rules and Regulations.
- Political affairs prevailing in the country.
- Institutional support provided by Government.

The literature also indicates that strategies of supply chain are more or less consolidated in developed countries in large size organizations. But very few attempts to consolidate and integrate SCM have been made in MSME sector, especially the medium size companies, in India. Though an effort is made to cover the most of the available literature in this research, the focus is more on SCM literature related to medium size manufacturing companies in India and abroad.

CONTRIBUTION OF SMES TO THE ECONOMY

The large scale contribution of SMEs to the economy of any country was emphasized many authors and research reports. SMEs play a vital role and act as one of the major driving forces in the economic development of every nation. It is found that as on July 2006, there are around 14 crore SMEs in 130 countries employing 65 percent of total labor force. In India, too the SMEs, contribute to the overall growth of economy in terms of exports, industrial production, creation of employment, promotion of entrepreneurs segment. Some salient features of SMEs are found to be (Deloitte Report, 2009),

- As on 1007, there are around 1.3 million SMEs in India.
- The GDP share of SMEs is around 9%.
- They contribute to 80% of industrial sector employment.
- They contribute approximately 45% of production in India.
- A significant 40% is from SMEs in export trade of India.
- 41 million people are employed in SMEs, which is just next to agriculture employment in the country.
- The total number of products manufactured by SMEs is 8000 which includes all varies.

It is found that, in manufacturing sector, compared to larger enterprises, SMEs can produce raw material, other spare parts and components, and sub-assemblies at lesser prices and they become the dedicated supplier to large enterprises (Rajesh K. Singh et. al, 2008). The quality aspect plays a vital role for the large enterprises to accept these products. Medium enterprises play a vital role in the economy of any country. They have certain unique characteristics compared to large enterprises. In terms of Organizational Structure they are organic by nature than to a bureaucratic structure in bigger enterprises (Ghobadian & Gallear, 1996). The characteristics of medium enterprises make them more flexible towards environment (Levy, 1998).

Over the years, SMEs have shown a consistent growth rate, both during protected economy and open economy (Ghose, 2001; Dalu & Deshmukh, 2001). In the wake of Globalization and emergence of global opportunities, the pros and cons of how Indian SMEs can leverage their existing strengths and gain competitive advantage was studied (Bala, 2007). This study found that SMEs in manufacturing sector got multi-fold opportunities due to globalization for both intermediate and consumer products. On the other hand, the study also found that the transition phase for Indian SMEs to exploit global opportunities is not smooth.

CONCLUSION AND STRATEGIES FOR SMES

SMEs face several challenges in the implementation of SCM. But, inspite of this it is found that by redesigning their strategies they can gain the advantage through SCM practices. Large enterprises view SMEs as satellites that would rotate around them seeking revenue and possible profit (Raghunath, 2001). In the present era of globalization SMEs should possess the ability to get the organization to innovate quickly and produce an acceptable product and service to capture upcoming business opportunity.

SMEs were found to be working with a traditional and conservative approach by focusing only on localized benefits which restricted their growth and scope of business (Thakkar et al, 2008). It is suggested that SMEs can look at SCM as a tool to develop strategies for improving their performance and sustain in the competitive market. They should take the advantage of SCM for maintaining mutual and useful relationships with manufacturers and large companies to obtain the operating leverage. They can look at manufacturers to gain mutual advantage and profits. With the help of mutual coordination the SMEs can look at minimizing the waste and reducing the levels of inventory at various levels. Implementation of SCM can aid the SMEs in minimizing the fluctuations in demand by increasing the levels of appropriateness demand forecasting. SMEs can improve their cash and inventory flow with the help of SCM (Thakkar et al., 2008).

Indian SMEs are currently in a transition phase and they can get advantages of globalization through highly focused efforts in terms of top management commitment and government support for this transition to be effective and useful (Bala 2007). Indian SMEs need to have vertical integration with Transnational Companies and horizontal cooperation through clustering and networking among the equal level SMEs. The study identified seven competencies for the growth of SMEs. They are, Identification of Niches, Development of New Products from time to time, Optimization of work environment, Customer driven quality standards, Frequent New technology introduction, Optimization of decisions and decision-making using information, being pro-active to market dynamics (Singh, Garg & Deshmukh, 2008)

The challenges facing SMEs in Singapore was highlighted (Roland et. al, 2006) by developing an integrated strategic supply chain positioning methodology. This methodology helps the SMEs to identify and define their strengths, opportunities, weaknesses and threats and will enable them to sustain in the global competition. Implementation of Information Systems in SMEs in India was found to be helping them in managing the day-to-day operations management (Sharma & Bhagwat, 2006). But the study revealed that though the SMEs are aware of and understand the significance of implementing the information systems, it is still not implemented by them in a professional way to gain the highest possible business outcomes. The strategic and operational choices of supply chain practices between large scale firms and SMEs were compared and analyzed on different competitive priorities. SMEs were classified into four types as Efficiency, Coordination, Collaboration and Innovation SMEs, based on Strategic Focus and Supply Chain relationship position. The growth paths for these four types of SMEs were provided for moving from one type to the other through a planned change and improvement in operational capabilities (Hong & Jeong, 2006).

Allan Macpherson and Alisn Wilson (2003) conducted a study of manufacturing SMEs using a supply chain competence map to investigate the extent to which these companies demonstrate the competencies in supply chain. The findings of this study were that the SMEs can demonstrate higher levels of competencies in supply chain, if the management development support systems available in large scale organizations are created in SMEs too. It is found that the SME sector is more heterogeneous in supply chain competencies than the large scale sector and therefore the management is highly specific and many factors impact the supply chain management (Goss & Jones, 1997).

A study on multi-sector SMEs in Pakistan (Khurrum et. al, 2007) revealed that SMEs in that country export oriented SMEs have healthier trend towards growth and dynamism. The second set of them were those who sell their products to OEMs. The successful companies found to be having more new and existing customers and more products

A study of techniques adopted by different medium to small firms in manufacturing sector was studied (Higginson & Alam, 1997) in Ontario. This survey conducted on SCM studied the techniques in four broad areas which included, Cycle Time and Inventory Management, IT and IS, Quality Assurance and Organizational and relationship issues. Based on the results of this survey, a framework was developed which concluded that the following techniques are used to implement SCM in medium sized companies.

They are,

- Development of good relationships with the members of Supply Chain.
- Development of high quality products and services.
- Sharing of information across the supply chain entities.
- Top management's commitment towards SCM.
- Process support to reduce cycle times.
- Minimization of inventory levels.
- Optimum number of suppliers and distributors.

Scope for Further Study

The current study is based on existing literature. The scope of this study can be extended by conducting an empirical study to validate and correlate the findings of this study in terms of challenges faced by SMEs in different countries.

REFERENCES

Achanga, P., Shehab, E., Roy, R., & Nelder, G. (2006). Critical success factors for lean implementation within SMEs. *Journal of Manufacturing Technology Management*, *17*(4), 460–471. doi:10.1108/17410380610662889

Arend, R. J., & Wisner, J. D. (2005). Small business and supply chain management: Is there a fit? *Journal of Business Venturing*, *20*(3), 403–436. doi:10.1016/j.jbusvent.2003.11.003

Chopra, S., & Meindl, P. (2001). *Supply Chain Management: Strategy, Planning, and Operation* (2nd ed.). Englewood Cliffs, NJ: Prentice-Hall.

Macpherson, A., & Wilson, A. (2003). Enhancing SMEs' capability: Opportunities in supply chain relationships? *Journal of Small Business and Enterprise Development*, *10*(2), 167–179. doi:10.1108/14626000310473193

Snider, B., da Silveira, G. J. C., & Balakrishnan, J. (2009). ERP implementation at SMEs: Analysis of five Canandian Cases. *International Journal of Operations & Production Management*, *29*(1), 4–29. doi:10.1108/01443570910925343

Subrahmanya. (2007). Development strategies for Indian SMEs: Promoting linkages with global transnational corporations. *Management Research News, 30*(10), 762-774.

Dalu, R. S., & Deshmukh, S. G. (2001). SWOT analysis of small and medium scale industry: A case study. Productivity, 42(2).

Dangayach, G. S., & Deshmukh, S. G. (2004). Advanced manufacturing technology implementation: Evidence from Indian small and medium enterprises (SMEs). *Journal of Manufacturing Technology Management*, *16*(5), 483–496. doi:10.1108/17410380510600473

Ghobadian, A., & Gallear, D. N. (1996). Total quality management in SMEs. *Omega*, *24*(1), 83–106. doi:10.1016/0305-0483(95)00055-0

Goss, D., & Jone, R. (1997). Organization structure and SME training provision. *International Small Business Journal*, *10*(4), 13–25. doi:10.1177/026624269201000401

Ghose, A. K. (2001). SMEs and environment protection. Productivity, 42(2).

Archer, N., Wang, S., & Kang, C. (2008). Barriers to the adoption of online supply chain solutions in small and medium enterprises. *Supply Chain Management: An International Journal*, *13*(1), 73–82. doi:10.1108/13598540810850337

Chopra, S., Meindl, P., & Kalra, D.V. (2008). Supply Chain Management: Strategy, Planning, and Operations (3rd ed.). Pearson Education.

Christopher, M., & Holweg, M. (2011). Supply Chain 2.0: Managing supply chains in the era of turbulence. *International Journal of Physical Distribution & Logistics Management*, *41*(1), 63–82. doi:10.1108/09600031111101439

Deloitte Touche Tohmatsu India Private Limited. (2009). *Bringing value to the changing Indian Landscape*. Author.

Gunasekaran, A., Okko, P., Martikainen, T., & Yli-Olli, P. (1996). Improving productivity and quality in small and medium enterprises: Cases and analysis. *International Small Business Journal*, *15*(1), 59–72. doi:10.1177/0266242696151004

Higginson, J. K., & Alam, A. (1997). Supply Chain Management Techniques in Medium-to-Small Manufacturing Firms. *The International Journal of Logistics Management*, *8*(2), 19–32. doi:10.1108/09574099710805646

Hong, P., & Jeong, J. (2006). Supply chain management practices of SMEs: From a business growth perspective. *Journal of Enterprise Information Management, 19*(3), 292–302. doi:10.1108/17410390610658478

Khurrum,, S., Bhutta, M., Arif, I. R., & Asad, U. (2007). SCM practices and the health of SMEs in Pakistan. *Supply Chain Management: An International Journal, 12*(6), 412–422. doi:10.1108/13598540710826344

Koh, S. C. L., & Maguire, S. (2004). Identifying the adoption of e-business and knowledge management within SMEs. *Journal of Small Business and Enterprise Development, 11*(3), 338–348. doi:10.1108/14626000410551591

Kukalis, S. (1989). The relationship among firm characteristics and design of strategic planning systems in large organizations. *Journal of Management, 15*(4), 565–579. doi:10.1177/014920638901500406

Lambert, D., & Cooper, M. (2000). Issues in supply chain management. *Industrial Marketing Management, 29*(1), 65–83. doi:10.1016/S0019-8501(99)00113-3

Lamming, R., Johnsen, T., Zheng, J., & Harland, C. (2000). An initial classification of supply networks. *International Journal of Operations & Production Management, 20*(6), 675–691. doi:10.1108/01443570010321667

Lenny, S. C. (2007). The impact of supply chain management practices on performance of SMEs. *Industrial Management & Data Systems, 107*(1), 103–124. doi:10.1108/02635570710719089

Levy, M., & Powell, P. (1998). SME flexibility and the role of information systems. *Small Business Economics, 11*(2), 183–196. doi:10.1023/A:1007912714741

Lim, R. Y. G., Baines, T., Tjahjono, B., & Chandraprakaikul, W. (2006). Integrated strategic supply chain positioning for SMEs: An empirical study. *The International Journal of Logistics Management, 17*(2), 260–276. doi:10.1108/09574090610689989

Lin, F., & Shaw, M. J. (1998). Reengineering the order fulfillment process in supply chain networks. *International Journal of Flexible Manufacturing Systems, 10*(3), 197–229. doi:10.1023/A:1008069816606

Maguire, S., & Magrys, A. (2001). The potential use of IT to gain competitive advantage for small and medium-sized enterprises. In *Proceedings of the 9th Annual High Technology Small Firms Conference*. Manchester Business School.

Mangan, J., & Christopher, M. (2005). Management development and supply chain manager of the future. *The International Journal of Logistics Management, 16*(2), 178–191. doi:10.1108/09574090510634494

McAdam, R. (2002). Large scale innovation – reengineering methodology in SMEs: Positivistic and phenomenological approaches. *International Small Business Journal, 20*(1), 33–50. doi:10.1177/0266242602201004

McCartan-Quinn, D., & Carson, D. (2003). Issues which impact upon marketing in the small firm. *Small Business Economics, 21*(2), 201–213. doi:10.1023/A:1025070107609

Mezgar, I., Kovacs, G., & Paganelli, P. (2000). Co-operative production planning for small- and medium-sized enterprises. *International Journal of Production Economics, 64*(1-3), 37–48. doi:10.1016/S0925-5273(99)00032-8

Monczka, R.M. & Morgan, J. (1997, January). What's wrong with supply chain management?. *Purchasing*, 69-72.

Quayle, M. (2003). A study of supply chain management practice in UK industrial SMEs. *Supply Chain Management: An International Journal, 8*(1), 79–86. doi:10.1108/13598540310463387

Rahman, S. (2001). A comparative study of TQM practice and organizational performance of SMEs with and without ISO 9000 certification. *International Journal of Quality & Reliability Management, 18*(1), 35–49. doi:10.1108/02656710110364486

Report of the Symposium on Supply Chain Management for Small and Medium Enterprises, Taipei, Republic of China. (2001). *Asian Cases on Supply Chain Management in SMEs.*

Sharma, M. K., & Bhagwat, R. (2006). Practice of information systems: Evidence from select Indian SMEs. *Journal of Manufacturing Technology Management, 17*(2), 199–223. doi:10.1108/17410380610642278

Simchi-Levi, D., Kaminsky, P. & Simchi-Levi, E. (2000). *Designing and Managing the Supply Chain: Concepts, Strategies, and Case Studies.* Irwin McGraw-Hill.

Singh, R. K., Garg, S., & Deshmukh, S. G. (2004). Competitiveness of small and medium enterprises, case of an Indian auto component manufacturing organization. *IIMB Management Review*, (December), 94–102.

Singh, R. K., & Garg, S. K. (2008). Strategy development by SMEs for competitiveness: A review", Bechmarking. *International Journal (Toronto, Ont.), 15*(5), 525–547.

Singh, R. K., Garg, S. K., & Deshmukh, S. G. (2008). Competency and Performance analysis of Indian SMEs and large organizations: An exploratory study. *Competitiveness Review: An International Business Journal, 18*(4), 308–321. doi:10.1108/10595420810920798

Stock, J. R., & Boyer, S. L. (2009). Developing a consensus definition of supply chain management; a qualitative study. *International Journal of Physical Distribution & Logistics Management, 39*(8), 690–711. doi:10.1108/09600030910996323

Thakkar, J., Kanda, A., & Deshmukh, S. G. (2008). A conceptual role interaction model for supply chain management in SMEs. *Journal of Small Business and Enterprise Development, 15*(1), 74–95. doi:10.1108/14626000810850856

Thakkar, J., Kanda, A., & Deshmukh, S. G. (2008). Supply chain management in SMEs: Development of constructs and propositions. *Asia Pacific Journal of Marketing and Logistics, 20*(1), 97–13. doi:10.1108/13555850810844896

Thakkar, J., Kanda, A., & Deshmukh, S. G. (2012). Supply chain issues in Indian manufacturing SMEs: Insights from six case studies. *Journal of Manufacturing Technology Management, 23*(5), 634–664. doi:10.1108/17410381211234444

The smarter supply chain of the future. (2009). IBM.

Todd, P. R., & Javalgi, R. G. (2007). Internalization of SMEs in India: Fostering entrepreneurship by leveraging information technology. *International Journal of Emerging Markets*, *2*(2), 166–180. doi:10.1108/17468800710739234

Valand, T. I., & Heides, M. (2007). Can the SME survive the supply chain challenge? *Supply Chain Management: An International Journal*, *12*(1), 20–31. doi:10.1108/13598540710724374

Wagner, B. A., Fillis, I., & Johansson, U. (2003). E-business and e-supply in small and medium sized businesses. *Supply Chain Management: An International Journal*, *8*(4), 343–354. doi:10.1108/13598540310490107

Westhead, P., & Storey, D. (1996, July 01). Management training and small firm performance: Why is the link so weak? *International Small Business Journal*, *14*(4), 13–24. doi:10.1177/0266242696144001

Wyer, P., & Mason, J. (1998). An organizational learning perspective to enhancing understanding of people management in small businesses. *International Journal of Entrepreneurial Behavior & Research*, *4*(2), 112–128. doi:10.1108/13552559810224576

Zahedirad & Shivaraj. (2011, October-December). Supply Chain: Barriers and Benefits Indian SMEs. *SCMS Journal of Indian Management*, 11-30.

Chapter 10
Government's Role in Setting Optimal Policies for Green Supply Chain

Kalicharan Sabat
Vignana Jyothi Institute of Management, India

ABSTRACT

In the last two decades lot of research was done in the area of Green design and Green supply chain management. But very few researchers have really tried to study the government's role in the implementation of Green designs and Green supply chain in any nation. In this paper simple general equilibrium model is used to analyze the benefits from taxes, and subsidies by government for green disposal efforts such as waste reduction, waste recycle, remanufacturing, and disposal on both the manufacturers and household consumers.

INTRODUCTION

In every year since 1993 has been in the top 20 warmest years on record[1]. Environmental degradation and increasing resource depletion has compelled the governments across the globe to set policies on green practices such as Green design, Green procurement, Green manufacturing, Green packaging, Green distribution and Green disposal for all the manufacturing companies. Furthermore, with a global concern to environmental issues, various countries have framed stringent laws on disposal of goods, and are imposing green barriers while importing commodities from the global markets.

Fullerton and Wu (1998) in the paper "Policies for green design" initiated the discussion on how the government can use different mix of policy instruments to deal with various organizational and market failures in managing household waste. They suggested in the absence of a market for recycled goods, the mix of consumption tax, packaging tax and recyclability subsidy on producers of different industrial and household products can lead to minimizing the amount of waste generated and maximization of reuse and recycling of material. Similarly, if household and industrial waste is collected free of charge, then the combined effect of consumption tax, packaging tax and recyclability subsidy on producers may lead to

DOI: 10.4018/978-1-4666-9795-9.ch010

Figure 1. Government policies for manufacturers and consumers at different stages of supply chain

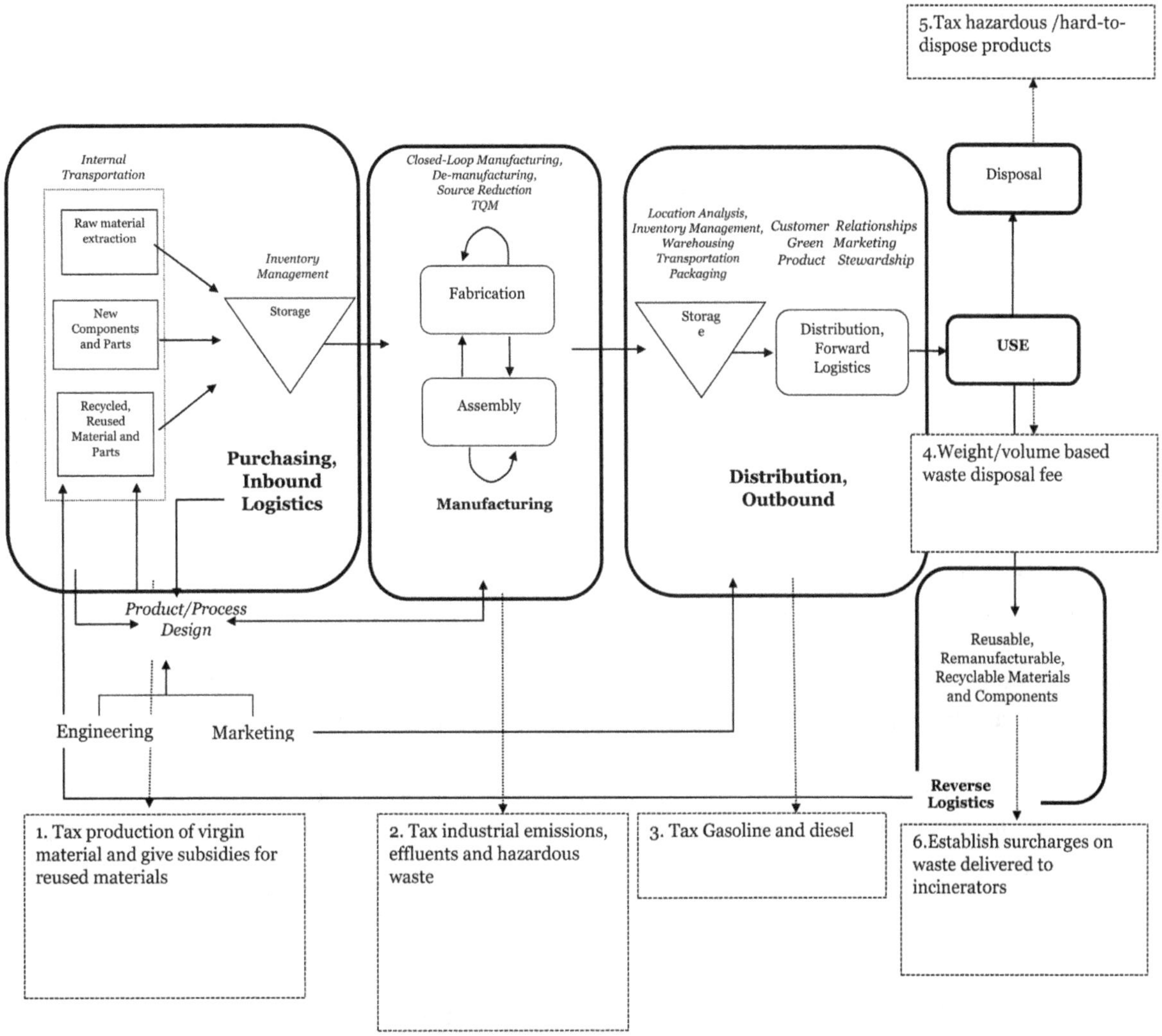

optimize the amount of waste in the entire product-life-cycle starting from design, to production, packaging, sales, use and disposal. Figure-1 shows how the government policies can target different stages in the product life cycle. Policies that affect product design will also affect product disposal and vice versa.

BACKGROUND

The concept of Green Supply chain Management was first put forward in US Michigan State University in 1996 (Zhou, 2009). Since then the scholarly interest in the area of Green Supply chain Management has rapidly developed across the world. Benita M. Beamon, (1999), in her Paper "Designing the green supply chain, Logistics Information Management", gave a new aspect to the Supply chain management by taking environmental challenges to production and manufacturing into thought. She insisted on reducing waste and minimizing the use of resources at each stage of the supply chain. Also to lower the environmental damage all the supply chain members should examine their "ecological footprint" and make Green supply chain as a practice for lowering the ecologic footprint of supply chains (van Hoek, 1999).

Organizations have started viewing many of these environmental programs like technological and organizational development projects, as possible alternatives for gaining a competitive advantage (Sarkis, 2002). Subhajit Mazumder and Anand Chatterjee in the paper "Green Supply Chain as a competitive advantage enabler in the FMCG sector" have said that cost strategy (low cost) and Evolving strategy (focus) are now out dated and for sustaining in the aggressive market they need to improve the supply chain efficiency and should become more environmentally conscious. A number of green practices were suggested by Srivastava (2007) likes green design, reverse logistics, product recovery, logistic design, solid waste management and so on. Sanjeev Swami and Janat Shah (2011) in the paper "Channel Coordination in Green Supply Chain Management: The Case of Package Size and Shelf-Space Allocation" addressed the benefits of vertical coordination in a green supply chain between the players at the upstream and downstream levels of the chain and to reduce the pack size of products that helps in saving environmentally costly transportation cost.

Aref A. Hervani, Marilyn M. Helms, Joseph Sarkis, (2005), in their paper "Performance measurement for green supply chain management", tried to design a Performance measuring system for GrSCM. In the paper they said in supply chains with numerous vendors, producers, distributors, and retailers the Green performance measurement is very challenging because it is difficult to attribute performance results to any one particular entity within the chain. And the problem gets amplified if the member organizations don't follow any standardized data, poor technological Integration among partners, cultural dissimilarities, lack of understanding of organizational plan, differences upon metrics, or unclear understanding of the needs for inter-organizational performance Measurement, etc. They suggested Environmental performance indicators can be used in evaluating the environmental performance of activities, processes, hardware and services.

GREEN INITIATIVES BY DIFFERENT GOVERNMENTS

As part of its commitment to improve the environment and the quality of life of its citizens, the Canadian government on April 1, 2006 made the policy of green procurement mandatory for all its government purchases.[2] On November 26, 2008 United kingdom (UK) government enacted the climate change act to improve carbon management and help the transition towards a low carbon economy in the UK.[3] The Japanese government is planning to introduce a new tax to curb carbon emissions from fiscal year 2016. Similarly in 2012 china announced increases in resource taxes on six minerals, such as tin, iron ore, and in February 2013, the US Department of Energy announced USD150 million in Advanced Energy Manufacturing Tax Credits for clean energy and energy efficiency manufacturing projects across the US.[4] Also the European Union (EU) too has waste management policies with aim to reduce the environmental and health impacts of waste and improve Europe's resource efficiency.[5] With most governments from the developed nations actively trying different green practices, the impacts from their efforts need to be evaluated.

A SIMPLE MODEL FOR ANALYSING WASTE REDUCTION AND WASTE REUSES EFFORT

The model is derived from the model suggested by Fullerton and Wu (1998) with the assumption that household consumer can also dynamically support the government's efforts in reducing waste. It is assumed that all the different manufacturers produce identical goods and all the household consumption is identical. A representative firm produces a consumable goods with the cost function $c(q, e)$, where $q \geq 0$ is the quantity of goods produced and $e \geq 0$ is the effort put by the firm to reduce waste at the source, such as reduced use of virgin toxic materials, minimizing material use for production, economic packaging, reverse logistics, etc. Any household in a nation has the utility function $u(q, a) - (m + d)$, where $q \geq 0$ is the quantity of goods produced and $a \geq 0$ is the amount of effort by the household in reducing waste, and 'm' is the monetary cost incurred in the consumption of q and 'd' is the cost incurred while disposing the waste. Household cost incurred in the disposal of waste (d) may include reuse, remanufacture, and recycle of products.

The amount of waste needed to be disposed after consumption will depend on q, e, a, and may be denoted by $w(q, e, a)$. Disposing waste may incurs two different types of cost. The first is the tangible cost of collecting, transporting, shorting, recycling and reuse of waste, denoted by "T" per unit of waste. The second type of cost is the intangible cost on environment and society, and is denoted by "I" per unit of waste. All the above functions are twice differentiable, so as to satisfy $cq > 0$, $ce > 0$, $uq > 0$, $ue < 0$, $wq > 0$, $we < 0$, $wa < 0$ and guarantying an unique solution to the problem.

The policy instruments available to the government are 1) unit consumption tax (t) 2) unit subsidy on waste disposal (λ), 3) unit subsidy on manufacturers waste reduction effort (γ), and 4) Governments direct investment for waste reduction and recycling facilities (θ).

Optimized Social Impact and Decentralizing Efforts

Due to quasi-liner preferences, the first optimal solution for the above model is maximizing household utility, less the cost of production, and cost of waste disposal by household, manufacturer, and the government. Then the objective function becomes $u(q, a) - c(q, e) - (T + I)\, w(q, e, a)$ and the first order condition for interior maximum are

$$uq - cq - (T + I)wq = 0, \tag{1}$$

$$- c_e - (T + I)we = 0, \tag{2}$$

$$ua - (T + I)wa = 0 \tag{3}$$

Each of the above conditions equates marginal social costs to marginal social benefits.

The following are the outcomes from decentralized market efforts. Let's assume the price for a unit consumable good is 'p' and the price of waste disposal per unit waste is 'χ'. If the above mentioned subsidies and taxes are implemented on household consumers, the objective function becomes $u(q, a) - (p + t)q - (\chi + \lambda) w(q, e, a)$. The first order conditions for maximum household consumer's utility are

$$u_q - (p + t) - (\chi + \lambda)w_q = 0 \quad (4)$$

$$u_a - (\chi + \lambda)w_a = 0 \quad (5)$$

From the given price and government policy variables, the producer's profit function may be derived as $pq - c(q, e) + (\gamma + \theta)e$ and then the first order conditions for profit maximization are

$$p - c_q = 0 \quad (6)$$

$$- c_e + \gamma + \theta = 0 \quad (7)$$

If Optimal Policies for Passive Household Waste Reduction Effort – a = 0

If the household consumers are not pro-active with respect to waste reduction effort, then the equations *(3)* and *(5)* are not applicable. Now if we compare the remaining equations for social optimal results and decentralized outcomes, we can easily find that they are equal if and only if $t + (\chi + \lambda) w_q = (T + I) w_q$ and $\gamma + \theta = - (T + I) w_e$. These conditions may lead to a series of optimal policies as the three parameters to be considered, t, λ, $\gamma + \theta$ are needed to be calculated using two equations.

One of the possible optimal government policy may be obtained by setting $t = 0$, $\lambda = T + I - \chi$ and $\gamma + \theta = - (T + I)w_e$. In this case the combined effort of unit subsidy on household waste disposal and a unit subsidy on the manufacturers and governments waste reduction effort can achieve the social optimal. If the waste disposal service to the household is provided at marginal cost so as $\chi = T$, then the unit tax for household waste disposal will be equal to the pre unit social and environmental cost of waste, ie., Pigouvian tax[6]. Instead suppose the household waste disposal is provided free of charge then, $\chi + \lambda = 0$, is possible because the problem of illegal waste disposal which is very common in most of the counties can be controlled. Once the illegal waste disposal problem is curbed then the remaining policy instruments can be implemented to achieve social optimum. That is, $t = (T + I)w_q$, $\gamma + \theta = - (T + I)w_e$ along with $\lambda = - \chi$ may also lead to social optimal. Again we can use consumption tax and waste reduction subsidy on manufacturers in the Pigouvian way.

If Optimal Policies for Passive Household Waste Reduction Effort - *a ≥ 0*

Once the household start active waste reduction efforts, then the comparison for social optimal conditions in the *equations (1-3)* and the decentralized efforts in the *equations (4-7)* will lead to three equations:

$$t + (\chi + \lambda)w_q = (T + I)\, w_q \quad (8)$$

$$\gamma + \theta = - (T + I)\, w_e \quad (9)$$

$$(\chi + \lambda)w_a = (T + I)\ w_a \tag{10}$$

Now there will be only one unique combination of policy devices that implement the social optimal: $t = 0$, $\lambda = T + I - \chi$, and $\gamma + \theta = -(T + I)\ w_e$. The reason for such uniqueness in the optimal policy is purely because government has only three policy devices to control the three private variables.

The optimality of the above policy can be automatically explained. Free waste disposal (or a fixed payment for waste disposal) does not provide the household or manufacturer any incentives to make any waste reduction effort. Everyone in the society (manufacturer, households, institutions, etc.) expect some kind of return on investment (ROI) from their efforts for reducing or safe disposing the waste. This is why most governments are charging taxes at Pigouvian rates.

CONCLUSION

From the above discussions we can conclude that even in the absence of active involvement from household consumers on waste reduction efforts, the governments are capable in setting policy procedure. The reasons behind governments' flexibility in setting policy is the availability of three policy instruments $(t, \lambda, \gamma+\theta)$ to target the two private variable (q,e). The role of the waste disposal tax and subsidies becomes more significant only when the household consumers and the manufacturers actively start reducing waste, and reusing/ remanufacturing/ recycling materials.

REFERENCES

Beamon, B. M. (1999). Designing the green supply chain. *Logistics Information Management*, *12*(4), 332–342. doi:10.1108/09576059910284159

Elovitz & Kenneth, M. (2010). Green design: Risk or reward? *Consulting-Specifying Engineer*, 47.

Fullerton, D., & Kinnaman, T. C. (1995). Garbage, Recycling, and Illicit burning or dumping. *Journal of Environmental Economics and Management*, *29*(1), 78–91. doi:10.1006/jeem.1995.1032

Fullerton, D., & Wu, W. (1998). Policies for green design. *Journal of Environmental Economics and Management*, *36*(2), 131–148. doi:10.1006/jeem.1998.1044

Hervani, A. A., Helms, M. M., & Sarkis, J. (2005). Performance measurement for green supply chain management. *Benchmarking: An International Journal*, *12*(4), 330–353. doi:10.1108/14635770510609015

Kopicki, R., Berg, M. B., & Legg, L. (1993). *Reuse and Recycling – Reverse Logistics Opportunities*. Oak Brook, IL: Council of Logistics Management.

Sarkis, J. (2003). A strategic decision framework for green supply chain management. *Journal of Cleaner Production*, *11*(4), 397–409. doi:10.1016/S0959-6526(02)00062-8

Sarkis, J., & Cordeiro, J. (2001). An Empirical Evaluation of Environmental Efficiencies and Firm Performance: Pollution Prevention Versus End-of-Pipe Practice. *European Journal of Operational Research*, *135*(1), 102–113. doi:10.1016/S0377-2217(00)00306-4

Srivastava, S. K. (2007). Green Supply-Chain Management: A State-of-the-Art Literature Review. *International Journal of Management Reviews*, *9*(1), 53–80. doi:10.1111/j.1468-2370.2007.00202.x

Swami & Shah. (2011). *Channel Coordination in Green Supply Chain Management: The Case of Package Size and Shelf-Space Allocation.* IIMB, Supply Chain Management Centre, Working Paper No: 348

van Hoek, R. I. (1999). From reversed logistics to green supply chains", Supply Chain Management. *International Journal (Toronto, Ont.)*, *4*(3), 129–135.

Zhou. (2009). *Study on the Implementation of Green Supply Chain Management in Textile Enterprises.* Academic Press.

Zhu, Sarkis, & Geng. (2005). Green supply chain management in China: pressures, practices and performance. *International Journal of Operations & Production Management, 25*(5), 449 – 468.

Zhu, Q., Geng, Y., Fujita, T., & Hashimoto, S. (2010). Green supply chain management in leading manufacturers: Case studies in Japanese large companies. *Management Research Review*, *33*(4), 380–392. doi:10.1108/01409171011030471

ENDNOTES

1 Mark Kaufman (January 10, 2007), *Climate Experts Worry as 2006 Is Hottest Year on Record in U.S.*, The Washington Post, available at http://www.washingtonpost.com/wp-dyn/content/article/2007/01/09/AR2007010901949_pf.html accessed on July 10, 2013

2 Policy on Green Procurement, Public works and Government services Canada, available at http://www.tpsgc-pwgsc.gc.ca/ecologisation-greening/achats-procurement/politique-policy-eng.html

3 Green policy, Green grants machine, available at http://www.greengrantsmachine.co.uk/page.aspx?SP=5419

4 The KPMG Green Tax Index 2013 available at http://www.kpmg.com/Global/en/IssuesAndInsights/ArticlesPublications/green-tax/Documents/kpmg-green-tax-index-2013.pdf

5 Being wise with waste: the EU's approach to waste management, available at http://ec.europa.eu/environment/waste/pdf/WASTE%20BROCHURE.pdf

6 Pigouvian taxes are named after economist Arthur Pigou. William Baumol was instrumental in framing Pigou's work in modern economics. It's a special tax that is often levied on companies that pollute the environment or create excess social costs through business practices. In a true market economy, a Pigovian tax is the most efficient and effective way to correct excess social cost.

Chapter 11
Cold Chain Logistics in India: A Study

Swapnil Saurav
JDA Software, India

Ravi Potti
JDA Software, India

ABSTRACT

Cold Chain refers to the transportation of temperature sensitive products like perishable goods from the point of origin to point of consumption in the food supply chain, which keeps it fresh and edible for a much longer period than in normal conditions. Cold Chain helps in transporting seasonal products and also making it available throughout the year. Two main parts of cold chain are transportation and storage systems. The key Indian industries where cold chain logistics play very important role are fruits and vegetables, milk and milk products (ice cream), Poultry and processed meat, marine products, pharmaceutical (mainly vaccines) and chemicals. An efficient cold chain industry ensures availability of food products as well as prevents spoilage of medicines. Country like India, where infrastructure is one of the major challenges, cold chain plays a critical role. Analysis for this study shows that cold supply chain network does not differ significantly from products to products at least in Indian scenario. Some of the challenges to the growth of sector in India are high energy cost, power deficit, rising real estate cost, lack of logistical support and uneven distribution of capacity. All these challenges bring down the operating margin of a company and makes it not so attractive business sector. But during last couple of years there is a positive environment being created for this sector in India. The growth in organized retail, growing interest in horticultural crops, demand for cold chain logistics from Pharmaceutical industry and various initiatives by government are some of the reasons why there is a renewed interest in this sector especially by private sector players. This study, which is focused on Indian cold chain logistics, analyzes the industry on PEST (Political, Economy, Social and Technology) model and presents top 3 factors on each of these 4 parameters.

DOI: 10.4018/978-1-4666-9795-9.ch011

INTRODUCTION

The Cold Chain refers to the transportation of temperature sensitive products like perishable goods from the point of origin to point of consumption in the food supply chain. Cold chain process helps reducing spoilage, retains the quality of the harvested products and guarantees a cost efficient delivery to the consumer which can be transported over long distance and can be preserved beyond the life of the product. Different products require different temperature level maintenance to ensure their integrity throughout the travel process. For instance, the most common temperature standards are "banana" (13 °C), "chill" (2 °C), "frozen" (-18 °C) and "deep frozen" (-29 °C). Staying within this temperature is vital to the integrity of a shipment along the supply chain. Any deviations from the prescribed temperature limits will damage the goods in the transit. Figure 1 below shows a typical cold chain logistics infrastructure.

The complete Cold chain logistics system consists of:

- **Cold Storages:** This part of logistics consists of acquiring products from the point of origin and storing them in temperature controlled environment for the further transit to the point of consumption or point of sale.
- **Refrigerated Carriers:** Refrigerated Carriers form the primary distribution in a Cold Chain process. The main aim is continuous maintenance of in-transit required temperature. Failure in this part will be a major setback to the whole process.
- **Packaging:** It can be viewed as the Value added service in the Cold Chain process. It contains adhering to the various specifications and packaging the products.
- **Warehousing:** Warehousing forms the secondary distribution system of the Cold Chain process. After transportation of the products, they need to be stored in a cold storage unit from where they can deliver to the clients based on their requirements.
- **Management Information Systems (Traceability and Tracking etc.):** Throughout the transit of the goods, there should be continuous monitoring of the temperature and efficient supply chain management system should have tracking systems in place to trace the products and vehicles in transit to avoid any malfunctions. A company must maintain proper RFID (Radio - Frequency Identification) and Data logger systems to excel in this area.

BACKGROUND

India ranks first in milk production in the world (close to 17% of total production), ranks 2 in fruits & vegetables production and has substantial production of marine, meat & poultry products. India needs

Figure 1. Cold chain logistics infrastructure

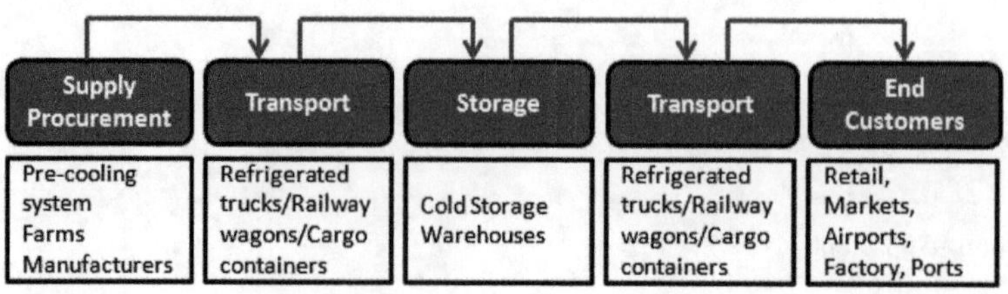

a fully developed cold chain sector. This study, which is based on the qualitative analysis of secondary data, has revealed that there is a tremendous scope for the development of cold chain facilities. Cold storage chains have been given infrastructure status (Budget 2011-12) so there is lot of focus on this industry now. Food processing minister, Mr Sharad Pawar had told Parliament on Aug 23, 2013, that inadequate storage facilities and unsafe storage of food grains resulted in wastage of fruits, grains and vegetables worth Rs 44,000 crore every year. It goes without saying that cold chains sector, if nurtured properly, can solve food problem of the India to some extent. For that first it needs to understand the current challenges that industry sector is facing and then work on the challenges.

Managing perishable items pose a significant challenge in the supply chain environment. Food stuffs, dairy products, green vegetables, fruits are all items of daily human consumption. These perishable items have a finite usable lifetime, and pass many hands during its movement from source to destination. The study of the supply chain of perishable products is particularly important because they account for more than 50% of supermarket sales, and the potential for extra profit from managing these items has been estimated at 15%. This thesis evaluates the channel dynamics and builds decision making models for supply chain of perishable items.

A perishable supply chain requires much attention to detail if the perishable goods are to get to the processing plant or to the end user as fresh as possible. There is a need of end-to-end refrigerated storage and transportation solutions for temperature-sensitive consumables. Demand for such a system would come from companies in Food, Retail, FMCG and Pharmaceuticals - to preserve product value across the supply chain. Some estimates say that the post-harvest losses of fruits and vegetables alone in India is more than the total production of fresh fruits and vegetables in Great Britain (Meera & Rajan 2007). According to a survey conducted by Corporate Catalyst India, a leading consulting firm, "Value addition of food products is expected to increase from 8% to 35% and that of fruits and vegetable processing from the current 2% to 25% by the end of 2025". The survey further reports that the dairy sector, which currently comprises the highest share of the processed food market, will experience marked growth (Manjunath 2013). Cold Chain logistics players in India are expected to deliver higher quality, increase profitability, ensure consumer safety and protect brand equity. The integrated solution is required at all the point in supply chain like production, storage and distribution of perishable products that require precise temperature control to retain product characteristics, active ingredients, freshness and nutritive values for longer durations (Mohapatra 2013). Cold Chain logistics in India is in nascent stage but increasingly demanding (Iyad 2013). Domestic volumes are increasing and also there is lot more international exposure now. Industry is fragmented and disjointed.

MAIN FOCUS OF THE CHAPTER

Problems

This paper attempts to bring out challenges faced by the retail, pharmaceutical and other industries which can be solved by cold chains and also look at the challenges and opportunities for independent cold chain service providers. Analysis is presented the holistic view on the subject by including the Political factor (how government regulations impact the sector), Economy (how business strategy impacts the sector), Society (challenges and benefits posed to the society) and Technology (technology usage and getting benefits from it). This study is based on studied opinions, published inputs, expert views available in public domain and on authors own qualitative analysis.

Inconsistent standards in different sections of the cold chain could lead to damage of food, either by shock or by undue temperature variations. This degrades food quality due to chemical reactions which are triggered off, which can otherwise be mitigated by low temperatures. The cold chain has a critical role to play in India as two-thirds of the country's population is dependent on agriculture for its livelihood. Of the close to 130 million tones of fruits and vegetables that the country produces, nearly 40 percent gets wasted (Gabriel, Bitran, Gurumurthio & Shiou 2007). More than 10 percent of annual milk production in the country is lost due to inadequate storage facilities (Forecast & Opportunities, 2013).

The paper addressed below questions in the research:

- Challenges faced by industries which use cold chain services and independent cold chain service providers.
- How can national and regional fragmented players can come together to create win-win situation for all the parties involved.
- Adaption of strategy and technology to Indian context.
- Anticipate road ahead for Cold Chain logistics in India.

Cold chain systems can be of strategic importance to companies since brand integrity, customer confidence, market share and profit are all at risk. This report will prove useful for the international and local companies looking to start cold chain operations in India. This can also help the government in designing much needed incentive programs for the sector which will further fuel the growth.

The research for this paper includes looking at various case studies available for Indian context and derived at the challenges for different industries and looked at key players in cold chain segment. The research also covered companies' documents and their strategies and based on these, research paper presents its observations and recommendations.

The research has divided the study group into two parts: companies using cold chain services and cold chain logistics providers. To identify challenges for industries using cold chain services are as shown in Table 1.

The total market value of Indian cold chain industry is expected to reach Rs 640 billion (approx. $310 billion) by end by 2017 (Aneesh & Arnav 2011). Some of the major industries that will fuel this growth

Table 1. Analytical steps for challenges in industries

Step 1	Identify top industries which need cold chain in India. This will be from various marketing research document
Step 2	Identify companies within each of these industries based on revenue size and presence in India
Step 3	Read through their Annual reports (if available), and other company reports.
Step 4	Looked for supply chain related post/comments by leading magazines, top executives comments/interviews and solution providers' press release.
Step 5	Talk to industry experts from JDA who work with clients on solving their problems
Step 6	Data collection- Data about retailers cold chain capabilities, its challenges and future strategies
Step 7	Present the Data Analysis

Table 2. Reasons for fueling growth in India

Industry	Growth Factor	Companies
Dairy & Ice Cream	India ranks first in milk production in the world (close to 17% of total production).	Amul, Heritage Foods, Britannia, Mother Dairy, Vadilal, Hatsun Agro Ltd., Parag Foods
Frozen Food	India is world's largest producer of fresh fruits and vegetables, milk, spices and select fresh meats and world's largest producer of several dry fruits, agriculture-based textile raw materials, roots and tuber crops and eggs.	McCain, Al Kabeer Group, Amigo Foods, Agri Freeze Foods, Capricorn Food Products India Ltd.
Fruits & Vegetable	India ranks 2 in fruits & vegetables production	Reliance Fresh, More, Vijetha (Hyderabad), Daulat Agro, Sudhir Traders
Seeds	India is largest producer of several staples such as millets and castor oil seed.	Nuziveedu Seeds, Tulasi Seeds, National Seeds Corporation Ltd.
Pharmaceuticals	The Indian vaccine market is somewhere in the range of Rs 10 billion, and is growing at 25-30 per cent year-on year (y-o-y). These vaccines require temperature control right from the manufacturing point to the consumption point, which makes cold-chain management imperative (Anil 2010).	Abbott, Aventis, Novartis, Ranbaxy, UCB India, MSD India Pharmaceuticals, Sanofi
Poultry & Sea Food	India is amongst world's largest producers of livestock and poultry meat, with one of the fastest growth rates, as of 2011. Aquaculture and catch fishery is amongst the fastest growing industries in India. Between 1990 and 2010, Indian fish capture harvest doubled, while aquaculture harvest tripled. India exported 600,000 metric tons of fish products to nearly half of all the world's countries.	Primeline Products International, Mathew & Thomas Global Traders, Seafood Express, Amigo Foods, Ocean Marine Foods, Fresco Seafoods

are given below in Table 2 along with the reasons. Table 2 also lists major companies under each of the industries, which were selected for the study. Companies were selected based on their revenue size.

Analysis is presented based on both primary and secondary data. The process followed for the research included interviewing senior employees of Reliance Fresh and Vijetha supermarket and read through various articles available in public domain. Then dividing the information from each company into 3 groups — current capabilities of the companies, growth drivers and companies' strategy related to cold supply chain logistics for future. Then grouping the data based on points of discussion headers of Political, Economic, Sociological, Technological (PEST) model. Since Cold Chain logistics is an industry which is dependent on government regulations, and other factors like economy of the company, social and technological factors, using PEST model for the analysis will correctly bring out the current challenges of the industry. This process is very similar process for identifying the challenges with cold chain logistics service providers, as shown in Table 2.

India has about 250 reefer transport operators (this includes independent firms) that transport perishable products. Of the estimated 25,000 vehicles in use, 80% transport dairy products (wet milk); only 5,000 refrigerated transport vehicles are available for all other commodities (Manjunath 2013). The research referred to the companies listed in the Table 3 below for the study. The research included looking at various company articles, news items and executive interview and put the cold chain related information under 3 different types - current capabilities of the companies, growth drivers and companies' strategy related to cold supply chain logistics for future and dividing all the information under PEST format.

The paper analyzed the data and came up with top factors against each of PEST factors and relevant analysis is presented in the sections to follow.

Table 3. Reasons for fueling growth in India

Players	Key Customers	Service	Operation
Gati (Kausar India Limited)	mainly serving Meat, FMCG and Fruits and Vegetables Industry	Integrated Logistics	Pan India
Snowman (Gateway Distripacks)	HLL, Amalgam, Amul, Mother Dairy, Baskin Robins, Mars etc	Integrated Logistics	National Player, 16 Cold Storages across India, 150+ Reefer trucks, runs specified reefer truck routes
Frick India	HLL, Allana, Al Kabeer	Refrigeration equipment and storage services	2-3 cold storages, centered around Delhi
Chaitanya Cold Storage	Joy ice creams, Farm Suzanne, McCain Foods, Domino's, Pizza Hut, Vadilal	Storage and Transportation	1 cold storage in Bangalore and 10-15 reefer trucks around Bangalore
Fresh Express	Nestle, Cadbury's	Storage and Transportation	2 cold storages in western India, 10-15 reefer trucks
RK Foodland	McDonald's, Hotels and Restaurants	Integrated Logistics	200+ Reefer trucks, Nationally integrated service provider in times to come
Western	F&B business	Storage and Refrigeration equipment	1 cold storage in Mumbai, 1 in Chennai
Kelvin Cold Storage	Processed food, dairy, meat, poultry, FMCG, QSR, Pharma	storage and distribution	All over India
Other local players covered in the study: Delhi Baroda Road Carrier (DBRC), Crystal Roadways, Evergreen, Trans Gulf			

Requirement in perishable goods supply chain are:

- **Increased Efficiency:** Faster transit and turnaround times.
- **A Broad Portfolio of Services:** To address every client's needs.
- **Improved Transparency:** Visibility into the logistics chain.
- **Better Tracking:** To have real time information.
- **Prioritized Treatment:** Since perishable goods are time sensitive, it needs prioritized handling at all times.

Solutions and Recommendations

Based on the analysis the current landscape of Cold Chains in India has been classified using PEST (Political, Economic, Sociological and Technological) methodology. Findings are explained further in the sections below.

A. The Current Landscape of Cold Chains in India

Research findings have been grouped in to the PEST model as discussed in the methodology. Table 4 lists the current challenges of the sector on Political, Economic, Sociological and Technological parameters.

Political Challenges: It was found that cold chain industry in India is fragmented and characterized by very few integrated logistics management players. The underdeveloped cold logistics infrastructure

Table 4. Current challenges of cold chains in India

Political	Economic:
• Lack of government support • Lack of infrastructure • Rising food inflation rate	• Lack of logistical support • Rising Real Estate costs • High energy costs
Sociological: • Fragmented players • Lack of trained man-power • Overall a risky business	**Technological:** • Network Optimization • Out-dated current storage technologies • Old transportation technologies

adds to the problems. Government support has been very little. There is a lack of required frame work and guidelines for the prospective Agri Business investors from the government which is not helping the potential investors. Only in Budget 2011-12, cold storage chains have been given infrastructure status. More than half of the fruits and vegetable produced are wasted because of lack of cold logistics infrastructure. Cold logistics carry mostly the edible goods; rising food inflation and the high cost of supply chain, makes situation even worse.

Economic Challenges: Due to the fragmented industry setup of cold chain in India, companies are unable to cover the entire value chain right from procurement in remote areas to delivery centers in cities. Also compounding the problem is continuously increasing real estate costs which sometimes can be as high as 10 to 15% to setup a fully integrated, international class cold storage facility. This makes the sector unattractive for investors. Energy costs too are rising. Most of the areas do not get uninterrupted power supply due to which companies turn to other means of power supply which are not only expensive but also not environment friendly. For a typical company in cold chain industry in India, these energy costs form major chunk of operating expenses which can as high as 25-30% which is 3 times higher than the cold chain companies in the western countries.

Sociological Challenges: The small fragmented players which may own just one piece of the supply chain network like warehouse, transport, etc increases the overall cost of the system. Lack of trained man-power affects the collection of fresh food produce from the farms, loading and unloading of vaccines, etc. Coupled with this erratic power supply in the remote areas is posing even bigger challenge for storage and transportation of the fresh food produce. Cold storage remains, overall, a risky business. Unlike other sectors, the cold logistics is completely dependent on power as it requires power from the point of production to the point of consumption. Any variability in temperature, either increase or decrease, leads to complete wastage of goods.

Technological challenges: Industry is looking for an Integrated Network optimization method to management the capacity and brings in overall efficiency in the system. The users of the system wants better visibility and hence better predictability into cold logistics system which a can achieved by an integrated network optimization application. Another important challenge inferred from the study was currently companies have very old storage systems. Companies should look to invest in better technologies. Even the transportation technologies are old and companies face challenges such as lack of cold vans/carriages for long haul travel for temperature sensitive products. Coupled with this shortage is the bad infrastructure in terms of Roads etc is increasing the problem manifold.

As it's inferred from Table 4, most of the challenges mentioned are interlinked. If one factor improves, it will impact other factors positively. It's important that all the players involved in cold logistics come together as partners and try solving the problems.

Table 5. Growth drivers: cold chains in India

Political • Ease of FDI restrictions • Rise in export demand • Establishment of National Centre for Cold Chain Development	Economic: • Growing retail sector • Demand from pharma & processed food sector • Rise in nuclear family, demand for frozen foods
Sociological: • Entrepreneur Attitude • Funds available from Bank • Government schemes for setting up small sector businesses.	**Technological:** • Making "world class facility" • Focus on logistics models • Technology for Quality & Customer Service

B. The Growth Drivers of Cold Chains in India

In last couple of years, government is seen taking some positive steps to develop this sector. Based on the study, some of the important growth drivers has been listed in Table 5 under Political, Economic, Sociological and Technological factors.

Political growth factors: During the budget of 2011-12, Indian government gave nod for 100% FDI in the cold chain sector via automatic route. It also accorded project import status at a concessional customs duty of 5% with full exemption from service tax for the initial setting up and expansion of cold storage, cold room and processing units. And also refrigeration units required for the manufacture of refrigerated vans or trucks were fully exempted. There has been rise in export demand for processed & frozen food.

Economic growth factors: Growing demand from horticultural crops coupled with growth in pharmaceutical sector and processed food sector will in turn drive the growth in cold chain industry. Most of the current companies cater to potatoes and retail industries. However, growth in other sectors will attract more players and drive the growth in industry. There is a rise in nuclear families in India and more and more house-hold have both working parents. This has pushed the demand for frozen foods.

Sociological growth factors: The educational institutes and other social organizations like National Innovation Network (NIN) are building a culture of entrepreneurship. Funds are easily available from banks, venture capitalists and also there are government schemes which help in setting up small and medium scale industries. All these factors have resulted in serious thinking for the development of the sector by private organizations.

Technological growth factors: A multitude of changes have occurred in last 2 decades and it's been made possible with advancement of technology. According to our analysis, companies are looking to either expand/ renovate the existing setup or going for complete new infrastructure. Companies are using technology to make their infrastructure of world class level. Companies are demanding the logistics model in their transportation or warehouse systems. Companies are looking for technology that will enable them to deliver better customer service.

C. Future Growth Strategy of Cold Chains in India

Based on the current challenges that the industry is facing coupled with the growth drivers the study also looked at what future holds for cold chain logistics in India. The study applied the growth factors identified in table 6 to the company's long term strategy and analyzed what various players in the in-

Table 6. Growth drivers: cold chains in India

Political	Economy:
• Public Private Partnership • Incentives for investment	• Modernizing Cold Chain • Transformation of 3PLs • Talent/Skills development
Sociological: • Demand for operations courses at universities • Change Management.	**Technological Strategy:** • Integrating Information • Specialized technologies introduction • World class technology

dustry can do to improve the current scenario. Findings are listed in Table 7 below under Business and Technology strategies.

Political strategies: With various growth drivers in place, Cold Chain Logistics Industry is seeing increase in government recognition and support. In future, the sector will see public private partnerships (PPPs) and government will give incentives to make this sector lucrative for investment by private players.

Economic strategies: As existing warehouse and transport system gets old and the demand of perishable goods continue to rise, cold chain will get more and more modernizing. Indian companies have started incorporating cold chain logistics into its business strategy. Many Indian companies are working to modernize their cold chain setup to take on MNCs. There is an emergence and acceptance of 3PL in Indian market and this will lead to transformation of these 3PLs to an integrated logistics service providers in near future. As skills and talent in this sector is limited, companies will invest in training and development of their employees.

Sociological strategies: Skill/Talent Developments through increased focus on supply chain management/Operations management by the Indian universities, resulting in a significantly higher availability of resources. There will be requirement of change management experts to handle the dynamics of the industry.

Technological growth factors: Push from business strategy can be seen in the ways companies are aligning its technology strategy. Companies want better visibility so they are putting in system and applications in place. Systems will help in advance planning to negate some of the challenges faced due to poor infrastructure. With FDI allowed in cold chain logistics, MNCs and Indian companies are partnering to bring latest technologies to India. New technologies like Mobile Racking are enabling companies to change its business model. Retailers can outsource the transport and warehousing processes required to bring the produce from the farm to the retailer's distribution centre (DC) and still have a complete control over the goods using information technology. More and more Wholesalers / independent service providers will invest in the pre-DC processes (and in cold chain equipment such as pre-coolers and reefer trucks) and modernizing & reinventing Cold Supply Chain.

FUTURE RESEARCH DIRECTIONS

The Indian Cold Chain industry is poised to grow and is expected to offer significant opportunities for National and International cold chain companies in the near future. Areas of future research can be to assess the Indian market and present a study of global best practices that can be implemented in India and also focus on the areas where innovation is required. Companies like Ingersoll Rand, Carrier, and

DuPont have already established their offices in India. It would be interesting to present case study on their Indian operations.

Each of the perishable goods requires unique supply chain approach. Future research will look at the supply chain for each of the perishable goods such as- Ice Cream, Frozen Food, Fruits & Vegetable, Seeds, Pharmaceuticals and Poultry & Sea Food and come up with supply chain design of these products.

Another aspect for research, in context to Indian cold chain logistics scenario, would be to study the impact of Foreign Direct Investment (FDI) on the cold supply chain sector. Assuming that, the FDI inflow will impact the sector in positive way.

From past few months the focus of the industry players has shifted from increasing the production to better manage storage and transportation. Research could be taken to analyze Demand and Supply. With increased demand of cold chain, it will also add pressure on the players to improve their performance to reduce obsolete inventory and shrinkages. Strategic competitive analysis could be done to show the capabilities of each players and group them based on their strength of handling particular type of goods.

CONCLUSION

The cold chain industry in India, over the last few years, has witnessed various positive changes. The India industry, according to reports, is anticipated to grow at a CAGR of 28.7% during 2012-2017, which will make the market reach US$ 11.6 Billion by 2017. It becomes important for all those who are interested in this growth market to understand the various factors impacting the industry and tailor their offerings accordingly. This report will give important insight to the readers.

In the long term cold chain companies which move to palletisation will reduce overall operating costs and gain efficiencies. Improvement in the government regulation and technology enhancement will also drive growth in the cold chain industry. Growth in various industries like pharmaceuticals, retail and processes food sector will in turn fuel the growth in the cold chain sector in India.

The future cold chain in India will be much different than what is today in terms of business, technology and functional aspects. Government, slowly but steadily is creating a positive environment for the industry. That why companies are adopting a broad based and long-term approach in investing in the sector. There will be many public private partnerships (PPP) in this sector which will result in overall improvement in the sector like setting up of fast track corridor for cold chain logistics, railway cold chain network, etc. With such renewed interest in the sector, cold chain logistics will be among the most vibrant industry sector in India.

REFERENCES

Admin. (2013). Cold Chain in India: Market Opportunities & Challenges. *AFRIS Blog*.

Aneesh, M., & Sinha, A. (2011). *Cold chain: Finally warming up to India?* Infrastructure Today.

Anil, G. (2010). *A Survey of Information Technologies in Logistics Management*. University of Texas Research Paper.

Eric, L., & Abhijeet, K. (2003). *Pharmaceutical Distribution in India*. Bio Pharm International.

Forecast & Opportunities. (2013). *Cold Chain Market in India*. ADS Reports.

Gabriel, Bitran, Suri, Gurumurthio, & Shiou. (2007). *Emerging Trends in Supply Chain Governance*. MIT Sloan School of Management.

Iyad, K. (2013). *20/20 Vision - Logistics at the end of 2020*. Paper presented at the 13th Annual Logistics CIO and Supply Chain Technology Forum.

Manjunath, M S. (2013). *India centric challenges*. Ingersoll Rand International (India) Ltd.

Meera, K., & Rajan, H. (2007). *Machine to Machine: A fresh approach to profits from perishables*. Infosys Whitepaper.

Mohapatra, B. B. (2013). *Indian Vaccine Market and Ensuring Quality from Production to Distribution*. Paper presented at the 2nd Annual Cold Chain India Conference.

Perspectives, A. (2007). *The Sun Rises on the Cold Chain Industry: The Impact of Organized Retail*. Retrieved from consultavalon.com

Rennie, S. (2011). *Country Report: India's cold chain industry*. Department of Commerce.

Rohit, Joshi, Kumar, Banwet, & Shankar. (2009). Indian cold chain: modeling the inhibitors. *British Food Journal, 111*.

Salin & Nayga, Jr. (2003). A cold chain network for food exports to developing countries. *International Journal of Physical Distribution & Logistics Management, 33*.

Sivakumar & Patel. (2013). *Cold chain industry-India*. International Exhibition & Conference on Cold Chain Industry.

World Economic Forum. (2013). *Outlook on the Logistics & Supply Chain Industry*. Global Agenda Council on Logistics & Supply Chain Systems.

ADDITIONAL READING

Aruoma, O. I. (2006). The impact of food regulation on the food supply Chain. *Toxicology*, *221*(1), 119–127. doi:10.1016/j.tox.2005.12.024 PMID:16483706

Li Dongjiu., Li Hao., Zhang Qingnian & Wang, Jiali. (2012). "Transport Hub-and-spoke Network Optimization Model Construction of Pharmaceuticals Cold-chain Logistics" e-Business Engineering (ICEBE).

Fearne, A., Hornibrook, S., & Dedman, S. (2001). *The management of perceived risk in the food supply chain: a comparative study of retailer led beef quality assurance schemes in Germany and Italy*. The Internaional Food and Agribusiness Management Review.

Guangshui, Ma & Hongjun, Guan. (2009). "The application research of cold-chain logistics delivery schedule based on JIT", Industrial Mechatronics and Automation, 2009. ICIMA 2009.

Lisa, K. (2013). *"Use of cold chains for reducing food losses in developing countries", The Postharvest Education Foundation*. PEF.

Liu, B. L. (2013). *Food preservation and cold chain*. Chemical Industry Publishers of China.

Liu, M., & Yue, T. (2011). *"The cold chain logistics performance evaluation on sideline products based on data envelopment analysis", Product Innovation Management*. ICPIM.

Luis Ruiz-Garcia1 & Loredana, Lunadei. (2010). "Monitoring Cold Chain Logistics by Means of RFID", Sustainable Radio Frequency Identification Solutions.

Oannis, Manikas & Leon, A. Terry. (2009). "A case study assessment of the operational performance of a multiple fresh produce distribution centre in the UK", British Food Journal Volume: 111.

Rai, S. K., Singh, V., & Vaish, A. (2013). *"Perishable Food Inventory Management: A Retailer's Persoective", International Journal of Business and General Management*. IJBGM.

Shouping, Gui., Wenjie, Feng & Zhiyong, Zhang. (2010). "Research on Multi-echelon inventory model of cold-chain logistics on random demand", Logistics Systems and Intelligent Management.

Verena, Raab., Brigitte, Petersen & Judith, Kreyenschmidt. (2011). "Temperature monitoring in meat supply chains", British Food Journal Volume: 113.

Xu, Xiaohong., Lan, Hongjie & Ruijiang, Wang. (2010). "Identification of critical control points of the food cold chain logistic process", Logistics Systems and Intelligent Management

Ye, Ren., Qizhou, Hu & Xuncheng, Huang. (2011). "Researching on benefit assignment of logistics system in cold chain for food", Emergency Management and Management Sciences (ICEMMS).

KEY TERMS AND DEFINITIONS

Air Cargo Containers: Containers designed to conform to the inside of an aircraft. There are many shapes and sizes of containers. Air cargo containers fall into three categories: 1) air cargo pallets 2) lower deck containers 3) box type containers.

Barrier to Entry: Factors that prevent companies from entering into a particular market, such as high initial investment in equipment.

Base Demand: The percentage of a company's demand that is derived from continuing contracts and/or existing customers. Because this demand is well known and recurring, it becomes the basis of management's plans. Synonym: *Baseload Demand.*

Bottleneck: A constraint, obstacle or planned control that limits throughput or the utilization of capacity.

Break-Even Point: The level of production or the volume of sales at which operations are neither profitable nor unprofitable. The break-even point is the intersection of the total revenue and total cost curves.

Bulk Storage: The process of housing or storing materials and packages in larger quantities, generally using the original packaging or shipping containers or boxes.

Capacity Planning: Assuring that needed resources (e.g., manufacturing capacity, distribution center capacity, transportation vehicles, etc.) will be available at the right time and place to meet logistics and supply chain needs.

CAPEX: A term used to describe the monetary requirements (CAPital EXPenditure) of an initial investment in new machines or equipment.

Cold Chain: A cold chain is a temperature-controlled supply chain.

Consortium: A group of companies that work together to jointly produce a product, service, or project.

Consumer Packaged Goods (CPG): Consumable goods such as food and beverages, footwear and apparel, tobacco, and cleaning products. In general, CPGs are things that get used up and have to be replaced frequently, in contrast to items that people usually keep for a long time, such as cars and furniture.

Cool Cargo: Temperature sensitive products, during transport, are called cool cargo.

Dedicated Contract Carriage: A third-party service that dedicates equipment (vehicles) and drivers to a single customer for its exclusive use on a contractual basis.

Domestic Trunk Line Carrier: An air carrier classification for carriers that operate between major population centers. These carriers are now classified as major carriers.

Duty Free Zone (DFZ): An area where goods or cargo can be stored without paying import customs duties while awaiting manufacturing or future transport.

Fourth-Party Logistics (4PL): Differs from third party logistics in the following ways; 1)4PL organization is often a separate entity established as a joint venture or long-term contract between a primary client and one or more partners; 2) 4PL organization acts as a single interface between the client and multiple logistics service providers; 3) All aspects (ideally) of the client's supply chain are managed by the 4PL organization; and, 4) It is possible for a major third-party logistics provider to form a 4PL organization within its existing structure. The term was registered by Accenture as a trademark in 1996 and defined as "A supply chain integrator that assembles and manages the resources, capabilities, and technology of its own organization with those of complementary service providers to deliver a comprehensive supply chain solution.", but is no longer registered.

Integrated Carrier: A company that offers a blend of transportation services such as land, sea and air carriage, freight forwarding, and ground handling.

Integrated Logistics: A comprehensive, system-wide view of the entire supply chain as a single process, from raw materials supply through finished goods distribution. All functions that make up the supply chain are managed as a single entity, rather than managing individual functions separately.

Obsolete Inventory: Inventory for which there is no forecast demand expected. A condition of being out of date. A loss of value occasioned by new developments that place the older property at a competitive disadvantage.

Outbound Logistics: The process related to the movement and storage of products from the end of the production line to the end user.

Pallet: The platform which cartons are stacked on and then used for shipment or movement as a group. Pallets may be made of wood or composite materials.

Radio Frequency Identification (RFID): The use of radio frequency technology including RFID tags and tag readers to identify objects. Objects may include virtually anything physical, such as equipment, pallets of stock, or even individual units of product.

Reefer: A term used for refrigerated vehicles.

Refrigerated Carriers: Truckload carriers designed to keep perishables good refrigerated. The food industry typically uses this type of carrier.

Shelf Life: The amount of time an item may be held in inventory before it becomes unusable. Shelf life is a consideration for food and drugs which deteriorate over time, and for high tech products which become obsolete quickly.

Shrinkage: Reductions of actual quantities of items in stock, in process, or in transit. The loss may be caused by scrap, theft, deterioration, evaporation, etc.

Third-Party Logistics (3PL): Outsourcing all or much of a company's logistics operations to a specialized company. The term "3PL" was first used in the early 1970s to identify intermodal marketing companies (IMCs) in transportation contracts.

Transportation Mode: The method of transportation: land, sea, or air shipment.

Unbroken Cold Chain: An unbroken cold chain is an uninterrupted series of storage and distribution activities which maintain a given temperature range. It is used to help extend and ensure the shelf life of products such as fresh agricultural produce, seafood, frozen food, photographic film, chemicals and pharmaceutical drugs.

Warehouse: Storage place for products. Principal warehouse activities include receipt of product, storage, shipment, and order picking.

Chapter 12
Service Quality in Supply Chain: A Case of Indian Automotive Industry

Asad Ullah
Aligarh Muslim University, India

Mohd. Adil
National Institute of Technology Hamirpur, India

ABSTRACT

In manufacturing industries or the supply chain, very few attempts have been made in the past to measure service quality or to test the linkages between service quality with supplier's satisfaction and loyalty. Thus, to fill the existing void in the literature, an attempt has been made to conceptualise a model proposing linkages of service quality with supplier's satisfaction and loyalty in the supply chain of automotive industry. The research design for this paper includes a combination of literature review and a survey of 220 and 170 practitioners from four and two automobile manufacturing companies located in Northern India and Western India, respectively. Structural Equation Modelling Techniques (SEM) has been used to examine the relationships among the variables and to test the goodness fit of the proposed model. Set of propositions related to model was empirically tested at supplier-manufacturer dyad and with an optimum fit of model, found that the data fits the model.

INTRODUCTION

The economic growth of a nation largely depends on three sectors, namely agriculture, manufacturing and the service sectors. In the Indian context, higher dependency and correlation has been found between the manufacturing and the service sectors. As it has been very rightly observed by Gupta and Singh (2012) that "It is very easy to calculate the loss due to poor sale but it is very difficult to calculate the loss due to poor service quality". One of the manufacturing sectors that is on the verge of a revolution and is predicted to become the global hub is the Indian automobile industry. This has happened as all the leading automobile manufacturers of the world have set manufacturing facilities to serve the local market and expand their reach globally. In order to accomplish their targets and to have better competitive advantage the Indian automobile companies have taken lead in adopting supply chain management

DOI: 10.4018/978-1-4666-9795-9.ch012

practices. The reason behind this as sighted by Christopher (1992), is that supply chains compete and not the companies. Also, it has been widely accepted and observed that increase in satisfaction of each member of supply chain can be brought about by only putting aside the traditional arms length relationship and developing closer partnership type arrangements. In developing such type of relationship, service quality has been recognised as an important tool. Also the relation of service quality with improved supply chain performance has been well accepted (Chow et al., 1994; Kearney, 1994; Mentzer et al., 1999, 2001; Pery and Sohal, 1999; Stanley and Wisner, 2002). In the context of supply chain, it is also evident that service quality not only has impact on suppliers, distributors, customers and employees but it also affects the overall business and growth of the organisation.

However, despite this universal recognition of the importance of service quality in supply chains, according to Nix (2001), this area has been little researched. Therefore, this paper attempts to address the need for more empirical based research into the experience of service quality between the business to business customers (manufacturer–supplier dyad) specifically in the context of Indian automobile industry.

The Indian Manufacturing Industry

The manufacturing industry in India has grown over the years and has been mechanized in such a way to enable economic growth, increase output and face firm competition from the global players. It also holds today the key to unlock many socioeconomic challenges of the Indian sub-continent. Studies conducted on the manufacturing industry have concluded that India has a working population of 75% (Umamaheshwari, 2014). Of these, only 600 million have acquired education till middle school level. Thus, in India, the manufacturing industry, is labor centric as it offers a good option for the semi-literate/ lesser educated employment seeking multitude. The manufacturing industry's success can be seen as one of the key components of growth in Asian economies. Studies have indicated that the Productivity of the manufacturing industry in India is approximately 1/5th of the productivity in the manufacturing industry of United States of America (Umamaheshwari, 2014). However, comparatively, performance of manufacturing industry is far different from service industry. In manufacturing industries the product delivered are tangible and the level of quality expected are very much based on the perceived level of expectations of the specific customers (Umamaheshwari, 2014). Umamaheshwari (2014) argues that the pursuit of the growth and success depends on the external service performance that comes from the quality of services that are provided internally within the organization by various distinct but linked units like design, operations, procurement, HR etc.

The Fast Growing Indian Auto Component Industry

The Indian Auto component Industry is undergoing raging changes and is constantly influenced by current economic, political and social conditions prevailing in the country. With unprecedented options available to customers, the demand of the customers is only spiraling –putting pressure on Tier 1 and 2 Suppliers. Borders only on the map, the customer value can hop from one supplier to another. This is forcing units to innovate not only their products but their production process as well (Umamaheshwari, 2014). This dimension dictates a self-imposed discipline amongst functions in an Enterprise. Openness, leadership styles –without upsetting the equations of individuals and teams are considered vital in daily management (Umamaheshwari, 2014).

The study is organised as follows: the next section provides an overview of perceived service quality, service quality dimensions, relationship among the variables under study. Subsequent section presents the research methodology including the sampling procedure and the composition of the questionnaire. The sections on analysis of data and the conclusions are followed by those on managerial contributions, limitations and future research.

LITERATURE REVIEW

Service quality issues have received growing attention from management and academic circles partially owing to proven and potential benefits (*viz.* enhances customer satisfaction and loyalty, attract new customer, word of mouth advertising, increases productivity, leads to higher market share etc.) and partially fuelled by Parasuraman, Zeithaml & Berry's seminal works in the 1980s (Parasuraman et al., 1985 and 1988; Yavas et al., 1997). Much of this focus, however, has been in developed countries (Herbig & Genestre, 1996) perhaps due to the prevailing sellers' market conditions in the emerging economies (Kassem, 1989; Firoz & Maghrabi, 1994; Yavas et al., 1997).

The concept of service quality is defined "as the degree of discrepancy between customer's normative expectations for service and their perception of service performance". However, a closer observation of literature suggests that service quality concept have been conceptualised differently in different service industries like retail, banking, hospitality and health care.

Considering service quality as of key strategic value in service sector, and particularly in retail, previous authors linked it to profitability (Lee and Hwan, 2005), customer retention (Rust and Zahorik, 1993), customer loyalty (Reichheld, 1996; Matos *et al.*, 2009; Chen et al., 2011), customer satisfaction (Dabholkar *et al.*, 2000; Yavas *et al.*, 2004; Matos *et al.*, 2009), word of mouth, complaint, recommendation and switch over (Yavas *et al.*, 2004). But the main emphasis has been on mapping the relationship between service quality and customer satisfaction (Moutinho and Goode, 1995; Sureshchandar *et al.*, 2002; Yavas *et al.*, 2004; Lee & Hwan, 2005; Rahman, 2005; Ladhari *et al.*, 2011); service quality and customer loyalty (Chao, 2008; Chen *et al.*, 2011); service quality, customer satisfaction and customer loyalty (Dabholkar *et al.*, 2000; Matos *et al.*, 2009); and service quality, customer satisfaction, complaint behaviour and commitment (Yavas *et al.*, 2004). However, in the context of manufacturing, it has been found that very few researchers have actually explored this concept particularly in supply chain management in an emerging economy like India. Further, the relationship of service quality with various outcomes such as supplier's satisfaction and supplier's loyalty subsequently impacting organizational performance and sustainable competitive advantage has also been less researched. Moreover, the conflicting empirical evidences (Bitner, 1990; Parasuraman *et al.*, 1988; Cronin & Taylor, 1992; Taylor & Baker, 1994) as to the direction of the causal link between service quality, satisfaction, and customer loyalty also highlight the need for this research as it could have an important managerial implications.

The present paper, thus, attempts to fill this gap by establishing linkages between service quality to suppliers satisfaction leading to supplier's loyalty. Also, it attempts to propose a comprehensive model (shown in Figure 1) that takes into account the interplay between these variables and empirically validates the same in Indian context. It is expected that apart from bridging the gap in existing literature, the present research would also provide valuable insights to academicians as well as automobile professionals in prioritizing service quality dimensions in their industry. The suggestions may help the management in developing appropriate promotional strategies by highlighting relevant aspects of service quality at and also retaining and attracting suppliers.

Figure 1. Structural model with constructs
***Key** OSQ= Overall Service Quality. TAN= Tangibles. REL= Reliability. ASU= Assurance. RES= Responsiveness. EMP= Empathy*

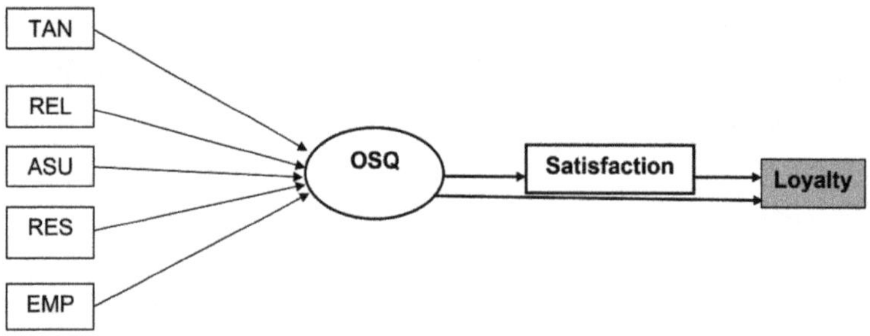

Dimensions of Service Quality

A great deal of research exists on service quality in the banking sector. To measure service quality and identify the dimensions that can be considered in evaluating the services, the most commonly used research instrument is SERVQUAL (Parasuraman *et al.*, 1988). This generic scale for measuring service quality in a variety of service sectors is used in most studies of service quality (Mels *et al.*, 1997; Othman & Owen, 2001; Lam, 2002; Zhou *et al.*, 2002; Chi Cui *et al.*, 2003; Zhou, 2004; Arasli *et al.*, 2005). But, Cronin & Taylor (1992) questioned the conceptual basis of the SERVQUAL scale and found it confusing with service satisfaction. Raising maximum criticism against expectation component of SERVQUAL, they opined that it should be discarded altogether and instead performance component alone be used. Besides theoretical arguments of Cronin and Taylor (1992), other researchers too (Babakus and Boller, 1992; Boulding *et al.*, 1993; Gotlieb, Grewal & Brown, 1994; Hartline and Ferrell, 1996; Jain & Gupta, 2004, Adil, 2011) provided considerable support in favour of the SERVPERF scale. Although alternative instruments are also available for specific use in different service sector such as banking sector (Avkiran, 1994; Bahia & Nantel, 2000; Aldlaigan & Buttle, 2002; Jabnoun & Al-Tamimi, 2003; Karapte *et al.*, 2005; Guo *et al.*, 2008), but they have not been used as extensively as SERVQUAL and SERVPERF (see Table 1).

Relationship among Study Variables

Relationship between Dimensions of Service Quality and Overall Service Quality

Initially, service quality used to be operationalized employing widely known 22-item SERVQUAL instrument developed by Parasuraman *et al.* (1988) comprising five-components (i.e. tangibles, reliability, responsiveness, assurance and empathy). However, Babakus & Boller (1992); Cronin & Taylor (1992, 1994); McAlexander *et al.* (1994) criticized SERVQUAL for its unfavourable psychometric properties and inefficiency in measuring customer expectations. In line with the suggestions of several researchers that the original SERVQUAL dimensions are directly linked to overall service quality (Kwan & Hee, 1994; Saurina & Coenders, 2002; Ijaz *et al.* 2011), the present study considered the following five hypotheses:

H01a: *Tangible* has a significant impact on overall service quality (OSQ).
H01b: *Reliability* has a significant impact on OSQ.

Table 1. Select studies on bank service quality from 1994 till 2011

Study	Country	Sample Size	Scale and Dimensions	α
Avkiran (1994)	Australia	791	17 items; P – E scores; five-point Likert scale 4 dimensions: *staff* conduct (7), *credibility* (3), *communication* (5), access to teller services (2)	.80- .88
Bahia & Nantel (2000)	Canada	115	31 items; P – E scores; seven-point Likert scale 6 dimensions: effectiveness and assurance (13), access (5), price (5), tangibles (4), service portfolio (2), reliability (2)	.78- .96
Aldaigan & Buttle (2002)	U.K.	975	21 items; P-only scores; seven-point Likert scale 4 dimensions: service system quality (11), behavioural service quality (5), machine service quality (2), service transactional accuracy (3)	.80-.93
Sureshchandar *et al.* (2002)	India	277	41 items; P-only scores; seven-point Likert scale 5 dimensions: core service (5); human element of service delivery (17), systematization of service delivery (6), tangibles of service (6), social responsibility (7)	.82-.96
Jabnoun & Al-Tamimi (2003)	UAE	462	22 items; P-only scores; seven-point Likert scale 3 dimensions: human skills (12), tangibles (5), empathy (5)	.77-.93
Jabnoun & Khalifa (2005)	UAE	115 Islamic and conventional banks	29 items; P-only scores; seven-point Likert scale 4 dimensions: personal skills (12), reliability (5), image (6), value (6)	.85-.94
Karatepe *et al.* (2005)	Cyprus	1,220	20 items; P-only scores; five-point Likert scale 4 dimensions: service environment (4), interaction quality (7), empathy (5), reliability (4)	.81-.92
Gao *et al.* (2008)	China	259 corporate	20 items; P - E scores; seven-point Likert scale 2 higher order constructs and 4 second order constructs: functional quality composed of reliability (3) and human capital (7); and technical quality including communication (6) and technology (4)	.88-.93
Nam (2008)	US & S. Korea	129 students	23 items; P-only scores; seven-point Likert scale 4 dimensions: helpful employees, functionality, appearance and trustworthiness	.72-.85
Adil (2011)	India	234	22 items; P-only scores; seven-point Likert scale 5 dimensions: tangibles (4); reliability (5); assurance (4); responsiveness (4); empathy (5)	.73-.88
Adil & Khan (2011)	India	130	14 items; P-only scores; seven-point Likert scale 5 dimensions: tangibles (3); reliability (3); assurance (3); responsiveness (3); empathy (2)	.57-.63
Khan & Adil (2011)	India	274	22 items; P-only scores; seven-point Likert scale 5 dimensions: tangibles (4); reliability (5); assurance (4); responsiveness (4); empathy (5)	.73-.85

Adapted from: Ladhari *et al.* (2011)
Key: P-E: perceptions minus expectations; P-only: performance only; α: Cronbach's Alpha, No. of items are given in parantheses

$H_0$1c: *Assurance* has a significant impact on OSQ.
$H_0$1d: *Responsiveness* has a significant impact on OSQ.
$H_0$1e: *Empathy* has a significant impact on OSQ.

Direct Relationship between Service Quality and Overall Satisfaction

Service quality and customer satisfaction are two very closely related and sometimes even identified concepts. Customer satisfaction is a key factor in assessing the quality of a service. If it is difficult to define the quality of service; it is even harder to determine the level of customer satisfaction (Meidutė-Kavaliauskiene et al., 2014). Customer satisfaction is generally described as a post-purchase response that occurs as a result of evaluation of pre-purchase expectations and perceived performance (disconfirmation) (Oliver, 1980). Previous studies (Cronin & Taylor, 1992; Cronin *et al.*, 2000; Aldlaigan & Buttle, 2002; Brady *et al.*, 2005; Bei & Chiao, 2006; Kassim & Souiden, 2007) report a positive effect of service quality on satisfaction. Treating service quality as an antecedent and satisfaction as an outcome factor is relevant and meaningful from a managerial perspective as well. As Goode *et al.* (1996) cogently argue for marketers, customer satisfaction represents a primary goal to strive for and if they wish to increase suppliers' overall satisfaction; they must target service quality factors which have a bearing on satisfaction. Thus, the following hypothesis was formulated:

$H_0$2: 'OSQ' has significant relation with Supplier's Satisfaction (SS)

Indirect Relationship between Service Quality and Customer Loyalty

Dick & Basu (1994) have characterized loyalty as a strong relationship among individual attitudes and the repetition of a purchase which represents proportion, sequence and the probability of purchases done from the same provider. Loyalty in this study refers to the continuing patronage of a particular manufacturer by a supplier over time. The prime objective of a company's strategy should be loyalty. It has been previously reported that the increase in profit resulting from a 5 percent increase in retention varies between 25 and 85 percent (Ladhari *et al.*, 2011). Both direct and an indirect influence (through satisfaction) of service quality on loyalty has been confirmed empirically in several studies (Fornell, 1992; Boulding *et al.*, 1993; Bloemer *et al.*, 1998; Cronin *et al.*, 2000; Karatepe *et al.*, 2005; Ehigie, 2006; Baumann *et al.*, 2007). Hence, we formed the proposition:

$H_0$3:'SS' has significant relation with Supplier's loyalty (SL)

Supplier Integration (SI)

Supplier integration is supply chain integration upstream. It involves a relationship between the firm and the upstream suppliers (Vijayasarthi, 2010). With supplier integration, suppliers provide information and participation in making decisions (Peterson et al., 2005). Such effective relationships and communications possess a major significance in advanced firms since suppliers know the components supplied better than the firms (Jammernegg & Reiner, 2007; Luo, 2007; Lotfi et al., 2013).

Supplier integration refers to acquiring operational, technical and financial information with the suppliers (Swink et al., 2007; Lotfi et al., 2013). Manufacturers and suppliers may share information including production plans, demand forecasts and levels of inventory. This information sharing results in enhancing the product and production requirements and better utilizing the supplier's and factory's capabilities and structure of cost (Swink et al., 2007; Lotfi et al., 2013).

Service Quality in Supply Chains

According to Ramanigopal and Mani (2011), consumer's satisfaction has a direct relation with the quality of service i.e. good quality of service gives better customer satisfaction and vice –versa. In today's global market, individual firms do not compete as independent entities but as an integral part of supply chains. Therefore, service quality in supply chain is defined as "how well an organisation meets or exceeds the customer expectations in unidirectional or bidirectional for each element of a supply chain".

Service quality in supply chain can also be considered as a three level development work. At the first level comes service quality followed by supply chain management and the third level, the combination of above two i.e. service quality in supply chain.

SCM and Competitive Potential

The impact of SCM practices on operational and business performance of the organisations have been studied by a number of researchers for example, Kim and Oh (2005), while studying the manufacturers and suppliers relationships found that on the basis of bargaining power in relationships, each partners resource commitment may vary depending upon the activity such as quality improvement and new product development. Similarly, Salvador, Forza, Rungtnsanatham and Choi (2001), while studying the causes of attaining improved time performance by the organisations interacting with its suppliers and with customers across the supply chain suggested two findings:

First, when organisations interact with its suppliers and customers on quality management issues, time performance would improve indirectly.

Secondly, when it interacts on material flow management issues, the impact on time related performance would either be completely or partially mediated by the internal practices.

RESEARCH METHODOLOGY

Objectives

Primary objectives of the present study are listed below:

- To determine customer perceived service quality at Indian automotive industry.
- To ascertain the determinants of service quality at Indian automotive industry.
- To examine the contribution made by each dimension of service quality in predicting supplier's satisfaction.
- To fit the proposed model into empirical data in order to draw meaningful statistical inferences.

The Research Instrument

Irrespective of the kind of the service, consumers generally evaluate service quality on the basis of factors such as tangibles, reliability, responsiveness, assurance, and empathy proposed by Parasuraman *et al.* (1988). Through SERVQUAL scale, data are generated on both expectation and perception aspects of service quality, followed by computation of gap scores (Prakash, 2011). While on the other hand, another group of scholars advocated that service quality can best be measured on the basis of perceptions only (Brady and Cronin,2001; Cronin and Taylor, 1994; Cronin and Taylor, 1994; Jain and Gupta, 2004; Kumar & Gulati, 2010; Adil and Ansari, 2012; Adil, 2013a) as mapping expectation brings in inefficiency and redundancy in the service quality instrument (Babakus and Boller,1992; McDougall and Levesque, 1994; Prakash, 2011).Thus, on the basis of the evidences in support of perceptions only scale, SERVPERF instrument developed by Cronin &Taylor (1992) was adopted in the present study. The research instrument comprises of two sections, of which section I consist of 22-items related to service quality where respondents were required to indicate their level of agreement with the help of a seven-point

Likert scale while section II focuses on generating details regarding respondent's demographic profile. The research instrument was developed in English and given independently to two subject experts from the subject area to obtain feedback regarding the content and wording of the measurement items. As an additional precaution, the questionnaire was first pre-tested on a representative sample of 6 managers from automotive industry.

The Sampling Plan

As it was not feasible to cover each automobile manufacturing unit of the country, it was decided to focus on regions such as New Delhi and Gurgaon from Northern India while Pune from Western India. A sample frame comprising 34 companies registered with Society of Indian Automobile Manufacturers (SIAM) was generated with unique code assigned to each of them. Five companies from Northern India while three companies from Western India were randomly picked. The five companies, thus identified, were BMW, General Motors India Pvt. Ltd., Hero Moto Corp, Honda Cars, and Honda Motorcycle & Scooter India Pvt. Ltd. from Northern India while three companies from Western India were Bajaj Auto Ltd, Fiat India and Force Motors Ltd. Respondents were approached by following snowball sampling technique.

Data Collection

The data were collected from November, 2012 to April, 2013.The structured questionnaire was personally administered on roughly 245 and 192 respondents in Northern and Western India, respectively. After discarding the questionnaires with incomplete information, 220 responses from Northern India while 170 responses from Western India were found suitable for further analysis giving a satisfactory response rate of 89.79 percent and 88.54 percent, respectively.

ANALYSIS AND DISCUSSIONS

Exploratory Factor Analysis (EFA)

Factor analysis was conducted using principal component method with varimax rotation. Factors with eigen values greater than 1.0 and factor loadings that are equal to or greater than 0.50 were retained (Hair *et al.*, 2006). In all 18 items loaded cleanly onto the five dimensions of service quality as suggested by Cronin & Taylor (1992). Due to high cross loadings, four items were dropped.

Reliability

A scale with alpha value greater than 0.60 is considered to be reliable (Schuessler, 1971). As shown in Table 1, the reliability coefficients of service quality dimensions were ranging from 0.647 to 0.989 and thus, were within the acceptable range. According to Adil and Rashid (2013), the variables were also assessed with item reliability tests. The purpose of item reliability is to determine the amount of variance in an item owing to the underlying construct rather than to error. The coefficient of item reliability of at least 0.5 is considered to be evidence of reliability. As shown in Table 2, except a few, rest of the items exceeded the minimum recommended value.

Table 2. Results of exploratory factor analysis

Variable	Loading [a]	KMO	α [b]	Loading [c]	Mean	S.D.	TVE (%)
TANGIBLES		0.663	0.694		5.23	1.08	62.3%
T1	0.765			0.765			
T2	0.788			0.788			
T3	0.732			0.732			
T4	0.651			0.651			
RELIABILITY		0.664	0.613		4.33	1.23	57.9%
R1	0.643			0.803			
R2	0.513			0.671			
R3	0.493			0.647			
R4	0.727			0.803			
R5	0.305*			X			
ASSURANCE		0.710	0.639		4.79	0.97	69.1%
A1	0.801			0.801			
A2	0.801			0.801			
A3	0.752			0.752			
A4	0.696			0.696			
RESPONSIVENESS		0.687	0.779		5.32	1.05	46.3%
Rs1	0.687			0.792			
Rs2	0.610			0.805			
Rs3	0.648			0.727			
Rs4	0.457**			X			
EMPATHY		0.743	0.693		4.27	1.13	49.8%
Em1	0.320*			X			
Em2	0.665			0.678			
Em3	0.717			0.989			
Em4	0.402*			X			
Em5	0.790			0.880			

Key: *a* = factor loading of all items. *b* = Cronbach's alpha coefficient. *c* = factor loading of retained items only. X= item deleted

Conceptual Model

Researchers such as Gunasekaran *et al.* (2001), Seth *et al.* (2006c) had argued the use of service quality; hybrid of service quality and performance; performance metrics; buyer-supplier partnership level; and quality of documentation for analyzing performance of the supply chain while Brignall and Ballantine (1996) have explored customer satisfaction, employee satisfaction, training and development of employees. Similarly, in this study, the proposed model is based on the premise that sub-constructs of service quality have significant relationship with overall service quality (OSQ). Moreover, researchers, in the present study, have used customer satisfaction to conceptualize supplier satisfaction. Supplier satisfaction is supplier's overall or global judgment regarding the extent to which linkages developed between him and the focal organization matches his expectations (Prakash, 2011). Taking a cue from extant literature

Table 3. Results of item reliability

Variable	IR
TANGIBLES	
T1	0.585
T2	0.621
T3	0.536
T4	0.424
RELIABILITY	
R1	0.645
R2	0.450
R3	0.419
R4	0.645
R5	X
ASSURANCE	
A1	0.642
A2	0.642
A3	0.566
A4	0.484
RESPONSIVENESS	
Rs1	0.627
Rs2	0.648
Rs3	0.529
Rs4	X
EMPATHY	
Em1	X
Em2	0.460
Em3	0.978
Em4	X
Em5	0.774

in support of strong link between customer satisfaction and loyalty (Anderson and Sullivan, 1993; Innis and La Londe, 1994; Jones and Sasser, 1995), loyalty, in this research is conceptualized as supplier's repeat sale and recommendation of the organization (refer, Figure 1).

Structural Equation Modeling Technique (SEM)

As other multivariate techniques examine only a single relationship at a time, the technique of structural equation modeling (SEM) is highly recommended (Hair *et al.*, 2006). To corroborate with the models of other studies with similar variables such as Adil (2013b), the present study too, examined structure of interrelationships through a series of equation depicting all relationships among factors involved (see, Figure 1 and Figure 3). The relationship in the proposed research model was, thus, translated into a series of structural equations and fitness of model was tested through five main steps as shown below (see Figure 2):

The results suggest that Satisfaction mediates the relationship between OSQ and Loyalty. The results also suggest that OSQ has a trivial effect on Loyalty, when the effect of Satisfaction is controlled for. Overall, these results are in consistent with the general notion that service quality often do not affect suppliers' loyalty directly, but rather require some form of complex adaptive use to mediate their effect such as suppliers' satisfaction in the present case.

Table 4 contains an overall summary of the key fit statistics for the proposed model. Except, the AGFI value which was less than the desired minimum of 0.9 i.e. 0.875, rest of the indices fell within the permissible range. The GFI, CFI and IFI values were higher than required minimum of 0.9, CMIN/df = 3.693, RMSEA = 0.038. χ^2 values were significant at 2.980 and thus acceptable. The above results suggest that measurement model provides a reasonably good fit. Once the overall model was found acceptable, each of the constructs was evaluated separately by examining the indicator loadings for statistical significance. The results of the analysis show that all standardized factor loadings were over 0.5 (see Table 2) and significant at the p = 0.01 level.

Figure 2. Steps involved in Testing Hypothesized Model
Source: Adil, Akhtar and Khan (2013)

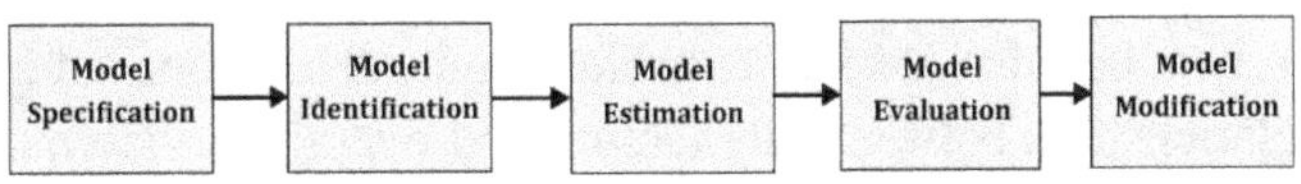

Figure 3. Structural model with main results

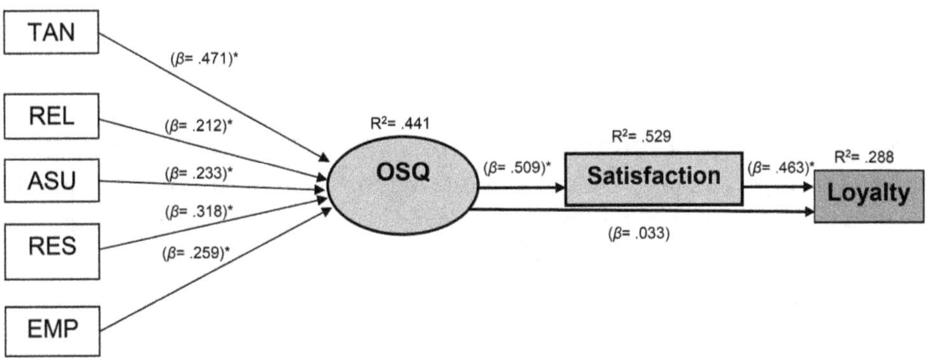

Table 4. Fit indices for SEM

Fit Indicators	CMIN/*df*	GFI	AGFI	CFI	IFI	RMSEA	χ2/*df*
Recommended Value	2.5 – 4.5	≥0.90	> 0.90	> 0.90	> 0.90	< 0.08	< 5.0
Observed Value	3.693	0.922	0.875	0.941	0.938	0.038	2.980

Source. Hair *et al.* (2006), Byrne (2001) and Kline (1998)

Key. GFI= Goodness of Fit Index. AGFI= Adjusted Goodness of Fit Index. CFI= Comparative Fit Index. IFI= Incremental Fit Index. RMSEA= Root Mean Square Error of Approximation

Thus, the data structure supported the conceptual model of the study. The standardized factor loadings as presented in Table 2 and the standardized path coefficients of the structural model in Figure 3 indicated that there existed strong evidence in support of the hypotheses considered for study.

MANAGERIAL IMPLICATIONS

This research contribution to the academic field is multi-fold. *Firstly*, this study contributes to leverage the existing gap in the contemporary literature in which research in the field of service quality in the manufacturing sector in general and automotive industry in particular, is scant. *Secondly*, this study conceptualized and validated a new service quality model specifically for the automotive industry, and this supports many other scholars' belief that the SERVQUAL model, although popularly used in various contexts, cannot be a one-size-fit-all model to all sectors and situations. The SQ model in this study contributes to enhance the understanding of service quality, as it is evidenced from the model's dimensions and attributes that the emphasis of overall service quality should not only be on the 'external outcome' that supplier concerns, but also the 'internal management'. These together construe the comprehensive SQ model and re-assure the belief that the quality of automotive in general should include not only quality of the service, but also those of employees. Besides this, the findings shed valuable insights on measures and critical underlying dimensions of service quality in the context of supply chain in automotive industry, specifically from an Indian perspective. These can be of immense use not only to academic researchers but also marketing professionals. The proposed model provides a deeper understanding of the relationships between key factors and overall service quality of supply chain in Indian automotive industry. This study also attempts to bridge the gap in extant literature on managers' perceptions of supply chain service quality in India.

Further, this study enriches and contributes to the current general literature on service quality and also serves as a preliminary framework for future studies to be engaged in the important sector of automotive. Hence, it is expected that this research would contribute to enhance the understanding of how service quality is assessed in automotive industry and how managers can improve on their service quality. Beside academic contributions, it is envisaged that this study makes several important contributions to the automotive industry. For the automotive industry, the model from this research can act as a measurement tool to benchmark their performance in relation to the industry. Automotive industry providers can act on the findings to realign their priorities in order to improve both overall service quality and supplier's satisfaction leading to supplier's loyalty. In addition, the SQ model could serve as a management tool for managers to identify areas of weaknesses and devise improvement plans to overcome them. This would help boost the company's competitiveness. Last but not least, the SQ model could also help in establishing common standards for overall service quality which the users can identify with. This will vastly improve communication and relationships between service providers and customers.

LIMITATIONS OF THE STUDY AND FUTURE RESEARCH

All researches have its limitations and this study is no exception. In designing the study, the researchers attempted to be as scientific as possible however, the present study nevertheless has the following limitations:

- Snowball sampling may have restricted the possibilities of taking a more homogenized sample in terms of occupation and other demographic factors.
- Research is based on particular geographic extent and thus results may not necessarily be generalizable.
- Study has not attempted to compare performance of automotive companies in terms of type of collaboration with service quality dimensions.
- Hypothesized model contains a limited number of variables.

Future research may be designed in a manner that addresses the limitation of the study pointed out above.

- Researchers should attempt to achieve larger representative sample which could provide more robust convergent reliability.
- More variables such as service charges, word of mouth, recommendation, customer trust etc. need to be studied.
- Future researchers could focus on diversifying the sample across different regions, ethnic groups.

REFERENCES

Adil, M. (2013a). The relationship between service quality and customer satisfaction in India's rural banking sector: An item analysis and factor-specific approach. *The Lahore Journal of Business*, *1*(2), 43–63.

Adil, M. (2013b). Modelling effect of perceived service quality dimensions on customer satisfaction in Indian bank settings. *International Journal of Services and Operations Management*, *15*(3), 358–373. doi:10.1504/IJSOM.2013.054447

Adil, M., Akhtar, A., & Khan, M. N. (2013). Refinement of internet banking service quality scale: A confirmatory factor analysis approach. *International Journal of Services and Operations Management*, *14*(3), 336–354. doi:10.1504/IJSOM.2013.052094

Adil, M., & Ansari, S. A. (2012). Service quality and customer loyalty at Indian rural banks: an empirical test of an integrative model. In *Proceedings of the National Conference on Contemporary Issues in Business, Management & Finance*. University of Kashmir.

Adil, M., & Rashid, I. (2013). Role Stress as a Barrier to Service Quality in Indian Banks: SEM Approach. *International Journal of Services and Sciences*, *15*(1), 40–57. doi:10.1504/IJSSCI.2013.057648

Chow, G., Heaver, T. D., & Henriksson, L. E. (1994). Logistics performance: Definition and measurement. *International Journal of Physical Distribution & Logistics Management*, *24*(1), 117–128. doi:10.1108/09600039410055981

Christopher, M. (1992). *Logistics & Supply Chain Management*. London: Pitman Publishing.

Grönroos, C. (1984). A service quality model and its marketing implications. *European Journal of Marketing*, *18*(4), 36–44. doi:10.1108/EUM0000000004784

Gupta, T. K., & Vikram, S. (2012). Service Quality in Supply Chain: A Review. *IACSIT International Journal of Engineering and Technology*, *2*(8), 1395–1398.

Kearney, A. T. (1994). Achieving customer satisfaction through logistics excellence. *Managing Service Quality*, *4*(2), 47–50. doi:10.1108/09604529410796224

Kim, B., & H. Oh, (2005). The impact of decision making sharing between supplier and manufacturer on their collaboration performance. *Supply Chain Management: An International Journal*, 223-36.

Mentzer, J. T., Flint, D. J., & Hult, G. T. (2001). Logistics service quality as a segment-customized process. *Journal of Marketing*, *65*(4), 82–104. doi:10.1509/jmkg.65.4.82.18390

Mentzer, J. T., Flint, D. J., & Kent, J. L. (1999). Developing a logistics service quality scale. *Journal of Business Logistics*, *20*(1), 9–32.

Nix, N. (2001). Customer service in supply chain management context. In Supply Chain Management. Sage.

Parasuraman, A., Zeithaml, V. A., & Berry, L. L. (1985). A conceptual model of service quality and its implications for future research. *Journal of Marketing*, *49*(3), 41–50. doi:10.2307/1251430

Pery, M., & Sohal, A. (1999). Improving service quality within the supply chain: An Australian study. *Total Quality Management*, *10*(4/5), 673–679. doi:10.1080/0954412997677

Stanley, L. L., & Wisner, J. D. (2002). "The determinants of service quality: Issues for purchasing", *European. Journal of Purchasing and Supply Management*, *8*(2), 97–109. doi:10.1016/S0969-7012(01)00009-0

Chapter 13
Food Cold Chain Management:
An Indian Perspective

Rajneesh Mahajan
Apeejay School of Management, India & Delhi Technological University, India

Suresh Garg
Delhi Technological University, India

P. B. Sharma
Delhi Technological University, India

ABSTRACT

Food cold chain management (FCCM) is a specialized kind of supply chain supported by the two different technologies first, refrigeration and second information technology. The refrigeration equipments like cold storages and cold storing devices. They can keep the food quality cum safety intact and enhance food items shelf life. The information technology enables the tracking and tracing through data loggers etc. Despite the huge fresh produce availability in India, food inflation and food security issues are major concerns for policy makers. The food wastage in India is close to US$ 8309 million annually (US$ 60.17) due to lack of processing and cold storage facilities. The aim of this paper is to discuss status of food cold storage in India. The chapter is based on extensive review of existing literature, studying novel cold storage status and presenting benefits of cold storage. The outcome of research will be that FCCM can play vital role in supplying quality, safe food products and instrumental in reduction of food wastages.

INTRODUCTION

The transportation and delivery of temperature-controlled eatable items is a fast-growing part of the overall food logistics sector. It is because of four major reasons such as volume growth of eatable items (Soon et al. 2012, Abadias et al., 2008); globalization of food logistics as global food foreign trade increases(Kittipanya-ngam et al. 2011, Acar et al., 2010); world trade organizations rigorous regulatory framework such as sanitary and phytosanitary measures (SPS), Codex standards etc. (Neeliah et al., 2013, European Commission, 2011) and consumer awareness and demand for global quality food products

DOI: 10.4018/978-1-4666-9795-9.ch013

(Mukherjee et al., 2012, Kearney, 2011). It will ensure by producers through managing transportation and delivery of temperature-controlled food. It is done either by companies' distribution system or by an integrated logistics provider who manages relationships with freight carriers, warehouses, brokers etc.

Food cold chain management (FCCM) is a specialized kind of supply chain supported by the two different technologies first, refrigeration and second information technology. The refrigeration equipments like cold storages, cold storing devices, refereed vehicles etc. can keep the food quality, safety intact and enhance its shelf life. The information technology enables the tracking and tracing through data loggers, radio frequency identification devices (RFID) for effective food supply chain management (FSCM). In the present chapter attention is paid to status of cold storage in India. Despite the huge fresh produce availability in India, food inflation and food security issues are major concerns for policy makers. The food spoilage and wastages to the scale of US$ 8309 million, is happing in India annually. FCCM can play vital role in supplying quality, safe food products and instrumental in reduction of food wastages. The objective of present chapter is to discuss status of food cold storage in India.

The structure of chapter is as follows: after introducing the subject matter, literature review on cold chain management is presented. The chapter briefly explains the need for cold storage. The present chapter also compares a global and Indian cold storage status and deal in cold storage management challenges and benefits. The chapter is concluded with key managerial insights and research contributions, limitations and future scope of research.

LITERATURE REVIEW ON COLD CHAIN MANAGEMENT

The review of literature on cold chain management is furnished to elaborate tasks and requirement for food supply chain management. The literature review pertaining to FCCM is discussed for the period of 2001-2014.

The operational CCM will be developed by performing mentioned tasks in table1. It may result in providing a much better and broad picture of Indian food supply chain on the world map. If current position of CCM and cold storage is improved then, it may be also instrumental in boosting domestic sales of processed food as well as its exports. It will support in tackling the issues of rising prices and food wastages. Rather it may shore up in creation of food surplus processed food which can be exported from India.

NEED FOR COLD STORAGES IN INDIA

India has total population 1.237 billion in 2012 to support. Indian gross domestic product (GDP) is US$1.842 trillion with GDP growth 3.2% in 2012. Indian economy is progressing with per capita GDP US $1489 in 2012. India has 60.5% land used for agricultural. India has the second largest population in the world to feed. Despite the huge fresh produce availability in India, food inflation and food security issues are major concerns for policy makers in the country. It affects the basic need for Indian citizens to have sufficient, healthy and affordable food. A national research study on quantitative assessment of harvest and post harvest losses for 46 agricultural produces in 106 randomly selected districts was done by central institute of post harvest engineering and technology (CIPHET) in 2010. The food wastage is

Table 1. Cold Chain Literature 2001-2013 and its tasks

Authors	Year	Contribution in Cold Chain Literature 2001-2014	Tasks for CCM
Omta et al.	2001	Cold chain to focus on the engineering facets narrows the view point on the chain structure that comprise the cold chain management (CCM) and the food businesses	Protector of Food
Rae-Smith and Ellinger's	2002	Business service offering the CCM and warehousing	Protector of Food
Victoria et al.	2003	CCM is a physical process that dictates the processed food supply chain management (PFSCM) of various food products	Logistics & FSCM
Tanner and Amos, Moureh and Flick, Wild et al., Punt and Huysamer, McMeekin et al. Bermejo et al. Rodríguez-	2003 2004 2005 2007 2006	Crucial to ensure that temperature of the distributing vans, temperature deviations may occur during distribution. Reports from literature indicate that deviations of 5 oC or more occur may lead to issues	CCM Ethics
Terpstra et al.	2005	It is required that consumers ought be aware about storage of fruits and vegetable (F&V), meat products or other perishables	CCM at consumers end
Arnold, Emond et al.	2008	Maintaining a required storage temperature for perishable foodstuffs is a key factor in preventing risks which can affect food safety and food quality	Quality and Food Safety ensured
Jedermann et al.	2009	There is a higher risk for perishable food of suffering damages even along with CCM. Excellent monitoring by advance IT tools can deliver perishable food in better manner.	Ensure compliance laid rules: CCM Ethics and Temperature as per product nature
Kuo et al.	2010	Temperature sensitive and perishable products (TSPPs) logistics, encourages and maintains food safety and quality	Logistics
Chen et al.	2011	Frozen food can provide people healthy, heterogeneous and safe foods. It suggests that consumers may have a great desire for frozen foods in the future too	Quality and food safety
Hafliðason et al.	2012	At the time of distribution there may be risks of temperature fluctuations occur, which can reduce the shelf life of the food products	CCM ethics and temperature as per product nature
Lorentz et al.	2013	Due to Globalization and retail revolution CCM is desired to protect food supply chains	Global prospective in CCM
Dutton, 2014	2014	Indian and Chinese cold chain facilities are particularly important because they supply raw materials to producer's worldwide. The challenge is to maintain temperature, identify the appropriate packaging and to monitor shipments all the way. It demands CCM becoming more sophisticated.	CCM act as protector in logistics

close to US dollars 8309 million annually (1US$@60.17INR) due to lack of processing and cold storages for fruit and vegetables. The results are illustrated in table 2.

Therefore, it is always desirable that India ought to save for the future and try to reduce waste of food products. One question is always haunting that how to eliminate wastages of food grains, fruits and vegetable? To prevent the wastage of agri produced and do make it available across all geographies and in all seasons, a well developed processed food sector supported by cold chain management can play vital role. It is because of a developed processed food sector coupled with cold storages would not only reduce wastage but also increase income of farmers.

Table 2. Percentage of losses estimated for major produces

Crop	Cumulative Wastage (%)
Cereals	3.9 – 6.0
Fisheries	2.9
Fruits and Vegetables	5.8-18.0
Meat	2.3
Milk	0.8
Oil seeds	6
Poultry	3.7
Pulses	4.3-6.1

Source: A study by CIPHET, 2010

GLOBAL STATUS OF COLD STORAGE

Total capacity of refrigerated warehousing is estimated around 458 million cubic meters globally in 2010, out of which 310 million cubic meters is public warehouses available on lease for food storage. It is observed that during last two years, approximately 192 million cubic meters of additional refrigerated warehouse capacity has been added. The additional capacity of 55 million cubic meters of public refrigerated warehouses was added since 2008. The increase of public refrigerated warehousing worldwide is 20% over the two years, or 10% compound annual growth rate.

The largest cold storage industry is in USA, India, and China as illustrated in Figure 1.

India and China are two large emerging markets demonstrating remarkable growth in cold storages. These two markets are contributing more than 68% of the growth capacity of cold storages globally. India and China have accounted for 133 million cubic meters (m3) of new capacity. In India and China, short-term capacity growth rates exceeded 100% in each year over the last two years, on a compound annual basis. Refrigerated warehousing space more than doubled in each year. The boom in capacity occurred while these emerging economies continued to grow, despite the financial crisis and recessions in Europe and the USA. The sustained growth in these economies is consistent with the observed rapid investment in new capacity over the last two years.

Figure 1. Global cold storage status
Source: GCSCR, 2010

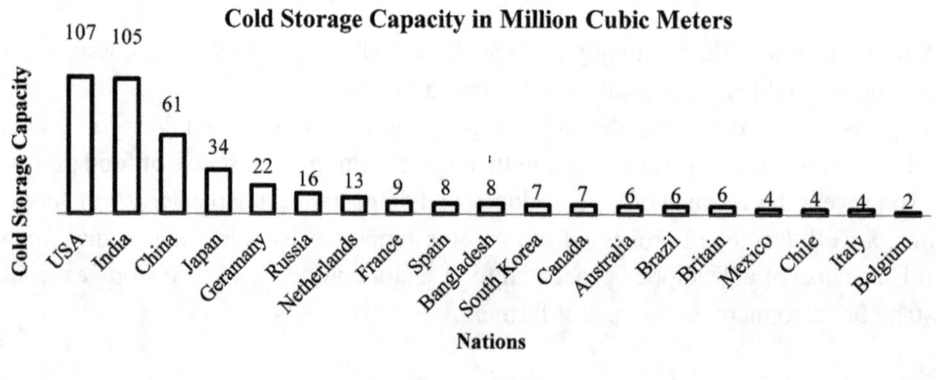

Although the emerging markets have growth in common, some distinctive features of the cold storage industry are worth noting. Indian warehousing is characterized by a large number of smaller warehouses and in terms of total capacity (at 105 million m3 as of 2010) is larger than China. China's growth has been rapid (14% compound annual growth) but more moderate than India. The total capacity in China of 61 million m3 includes approximately 7 million m3 of private warehouses. Private warehouses are 12% of the total capacity in China. Of the reported capacity of refrigerated warehouses in China, none are government-owned. In India, there are relatively few public refrigerated warehouses, less than half a million m3. Government-owned refrigerated warehouses comprise 4 million m3 or 4% of the capacity in India (GCSCR,2010). Cooperatively-owned warehouses are observed in India but not China. Cooperatives in India are classified as private facilities.

STATUS OF COLD STORAGE IN INDIA

India is the largest producer of fruits and second largest producer of vegetables. In spite of that per capita availability of fruits and vegetables is quite low because of post harvest losses. The quality of perishable food products may deteriorate if delivered later then the usable time. It is because of perishable nature of the produce. This type of produce requires a cold chain arrangement to maintain the quality and extend the shelf life if consumption is not meant immediately after harvest. In the absence of a cold storage and related cold chain facilities, the farmers are being forced to sell their produce immediately after harvest which results in glut situations and low price realisation (Roy 2008, Umali-Deininger & Sur, 2006).

A suitable agro climatic conditions and better availability of package of practices, the production is gradually rising. Although, there is a huge scope for enhancing the production of fresh F&V, but due to lack of cold storage and cold chain facilities are becoming major bottlenecks in tapping the potential. The cold storage facilities now available are mostly for a single commodity like potato, orange, apple, grapes, pomegranates, flowers, etc. which results in poor capacity utilisation (GCSCR,2010).

Food can be preserved by storage at low temperature. It may retard the growth of micro organisms. Micro organisms are the spoilage agents and consist of bacteria (Massilia et al. 2009). The low temperature may or may not wipe out spoilage agents whereas high temperature can be effective greatly to reduce activities of spoilage agents. It can provide a practical way of preserving perishable foods in a natural manner.

Cold storages can be utilized to store either a single food commodity or multiple food commodities. It is depending upon the business houses financial position, size of customer and their business activities. It can be planned to store the produce entirely owned by one person or on rental basis or in combination of the other parities. In Indian scenarios usually encourages cold storages where 70% of the capacity is available to farmers for storage on rentals and rest 30% on rent to traders. Financial viability of a unit depends upon the intended pattern of use and rental rate prevalent in an area. Considering 70:30 utilization of the capacity for rentals and own use or for traders is optimal and in terms of area is considered, it is approximately 5000 metric tonne (MT) capacity.

India had produced 147 million tonne vegetables in 2011 accounting for nearly 11% of the world's total production. India currently has a cold chain capacity of only 9 million tonne vegetables, leading to huge wastage annually. This has also impacted India's share in the global export of vegetables that at present is around 1.7% only.

The cold storage market is expected to grow at a CAGR of 27%, a giant leap from US $1781 million in 2009 to US$11345million by 2017 as per Associated Chambers of Commerce and Industry of India (Assocham)'s report (Business Standard, Economics Time, 2013).

TOTAL NUMBER OF COLD STORAGES IN INDIA

The location of cold storages in India is highly concentrated to a few states (names of states with number of cold storages) such as Uttar Pradesh (1589), Maharashtra (466), West Bengal(463), Punjab (422), Gujarat (398), Andhra Pradesh(290), Bihar (243), Haryana (244), Madhya Pradesh (197) etc., These states creates the hub of cold storages in India. Over 50 per cent of the cold storage capacity is confined to Uttar Pradesh, Maharashtra, West Bengal, Punjab and Gujarat among five states of India. On the other hand, majority of Indian states lack investments from government as well as private players are not forthcoming for investment in cold chain management (CCM) and storages in other parts of India. This has made presence of modern cold chain technology minimal in such parts of the country. Figure 2, radar graph is depicting the clear status and bifurcation of cold storage in India. The number of cold storage put together is five though three hundred eight one (5381) in India as per Agricultural Marketing Information Network(agmarknet), GOI, 2010.

Figure 2. Total number of cold storage in India
Source: www.agmarknet.nic.in, GOI, 2010

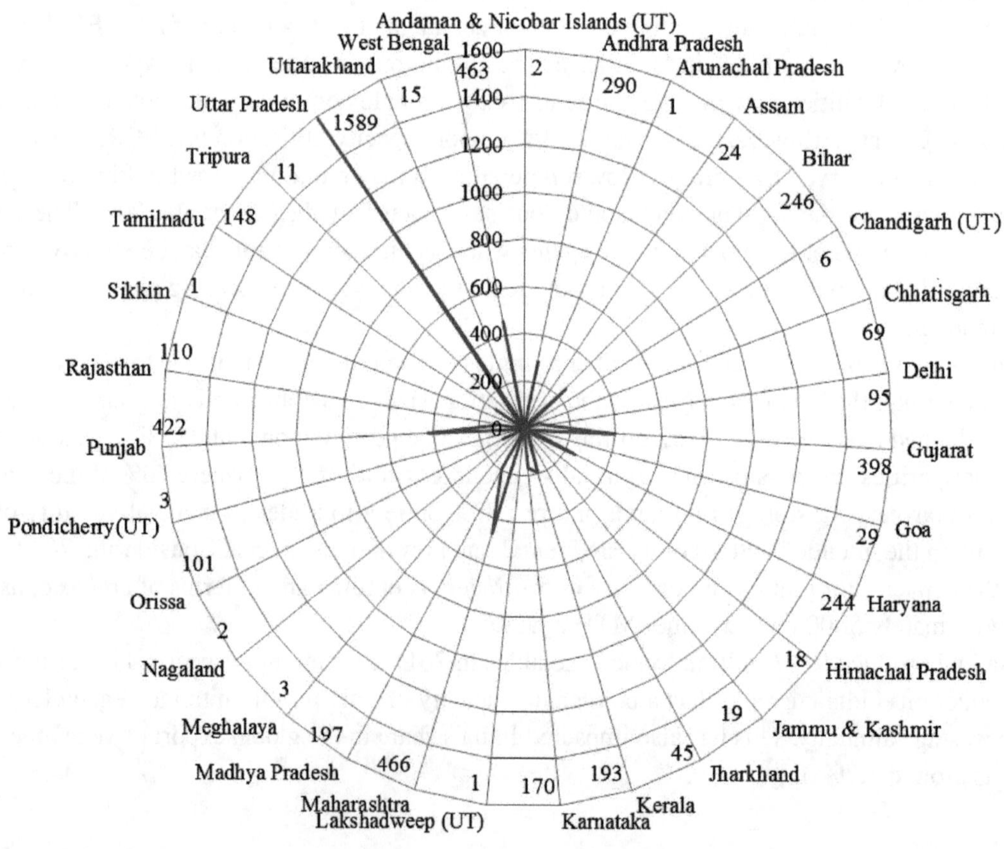

EXCLUSIVE POTATOES COLD STORAGE IN INDIA

The total vegetables sale is largely influenced by the potatoes in India. It is an important ingredient in Indian dishes and thus in demand round the year. Potato contains several phytochemicals such as phenolics, flavonoids, polyamines, and carotenoids, which are highly desirable in diet because of their beneficial effects on health of consumers. The concentration and stability of these constituents are affected by several factors such as genotype, agronomic factors, postharvest storage, cooking and processing conditions. The potatoes are stored and processed into a variety of products before consumption (Ezekiel et al., 2013). In India potatoes meant for household consumption are stored commercially for periods of 6-9 months in cold storage maintained at 2-4°C, relative humidity 90-95%.

There are 2862 exclusive cold storages are being used for storing the fresh potatoes at plus 2-4oC temperature in India. The location of cold storages in India is highly concentrated to a states like Uttar Pradesh (45), West Bengal (14), Punjab(12), Bihar (8) and Gujarat(7). They are top five states operating exclusive of cold storages for potatoes storage in India. Over 86 per cent of the cold storage capacity is confined to five states of India that is Uttar Pradesh with 45%, West Bengal with 14%, Punjab with 12%, Bihar with 8% and Gujarat with 7% contribution in total. The rest of 11 states namely Madhya Pradesh, Orissa, Rajasthan, Chhatisgarh, Karnataka, Jharkhand, Himachal Pradesh, Jammu & Kashmir, Tripura, Chandigarh contribute 14% in exclusive Potatoes cold storages in India as depicted in figure 3.

Figure 3. Exclusive potatoes cold storage in India
Source: www.agmarknet.nic.in, GOI, 2010

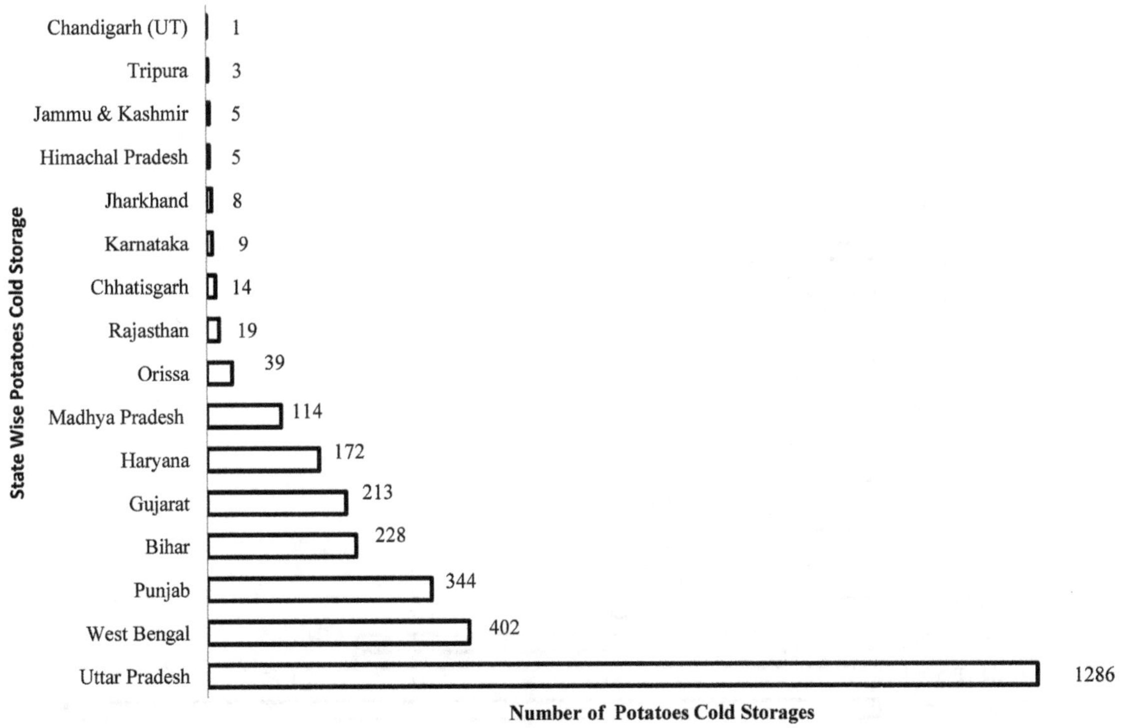

MULTI PURPOSE COLD STORAGES IN INDIA

There are total 1584 multipurpose cold storages are being utilized for storing the various product categories of fruits and vegetables in India. The location of multipurpose cold storages in India is highly concentrated to states such as Uttar Pradesh (294), Andhra Pradesh (212), Maharashtra (189), Gujarat (97) and Karnataka (96). They are top five states operating multipurpose of cold storages in India. Over 56 per cent of the cold storage capacity is confined to five states of India that is Uttar Pradesh with 19%, Andhra Pradesh with 13%, Maharashtra with 12%, Gujarat and Karnataka with 6% each contributing in total multipurpose cold storages in India. The rest of 23 states namely West Bengal, Delhi, Chhatisgarh, Jharkhand, Orissa, Assam, Bihar, Uttrakhand, Jammu & Kashmir, Kerala, Tripura, Himachal Pradesh, Chandigarh, Meghalaya, Nagaland, Arunachal Pradesh, Goa and Sikkim contributes 44% to multipurpose cold storages in India as per Agricultural Marketing Information Network(agmarknet), GOI, 2010 (figure 4).

Figure 4. Multi purpose cold storage in India
Source: www.agmarknet.nic.in, GOI, 2010

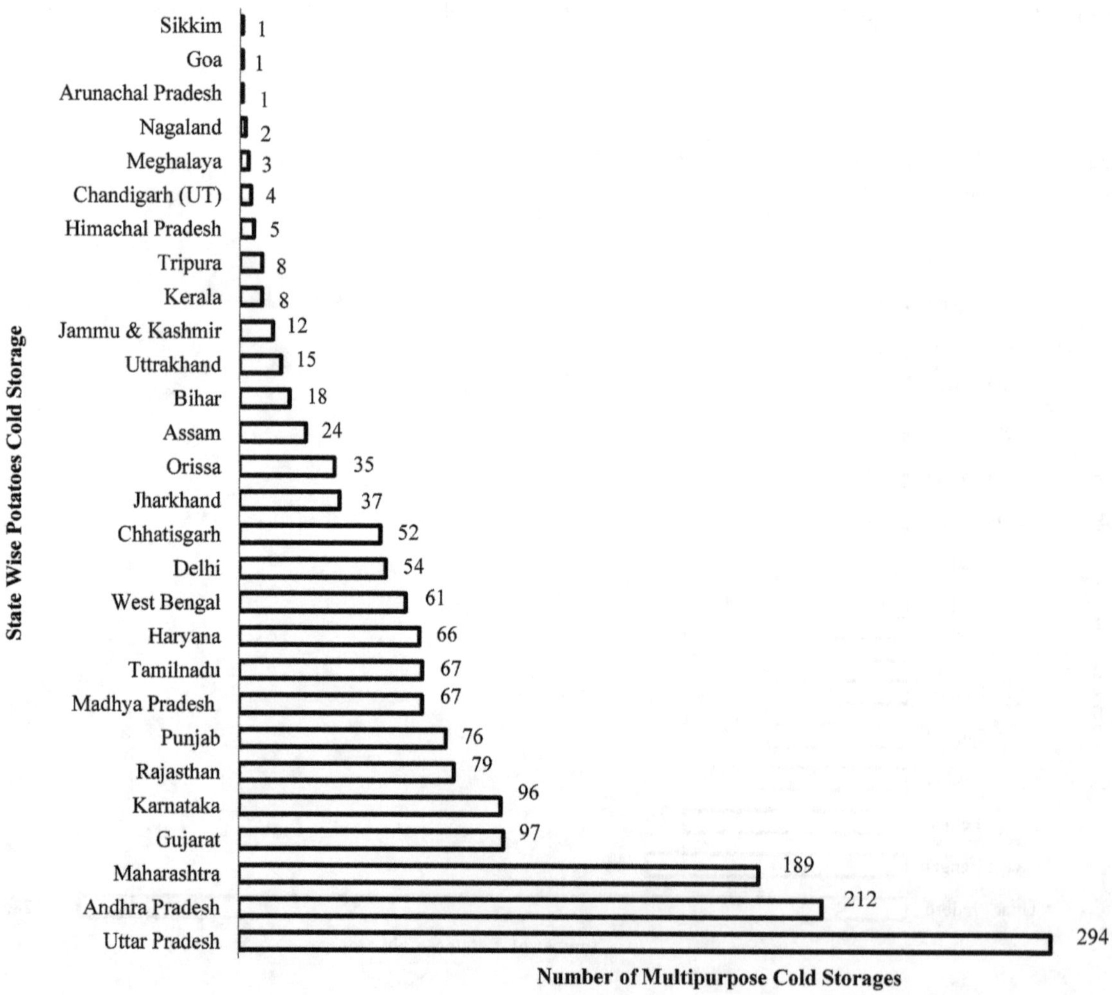

TYPE OF OWNERSHIP OF COLD STORAGE IN INDIA

All the cold storages are classified into three type of ownership in India. That is private ownership, cooperative ownership and public control. The private sector ownership of cold chain is 91%, followed by cooperative ownership 7% and rest 3% cold storages owned by public sector or Government of India. The maximum cold chain capacity is owned by private parties 100649004 cubic meters, cooperatives offer 4028520 cubic meters and public sector offer 460281 cubic meters capacity for storage of food products in India shown in figure 5.

PRODUCT WISE STATUS OF INDIAN COLD STORAGES

India has six types of cold storages utilized for storing potatoes, multipurpose, fruit and vegetable, meat and fish, milk & milk products and others. In absolute terms contribution in total number of cold storages in India for potato storage 2862, multipurpose 1584, fruit and vegetable 160, meat and fish 497, milk & milk products 191 and others 87. They offers respective capacity in cubic meters potatoes 79,233,158, multipurpose 24,272,033, fruit and vegetable 414,636, meat and fish 810,532, milk & milk products 293,389, others 114,053 as illustrated in figure 6 as per the as per Agricultural Marketing Information Network(agmarknet), GOI, 2010.

Figure 5. Types of ownership of cold storages in India
Source: www.agmarknet.nic.in, GOI, 2010

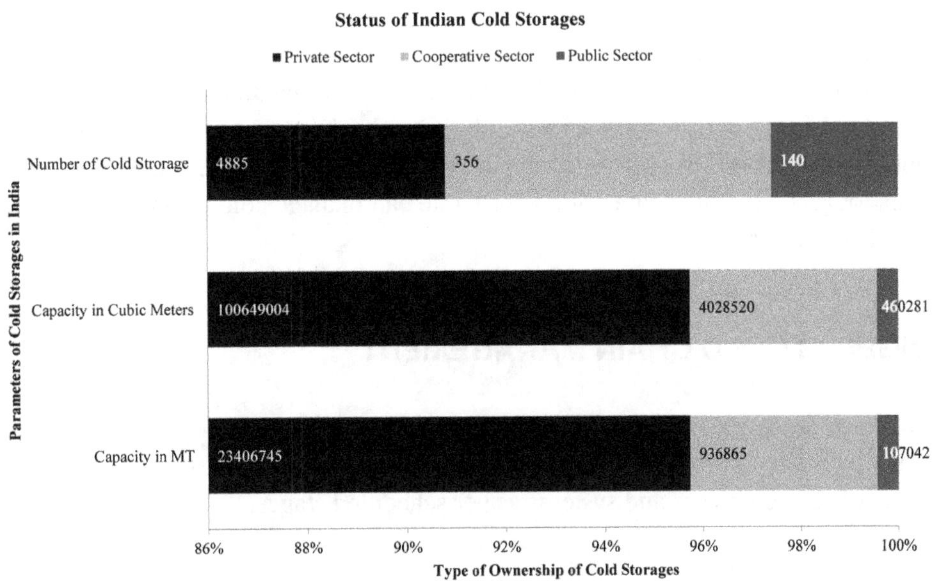

Figure 6. Product wise status of Indian cold storages
Source: www.agmarknet.nic.in, GOI, 2010

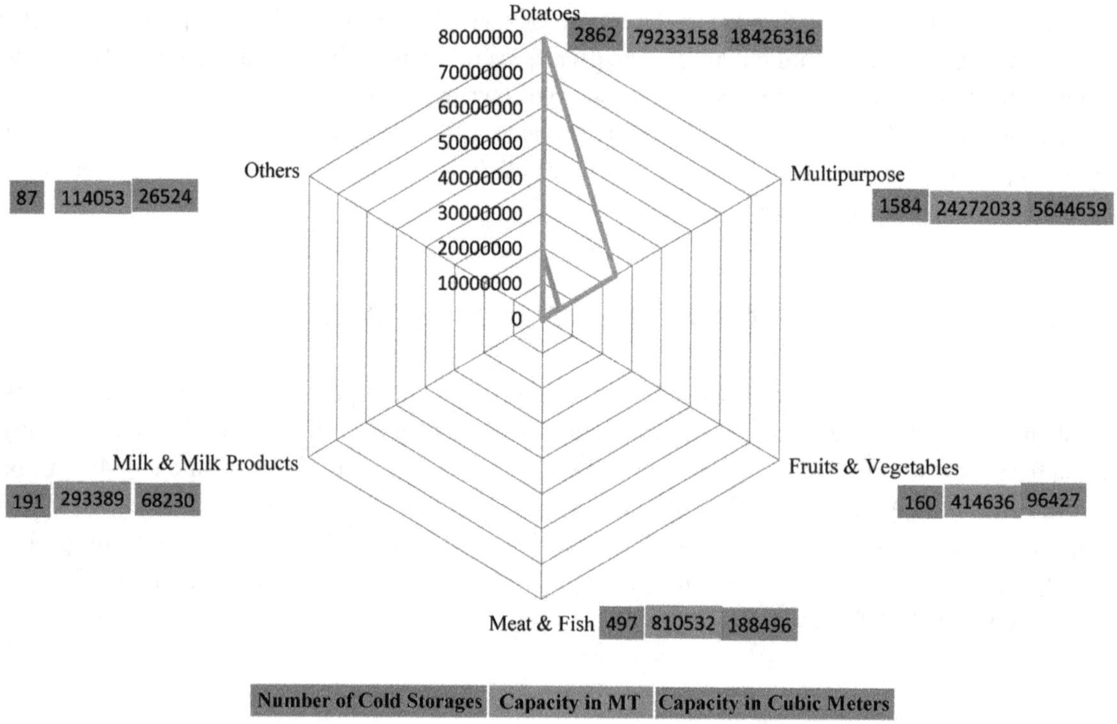

COLD CHAIN LOGISTICS

Transportation for environmentally sensitive products especially eatables, food safety, quality assurance (QA) and quality control (QC) and frequent inspections are major concerns for food processors and manufacturers globally. The food safety audits and inspections ought to be essential drill to ensure adherence of laid processes and systems in food processing and manufacturing. It is the most desired requirement of food safety certification across glob. It can be further ensured by recent advance information technology systems (ITS) such as traceability through radio frequency identifier device (RFID), bar codes, data loggers etc. This manner food corporate can manage cold chain logistics with ease and accuracy effectively (figure 7).

CHALLENGES IN COLD CHAIN MANAGEMENT

There are five broad challenges in path of cold chain management depicted in figure 8. These challenges ought to be met by the food processors and manufactures by ensuring a systematic documentation and implementation of all the process and systems as per scheduled stages.

Figure 7. Crucial cold chain logistics applications

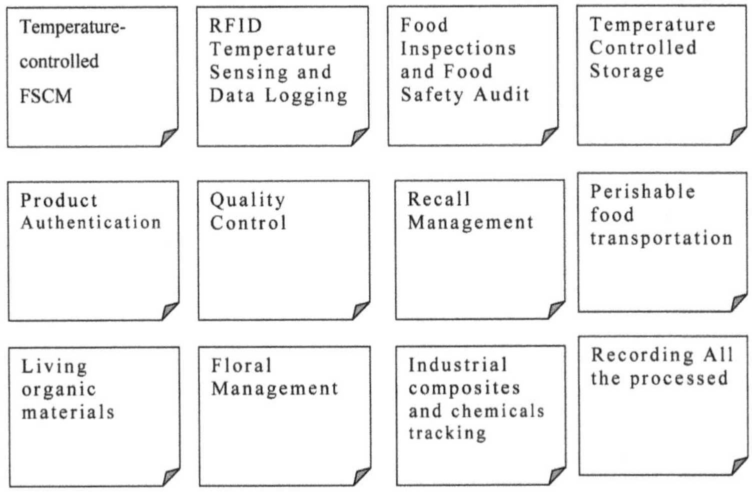

Figure 8. Challenges in cold chain management

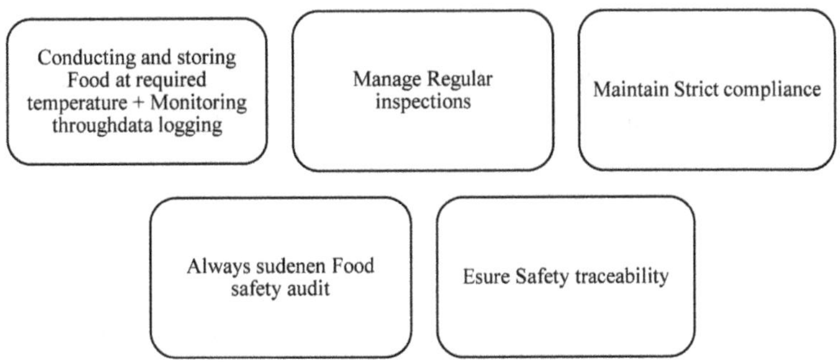

ADVANTAGES OF COLD CHAIN MANAGEMENT

The following is the mutually exclusive list of advantage of cold chain management and storage

- Ensures food safety and inspection process to check Hazards.
- Digital temperature and time tracking protects against CCM break.
- Provides food tractability identification supports inventory management and pilferages.
- Track and trace perishable food wastages can be eliminated to maximum extent.
- RFID is used for simple product and equipment identification supports in stock management as well as transit information.
- Cradle-to-grave product and equipment traceability, back track from source.
- Real-time safety is created.

DISCUSSIONS AND CONCLUSION

In this paper attention is paid to cold storages status of India. Despite the huge fresh produce availability in India, food inflation and food security issues are major concerns for policy makers. The central institute of post harvest engineering and technology (CIPHET) in 2010, done a survey in India on post harvest losses. They estimated that food wastage is close to US$ 9090 million annually due to lack of processing and cold storage facilities for food.

India currently has a cold chain storage capacity of about 9 million tonne vegetables. It is leading to huge wastage per year of food products. It has created an impact on India's share in the global export of F&V. Indian export of F&V is very low around 1.7% annually. The cold storage sector market was expected to grow at a CAGR of 27%, a giant leap from US $1781 million in 2009 to US$11345million by 2017 as per Associated Chambers of Commerce and Industry of India (ASSOCHAM)'s report (Business Standard, Economics Time, 2013). This magical figure is only possible if private business units are forth coming to invest in cold storage building and along with more GOI's policies towards CCM and cold storage development near farms.

CONTRIBUTION

This study provides support for the conceptual framework. It has added to the analysis and interpretation of status of cold chain storages and its requirements and benefits for PFS.

LIMITATIONS

The study uses and tests a conceptual framework extracted from literature in relation to the determinants. A desk study focusing on published literature and historical documentation.

MANAGERIAL IMPLICATIONS

The discussions and research observations of the present paper clearly indicate that in spite of the Indian Government efforts, the target beneficiaries such as farmers, processors, logistic service providers etc. are not to make use of Indian cold chain storages because of lack of awareness, expensive rentals of cold storage against the product sales realization as well as belief that CCM product have lesser nutrition value, technological constraints. It is identified that cold chain has huge potential for future growth and can support the goal of food wastage minimization and boost the processed food growth. Cold chain Managers can apply the learning's from present paper in further use in other sectors of food research. The cold chain managers can encourage the private participations from corporate to built and invest in developing and upgrading existing cold chain storage.

FUTURE SCOPE OF RESEARCH

Future research may be to compare the Indian cold storage status with other advance nations like USA, Germany etc. Second, this study focused only on cold storage status of India only, as India is huge country with many disparities in term of economics development, efforts should be preparing more cold storages near to farms and their number must be increased manifolds. An empirical research should be done to test the theory claims on cold chain.

REFERENCES

Abadias, M., Usall, J., Anguera, M., Solsona, C., & Viñas, I. (2008). Microbiological quality of fresh, minimally-processed fruit and vegetables, and sprouts from retail establishments. *International Journal of Food Microbiology*, *123*(1-2), 121–129. doi:10.1016/j.ijfoodmicro.2007.12.013 PMID:18237811

Acar, Y., Kadipasaoglu, S., & Schipperijn, P. (2010). A decision support framework for global supply chain modeling: An assessment of the impact of demand, supply and lead-time uncertainties on performance. *International Journal of Production Research*, *48*(11), 3245–3268. doi:10.1080/00207540902791769

Andrej, O., & Mojca, J. (2009). Maintaining a cold chain from purchase to the home and at home: Consumer opinions. *Food Control*, *20*(2), 167–172. doi:10.1016/j.foodcont.2008.03.010

Annual Report. (2011). *Ministry of Food Processing India (MOFPI)*. Retrieved on September 23, 2011 from http://mofpi.nic.in

Arnold, J. R., Stephen, C., & Lloyd, C. (2008). *Introduction to Materials Management*. Pearson Education International.

Business Standard. (2013). *Newspaper article*. Retrieved on February 7, 2013 from http://www.business-standard.com/article/companies/cold-chain-industry-to-touch-rs-64-000-cr-by-2017-assocham-113020600684_1.html

Chen, K. Y., & Shaw, Y. C. (2011). Applying back propagation network to cold chain temperature monitoring, Advanced Engineering Informatics. *Advanced Engineering Informatics*, *25*(1), 1–22. doi:10.1016/j.aei.2010.05.003

Cold Storage Capacity. (2013). Retrieved on June 02, 2013, http://articles.economictimes.indiatimes.com/2013-02-06/news/36949978_1_

Economics Times. (2013). *Newspaper article*. Retrieved on February 2013, available at http://articles.economictimes.indiatimes.com/keyword/cold-chain

Emond, J. P. (2008). The cold chain. In S.B. Miles, S.E. Sarma, & J.R. Williams (Eds.), RFID Technology and Applications. Cambridge University Press.

European Commission. (2011). *Rapid Alert System for Food and Feed*. Retrieved on August 15, 2013 from http://ec.europa.eu/food/food/rapidalert/archive

Ezekiel, R., Singh, N., Sharma, S., & Kaur, A. (2013). Beneficial phytochemicals in potato a review. *Food Research International*, *50*(1), 487–496. doi:10.1016/j.foodres.2011.04.025

Hafliðason, T., Ólafsdóttir, G., Bogason, S., & Stefánsson, G. (2012). Criteria for temperature alerts in cod supply chains. *International Journal of Physical Distribution & Logistics Management*, *42*(4), 355–371. doi:10.1108/09600031211231335

Jedermann, R., Garcia, L.R., & Lang, W. (2009). Spatial temperature profiling by semi-passive RFID loggers for perishable food transportation. *Computers and Electronics in Agriculture, 65*(2), 145–154.

Kittipanya-ngam, P., Shi, Y., & Gregory, M. J. (2011). Exploring geographical dispersion in Thailand-based food supply chain (FSC), Benchmarking. *An International Journal*, *8*(6), 802–833.

Mahajan, R., Garg, S., & Sharma, P. B. (2011). Indian frozen peas market: A case study on FPIL. *International Journal of Globalisation and Small Business*, *4*(2), 154–169. doi:10.1504/IJGSB.2011.042250

Massilia, R. M. R., Melgar, J. M., Fortuny, R. S., & Belloso, O. M. (2009). Control of Pathogenic and Spoilage Microorganisms in Fresh-cut Fruits and Fruit Juices by Traditional and Alternative Natural Antimicrobials. *Comprehensive Reviews in Food Science and Food Safety*, *8*(3), 157–180. doi:10.1111/j.1541-4337.2009.00076.x

McMeekin, T., Baranyi, J., Bowman, J., Dalgaard, P., Kirk, M., Ross, T., & Zwietering, M. H. et al. (2006). Information systems in food safety management. *International Journal of Food Microbiology*, *112*(3), 181–194. doi:10.1016/j.ijfoodmicro.2006.04.048 PMID:16934895

McMeekin, T. A., Smale, N., Jenson, I., Ross, T., & Tanner, D. (2006b). Combining microbial growth models with near real-time temperature monitoring technologies to estimate the shelf life and safety of foods during processing and distribution. In *Proceedings of the 2nd International Workshop Cold Chain-Management*. University of Bonn.

Moureh, J., & Flick, D. (2004). Airflow pattern and temperature distribution in a typical refrigerated Truck configuration loaded with pallets. *International Journal of Refrigeration*, *27*(5), 464–474. doi:10.1016/j.ijrefrig.2004.03.003

National Centre for Cold Chain Development. (2012). Retrieved on May 02, 2012, from http://nhm.dacnet.nic.in/NCCD/AboutUs.html

Neeliah, S. A., Neeliah, H., & Goburdhun, D. (2013). Assessing the relevance of EU SPS measures to the food export sector: Evidence from a developing agro-food exporting country. *Food Policy*, *41*(1), 53–62. doi:10.1016/j.foodpol.2013.04.002

Omta, S. W. F., Trienekens, J. H., & Beers, G. (2001). Chain and network science: A research framework. *Journal on Chain and Network Science*, *1*(1), 1–6. doi:10.3920/JCNS2001.x001

Punt, H., & Huysamer, M. (2005). Supply chain technology and assessment—temperature variances in a 12 m integral reefer container carrying plums under a dual temperature shipping regime. *Acta Horticulturae*, *687*(687), 289–296. doi:10.17660/ActaHortic.2005.687.35

Rae-Smith, J. B., & Ellinger, A. E. (2002). Insights from the introduction of an online logistics service system, Supply Chain Management. *International Journal (Toronto, Ont.)*, *7*(1), 5–11.

Report on Cold Storage of India. (2013). *Agmark*. Retrieved on August 29, 2013, from http://agmarknet.nic.in/coldstorage.htm

Rodríguez-Bermejo, J., Barreiro, P., Robla, J. I., & Ruiz-Garcia, L. (2007). Thermal study of a transport container. *Journal of Food Engineering*, *80*(2), 517–527. doi:10.1016/j.jfoodeng.2006.06.010

Roy, D., & Thorat, A. (2008). Success in High Value Horticultural Export Markets for the Small Farmers: The Case of Mahagrapes in India. *World Development*, *36*(10), 1874–1890. doi:10.1016/j.worlddev.2007.09.009

Salin, V., & Nayga, R. M. Jr. (2003). A cold chain network for food exports to developing countries. *International Journal of Physical Distribution & Logistics Management*, *33*(10), 918–933. doi:10.1108/09600030310508717

Soon, J. M., Bharu, K., & Manning, L. (2012). Fresh produce-associated outbreaks: A call for HACCP on farms? *British Food Journal*, *114*(4), 553–597. doi:10.1108/00070701211219568

Srinivas, M.R., Balamuralidhara, V., & Kumar, T.M.P., & khattri,S. (2013). Cold Chain Management in Vaccines Logistics - A Review. *Journal for Clinical Studies*, *5*(3), 50–56.

Tanner, D., & Amos, N. D. (2003b). Modelling Product Quality Changes as a Result of Temperature Variability in Shipping Systems. *International Congress of Refrigeration*, Washington, DC.

Terpstra, M. J., Steenbekkers, L. P. A., de Maertelaere, N. C. M., & Nijhuis, S. (2005). Food storage and disposal:consumer practices and Knowledge. *British Food Journal*, *107*(7), 526–533. doi:10.1108/00070700510606918

Umali-Deininger, D., & Sur, M. (2006). *Food safety in a globalizing world: Opportunities and challenges for India*. Paper presented at the International Association of Agricultural Economists Conference, Gold Coast Australia.

Vasudeva, M. (2009). *Evolution of Bulk Handling from Conventional Bagged Food-grains handling and its socio-economic impact on India*. Retrieved on November 15, 2013, http://www.geaps.com/knowledge/proceedings/article.cfm

Wikipedia. (2008). *The Free Encyclopedia*. Wikimedia Foundation, Inc. Retrieved on November 14, http://en.wikipedia.org/wiki/Supply_chain

Wild, Y., Scharnow, R., & Rühmann, M. (2005). *Container Handbook, 3, Gesamtverband der Deutschen Versicherungswirtschaft e.v*. Berlin, Germany: GDV.

ADDITIONAL READING

Mahajan, R., Garg, S., & Sharma, P. B. (2011). Indian frozen peas market: A case study on FPIL. *International Journal of Globalization & Small Business*, *4*(2), 154–169. doi:10.1504/IJGSB.2011.042250

Mahajan, R., Garg, S., & Sharma, P. B. (2013). Frozen Corn Manufacturing and its Supply Chain: Case study using SAP-LAP approach. *Global Journal of Flexible Systems Management (Springer)*, *14*(3), 167–177. doi:10.1007/s40171-013-0040-y

KEY TERMS AND DEFINITIONS

Cold Chain Management: Cold chain management is practiced utilizing the refrigerated shippers, cold storages and refereed vans. They will maintain the specific temperature require for product storage. For Dairy and its culture products like lassi, curd etc. at +4 to +6 degree, ice creams at -22 to -24 degree, frozen items like vegetables, fruit and Meat at minus 18- 25 degree.

Food Cold Chain: It can be defined as food chains requires two kind of temperature support during storage. First, cold temperature varies between 0 to plus 8 degree and second, frozen temperature varies from minus 1 to 25 degree for storage and transportation of food products.

Food Quality and Safety: The food growers and processers ought to ensure the food quality and food safety. If food growers will use appropriate seed and farming techniques the crops quality will be maintained and food processors will deploy the safe food manufacturing practices through implementing hazard analysis and critical control points (HACCP).

Food Storage: It is age old practice to protect food for insects and rat bit for consumption after certain time and in off season.

Food Wastages: Food wastages can be defined as the food which is not used for human consumption. It can be classified into two categories, first food wastage in plate and second, food wastages during storage and transportation due to poor post harvesting practices.

Processed Food Supply Chain Management: Processed food supply chain carries processed food (PF) using its network that comprises of farmers, food processors, manufactures, carrying and forwarding agents (C&F's), distributors and retailers. There are two types of processed, dry and require temperature conditioning (minus or plus) for its storage.

Refrigerated Warehousing: The role of refrigerated warehousing will ensure the adequate temperature and atmospheric conditions to preserve the food from spoilage.

Chapter 14
Strategic Development of Responsible Warehousing with Safety Partnership

Lilis Surienty
Universiti Sains Malaysia, Malaysia

Hui-Nee Auyong
Universiti Sains Malaysia, Malaysia

Suhaiza Zailani
Universiti Malaya, Malaysia

ABSTRACT

Occupational safety and health (OSH) issues have become a major concern to many corporations in Malaysia since the enactment of the Occupational Safety and Health Act 1994. While safety management system has been researched, only limited literatures have given attention to warehousing. The development of effective partnerships between customers and vendors to identify opportunities for enhancing safety management system. Employee safety behaviour in the workplace is crucial in approaching towards safety compliance and safety performance indicators. The main objective was to test the safety behaviour of the workers. Data were collected throughout the warehousing processes of a multinational electronics manufacturer in Malaysia. A questionnaire has been filled-up by the warehousing employees. This study proposes the theory of planned behaviour were to explain the linkage between customer - vendor partnership and safety behaviour. Respondents were questioned partnership with its customers to systematically improve safety behaviour. Usually most of the occupational accident or disease took place in the material handling operations and activities with cargo interface. The findings of this study show that customer-vendor partnership has a significant and positive relationship on safety behaviour. This approach to be examined for improved safety ownership and the possibility that enhancements would become an essential part of the vendor's processes, systems, and culture. This customer-vendor partnership approach will develop safety improvements for next heights of success. SPSS was applied for processing the data. It was found that information dissemination and collaboration with the client are important. Employees should participate in the safety program as required by the clients so as to obtain the merits of high performance workforce.

DOI: 10.4018/978-1-4666-9795-9.ch014

INTRODUCTION

The effectiveness of a safety management system depends on to what extent employees understand and accept the safety management system. Only through a systematic safety management system that is accepted by all employees can an organisation effectively eliminate workplace injuries. Without knowing how well employees subscribe to the organisation's safety management system, the aim to eliminate or reduce occupational safety and health (OSH) related injuries will not successful. Occupational safety and health should be able to influence behavioural changes among employees and bring the effort of both employers and employees together to manage an organisation's safety management system. The employee safety behaviour also gives impacts to organisational safety performance. In this study which will be concentrated on customer-vendor relationship as main factor that will influence safety behaviour in the workplace.

According to the Labour Force Survey Report Malaysia (DOS, 2013), by 2012, they were 12.723 million employed persons. The number of employed persons data show that the manufacturing sector employs 2.228 million persons (17.5%), agriculture, forestry & fishing sector employs 1,602 million persons (12.6%), construction sector employs 1.164 million persons (9.1%), and Transport, Storage and Communications sector employs 0.624 million persons (4.9%) of the work force (Table 1).

According to the International Labour Organisation (ILO), every year over two million people die around the world due to work-related accidents or diseases. In Malaysia, the major policy concerning OSH is Occupational Safety and Health Act (OSHA) 1994. The percentage of accidents in manufacturing sector (2011: 29%) is high and the data shows manufacturing sector is among the most hazardous workplace. The percentages of accidents in Transportation and storage sector (2011: 6%) are higher than some other sectors (Table 2). The un-proportionately accident rate in the transportation & storage industry in contrast of the construction industry is probably due to the latter's better OS&H management as compared to the fragmented authorities in the transportation industry.

Theory of planned behaviour (TPB) is an extension of Theory of reasoned action (TRA) that it offers room to address those behaviours with regard to being under volitional control of the individual (Ajzen, 1985, 1991). In a Malaysian study on safety practices at engineering laboratories and workshops (Koo, et al., 2014) shown that three explanatory variables in the TPB can explain behavioural intention of practising safety with 52% of variance and safety knowledge as an extension construct of the TPB model with behavioural intention can explain the safety behaviour with the variance of 40%. The paper used the Theory of Planned Behaviour (TPB) as the theoretical basis to assess differential qualities of resources, capabilities and work processes. According to TPB, behavioural decision is made due to attitude, subjective norms and perceived behavioural control which influences the behaviour primarily through their effects on behavioural intention. Behavioural study is of equal importance where it needs to include the human decision makers (suppliers, customers and service providers) of multiple firms (CSCMP, 2008). Therefore, inclusion of customer perspective can provide empirical evidence to explain the customers' perception, attitude and response patterns. The paper will study the factors that influence the effectiveness of safety behaviour and its correlation with the accident rate in storage / warehousing. Moreover, the study will also provide some suggestions and critiques on the above objective. The main objective of this study to identify the perceived norms of customer - vendor partnership towards safety behaviour.

Table 1. Number and percentage distribution of employed persons by industry, Malaysia, 2010, 2011 and 2012

Industry	2010 ('000)	2010 (%)	2011 ('000)	2011 (%)	2012 ('000)	2012 (%)
Agriculture, forestry & fishing	1,614.9	*13.6*	1,410.0	*11.5*	1,601.7	*12.6*
Mining & Quarrying	57.2	*0.5*	76.0	*0.6*	80.6	*0.6*
Manufacturing	2,108.5	*17.7*	2,222.3	*18.1*	2,227.9	*17.5*
Electricity, gas, steam and air conditioning supply	55.5	*0.5*	51.6	*0.4*	62.1	*0.5*
Water supply; sewerage, waste management and remediation activities	66.7	*0.6*	70.8	*0.6*	80.4	*0.6*
Construction	1082.7	*9.1*	1,133.6	*9.2*	1,163.7	*9.1*
Services	6,914.2	*58.1*	7,095.6	*59.6*	7,310.2	*59.0*
- Wholesale and retail trade; repair of motor vehicles and motorcycles	1887.8	*15.9*	1,999.5	*16.3*	2,116.0	*16.6*
-Transportation & Storage	554.7	*4.7*	605.2	*4.9*	624.0	*4.9*
- Accommodation and food and beverage service activities	856.7	*7.2*	942.3	*7.7*	957.0	*7.5*
- Information and communication	178.9	*1.5*	207.6	*1.7*	209.2	*1.6*
- Financial and insurance/takaful activities	323.4	*2.7*	317.6	*2.6*	322.9	*2.5*
- Real estate activities	58.5	*0.5*	61.2	*0.5*	69.0	*0.5*
- Professional, scientific and technical activities	285.6	*2.4*	329.0	*2.7*	307.8	*2.4*
- Administrative and support service activities	359.2	*3.0*	448.9	*3.7*	530.9	*4.2*
- Public administration and defence; compulsory social security	787.7	*6.6*	749.0	*6.1*	697.6	*5.5*
- Education	779.3	*6.5*	785.0	*6.4*	786.2	*6.2*
- Human health and social work activities	280.0	*2.4*	382.5	*3.1*	414.8	*3.3*
-Arts, entertainment and recreation	91.6	*0.8*	86.4	*0.7*	84.4	*0.7*
- Other service activities	182.9	*1.5*	181.5	*1.5*	190.2	*1.5*
Activities of households as employers	285.4	*2.4*	222.5	*1.8*	194.6	*1.5*
Activities of extraterritorial organisations and bodies	2.5	*0.0*	2.1	*0.0*	2.1	*0.0*
Total	**11,899.5**		**12,284.4**		**12,723.2**	

Source: Labour Force Survey Report Malaysia 2012 (Department of Statistics)

VALUE-ADDED SAFETY PARTNERSHIPS

Safety is among the performance measure of logistics service providers (Mejza et al., 2003). More specifically, the increase of safety levels reduces the possibility of working accidents within the warehouse and, therefore, the predictions for compensation payments to employees (Panta & Vrysagotis, 2011). If the aforementioned reason focuses on cost reduction, the following reason (Waring & Lever, 2004) focuses on revenue increase. Alternatively, businesses with high safety levels, low number of accidents and safety promotion policies in work at the warehouse that have been proved in action, attract investment capital directed towards socially responsible companies that have a consequent increase of their financial

Table 2. Accidents reported by sector, 2009, 2010 and 2011

Industry	2009	2010	2011
Manufacturing	17,206	17,569	17,106
Trade/Wholesale & Retail	9,197	9,437	9,724
Construction	4,108	4,665	4,937
Transportation & Storage	3,690	3,642	3,851
Agriculture, forestry & fishing	2,696	2,561	2,410
Finance & Insurance	780	840	987
Electricity, gas, water & sanitary services	544	643	773
Mining & Quarrying	370	370	421
Other Services	16,595	17,912	19,668
Total	55,186	57,639	59,897

Source: SOCSO Annual Report 2011

revenues. Hence, many organisations are seeking to manage their logistics operations strategically to gain a competitive advantage (Hum, 2000).

In the case study of Renault, the role of the safety and health bodies in charge of safety at work can take the form of a partnership with the customer (EU-OSHA, 2000, p. 93). With regard to the customer – vendor relationship, the UK good neighbour scheme encourages organisations to share their expertise in managing safety and health with their vendors, neighbouring organisations and the wider community and foster closer partnerships, where knowledge acquired is likely to be of wider benefit to a vendor (EU-OSHA, 2000, p. 150). For instance, the GME Fair Trade Code of Conduct developed by a trade union and applicable to the textiles, clothing, leather and footwear (TCF) sector, states: "Vendors shall ensure that workers are included in all appropriate state or private security provisions, especially invalidity benefit, injury benefit, sickness benefit, old-age benefit/pension, maternity benefit and unemployment benefit (ILO, 2001, p. 30).

For the customer-vendor partnership, Weibert and Plunkett (2006) wrote a success story of Marshall Space Flight Centre (MSFC) due to customer's emphasis on safety. In addition, Egels-Zandén (2007) examined vendors' compliance with codes of conduct with Swedish (customers) toy retailers' codes of conduct in China, which prescribe that employees should have physical examinations at least once a year, and that employees have the right to health and safety education to avoid work-related accidents.

Consequently, the company may make a conscious decision to ride on a strategic partnership approach to safety. Logistics has become a strategic factor in creating competitive advantage in supply chains (Ramanathan & Ramanathan, 2013). Customers often want to have third-party logistics alternatives for safety and competitive reasons (Hertz & Alfredsson, 2003). The company may work with customer to improve its current processes and implement responsible new processes.

OUTCOMES OF SAFETY BEHAVIOUR

Employee safety behaviour has major roles in keeping a safer workplace and minimising the possibility of accident rates in the workplace. Compliance with safety rules and regulations is important in lowering the risk of accidents. Becker, Billings, Eveleth and Gilbert (1996) have studied the rationale for using commitment as a criterion in assessing the impact of organisational transformation on employee-organisation relations. The perception on contribution of safety behaviour will be measured in terms of safety internalisation.

METHODOLOGY

The study focuses at the warehousing activities in an electronics manufacturer in Kedah, Malaysia. The Group Company's began with the invention of glass crystal in Japan in 1941. The Company is a leading electronics vendor that manufacturers LCD and LSI components. Packaged product today is transported for shipment to East Asia. Since its Kedah plant establishments in 2005, this company has grown from one line double-up to two lines with the number of employees are at approximately 500 employees.

The customer-vendor partnership measures were adapted from Bhattacherjee (2000), Lim, Gan and Chang (2002), and Lin and Lee (2004). Likert-scale type format was applied for employee safety behaviour questions. Questions on demographic information used ordinal and nominal scales. The population of this study is employees who work in warehousing function in this company. Data was collected from warehousing activities in the Kedah plant. This study is primarily conducted to ascertain the criteria used to measure the safety behaviour, and the factors that contribute to safety excellence in warehousing. The study examines the influential factors or independent variables that may affect the safety outcome the dependent variable.

FINDINGS

The Cronbach's alpha in the study recorded 0.574, which is below 0.6 but above 0.5, therefore, the reliability may be acceptable for exploratory work (Field, 2000). The *mean* for "Clients influence me to adopt safety measures" was 3.25 with a Standard Deviation *(SD)* of 0.707, the *mean* for" I feel under pressure from the clients to adopt safety measures" was 3.25 with a *SD* of 0.886 and the *mean* for "My colleagues share with me business knowledge from clients" was 2.75 with a *SD* of 0.886 (Table 3).

The profile of the respondents consists of 100% of male respondents. Most of the respondent's age is from 26 to 30 years old at 62.5%. In terms of ethnic group, they are 62.5% Malay. Status of respondents' highlights 57.1% married. For the education level, mostly respondents have education until SPM (O-Level) level (87.5%). The highest percentage of respondent's position is in non-executive level that is 87.5%.

Table 3. Mean and standard deviation

Items	Mean	Standard Deviation
Clients influence me to adopt safety measures.	3.25	.707
I feel under pressure from the clients to adopt safety measures.	3.25	.886
My colleagues share with me business knowledge from clients.	2.75	.886

Table 4. Demographic profiles of the respondent

Attributes	Description	Percentage (%)
Gender	Male	100
	Female	0
Age	21-25	12.5
	26-30	62.5
	36-40	12.5
	41-45	12.5
Ethnic	Malay	62.5
	Indian	37.5
Marital Status	Single	42.9
	Married	57.1
Education	SPM	87.5
	STPM	12.5
Position	Non-Executive	87.5
	Supervisor	12.5
Years of Service	< 5 years	75
	6 to 10 years	25
OSH Training	No	0
	Yes	100
OSH Experience	< 1 years	50
	1 to 5 years	25
	6 to 10 years	25

The majority of the respondents have worked at the current company for less than five years (75%). All of the respondents had received occupational safety and health training within last three years. Overall, half of the respondents have less than one year occupational safety and health experience (50%), one to five years (25%) and six to ten years (25%). Table 4 shows the profile of the respondents.

Enter multiple regression using customer-vendor partnership explains a statistically significant portion of variance (R Square = 0.853) for safety behaviour at 0.05 level of significance. "I feel under pressure from the clients to adopt safety measures" is considered as the effective factor in predicting variance of the dependent variable (Beta=0.768). Table 5 shows the results of multiple regression.

DISCUSSION

Most of work undertaken at the warehouse involves unloading containers, warehouse storage and product transfer. One of the warehouse's goods involved receiving large loads of 30 kg box of chemical container. Through consultation, employees and management recognised that the job needed to be replaced to ensure the manual task risks associated with the goods were reduced via use of mechanical aids including forklift

Table 5. Multiple regression results

Items	*B*	Beta	t-value	*p*
Constant	4.129		5.771	0.004
Clients influence me to adopt safety measures.	-0.484	-0.484	-1. 936	0.125
I feel under pressure from the clients to adopt safety measures	0.613	0.768	3.726	0.020
My colleagues share with me business knowledge from clients.	-0.290	-3.364	-1.543	0.198
F = 7.708		*R =0 .923*		
Sig-F =0 .039		*R Square =0 .853*		

and pallet jack. There is a significant and positive relationship between customer – vendor partnership on employee safety behaviour. The effect of customer – vendor partnership would expect employees obey occupational safety and health programs leads to greater obedience to the laws. The customer – vendor partnership increases the likelihood of being safe workers, consistent with Weibert and Plunkett (2006) and Egels-Zandén (2007) who found that customer – vendor partnership increased the likelihood of safety behaviour. A possible explanation is that the respondent may feel that their customers are likely to have more information about occupational safety and health. This finding is consistent with previous studies and suggests that employee who have appropriate customer – vendor relationship are expected to enhance their safety behaviour.

CONCLUSION

The study attempted to extract the elements for excellence safety behaviour for companies to educate their employees in order to completely accept the safety program and reach world-class best practice. The study examined the direct relationship between customer – vendor relationship and safety behaviour. The safety and health result of the study was that warehousing safety program enables the company to find efficiencies throughout their warehouse operation and reduce costs. About the room for improvement, the study is involved only one company, hence the result does not represent the whole electronics sector.

Findings of the study provided important implications to the industry in their planning to improve the effectiveness of the safety program. The finding found a direct relationship between customer – vendor relationship and employee safety behaviour. The research provided some knowledge and insight on safe work culture and we have studied the factors that give impacts the effectiveness of safe work culture.

Hence, a proposal follows that OSH program should be using customer – vendor partnership. Employees can learn and adapt positive safety behaviour through customer – vendor partnership. In order to improve the workers' safety in the workplace, an effective occupational safety and health committee should be established. The safety and health committee's job is ensuring compliance with the Occupational Safety and Health Act which came into force in 1994 and promotes safe conduct of work. When employees see clearly the active involvement by their management to bring safety to higher levels then they will be motivated and prevent accidents from happening at all. Customer – vendor partnership will increase the level of employees' understanding towards their company's safety management and contributed to safer behaviour.

REFERENCES

Ajzen, I. (1985). From intentions to actions: A theory of planned behaviour. In J. Kuhl & J. Beckman (Eds.), *Action-Control: From cognition to behaviour* (pp. 11–39). Heidelberg, Germany: Springer. doi:10.1007/978-3-642-69746-3_2

Ajzen, I. (1991). The theory of planned behaviour. *Organizational Behavior and Human Decision Processes*, *50*(2), 179–211. doi:10.1016/0749-5978(91)90020-T

Becker, T., Billings, R., Eveleth, D., & Gilbert, N. (1996). Foci and Bases of Employee Commitment: Implications for Job Performance. *Academy of Management Journal*, *39*(2), 464–482. doi:10.2307/256788

Bhattacherjee, A. (2000). Acceptance of e-commerce services: The case of electronic brokerages. *Systems, Man and Cybernetics, Part A: Systems and Humans, IEEE Transactions on*, *30*(4), 411–420. doi:10.1109/3468.852435

CSCMP. (2008). *Supply chain management and logistics management definitions*. Retrieved 19 August from: http://cscmp.org/AboutCSCMP/Definitions/Definitions.asp

Department of Statistics (DOS). (2013). *Labour Force Survey Report Malaysia 2012*. Author.

Egels-Zandén, N. (2007). Suppliers' Compliance with MNCs' Codes of Conduct: Behind the Scenes at Chinese Toy Suppliers. *Journal of Business Ethics*, *75*(1), 45-62.

European Agency for Safety and Health at Work (EU-OSHA). (2000). *Occupational safety and health in marketing and procurement*. Luxembourg, Belgium: Office for Official Publications of the European Communities.

Field, A. (2000). *Discovering statistics using SPSS*. London: Sage Publications.

Hertz, S., & Alfredsson, M. (2003). Strategic development of third party logistics providers. *Industrial Marketing Management*, *32*(2), 139–149. doi:10.1016/S0019-8501(02)00228-6

Hum, S. H. (2000). A Hayes-Wheelwright framework approach for strategic management of third party logistics service. *Integrated Manufacturing Systems*, *11*(2), 132–137. doi:10.1108/09576060010314125

International Labour Office (ILO). (2001). *Self-regulation in the workplace: Codes of conduct, social labelling and socially responsible investment*. MCC Working Paper No. 1, Series on Management Systems and Corporate Citizenship, Geneva, Management and Corporate Citizenship Program, Job Creation and Enterprise Development Department.

Koo, K. E., Nurulazam, M. D. A., Rohaida, M. Z. S., Teo, T. G., & Salleh, Z. (2014). Examining The Potential of Safety Knowledge As Extension Construct For Theory of Planned Behaviour: Explaining Safety Practices of Young Adults At Engineering Laboratories And Workshops. *Procedia: Social and Behavioral Sciences*, *116*, 1513–1518. doi:10.1016/j.sbspro.2014.01.426

Lim, J., Gan, B., & Chang, T.-T. (2002). A Survey on NSS Adoption Intention. *Proceedings of the 35th Hawaii International Conference on System Sciences (HICSS)*. doi:10.1109/HICSS.2002.993888

Lin, H.-F., & Lee, G.-G. (2004). Perceptions of senior managers toward knowledge-sharing behaviour. *Management Decision*, *42*(1), 108–125. doi:10.1108/00251740410510181

Mejza, M., Barnard, R., Corsi, T. M., & Keane, T. (2003). Driver management practices of motor carriers with high compliance and safety performance. *Transportation Journal*, *42*, 16–29.

Panta, M., & Vrysagotis, V. (2011). A new managerial approach to safety at workplace in warehousing systems: A case study. *Advances in Management & Applied Economics*, *1*(3), 115–126.

Ramanathan, R., & Ramanathan, U. (2013). Adoptionof RFID technologies in UK logistics: Moderating roles of size, barcode experience and government support. *Expert Systems with Applications*, *39*(12), 11115–11124.

Social Security Organisation (SOCSO). (2012). *Annual Report 2011*. Author.

Waring, P., & Lever, J. (2004). The impact of Socially Responsible Investment on Human Resource Management: A conceptual Framework. *Journal of Business Ethics*, *52*(1), 99–108. doi:10.1023/B:BUSI.0000033110.21900.da

Weibert, M., & Plunkett, C. (2006). Motivational Safety System: One Organisation's Experience Moving toward World-Class Performance. *Professional Safety*, (February), 34–39.

Chapter 15
Performance Measurement:
Measuring Retail Supply Chain Performance

Neha Grover
University of Petroleum and Energy Studies, India

ABSTRACT

The purpose of this chapter is to study what supply chain managers of a retail supply chain measure. In other words aim here to identify the key indicators for measuring retail supply chain performance. A qualitative approach is adopted. Published literature from refereed journals on supply chain performance measurement has been considered. Besides the need for organizations to adopt a holistic approach, firms remain focused on traditional financial measures (gross profit margin, Interest coverage, and debt and equity ratios). The chapter identifies key indicators for performance measurement and classifies them into four major categories: transport optimization, information technology optimization, inventory optimization and resource optimization. These key indicators are arranged precisely for retail industry. From a supply chain perspective, the non financial measures such as on-time delivery, training of employees, warehouse layout, etc. are also important aspects of measuring supply chain performance. Further research can be carried out to validate the relevance and applicability of identified indicators. The study can be further conducted to measure the interrelationships between the KPIs and their impact on financial performance of the firm. In this chapter the author attempts to identify the performance indicators specifically for retail supply chain. The identified measures are further categorized based on its operations.

INTRODUCTION

With increasing competition and advancement in technologies, firms have been experiencing significant changes in the way of doing businesses. The term Supply Chain Management [SCM] was originally introduced by consultants in the early 1980s, and has subsequently gained tremendous attention. Since then, this paradigm has undergone huge developments nationally and internationally. It is important for companies to know what they are doing and where they can reach, and measurement is the first step that

DOI: 10.4018/978-1-4666-9795-9.ch015

leads to control and eventually to improvement. Thus, with the evolution of new concepts in managing supply chain, there has been a revolution in supply chain management practices. Researchers and practitioners have realized that dependence on financial indicators could be a hindrance in achieving competitive advantage. The bigger challenge for the managers is to identify and develop suitable performance measures and metrics to make right decision that would contribute to an improved organizational competitiveness.

Performance measurement has been defined and redefined over the years. It has its roots in early accounting system in the late Thirteenth century, when traders used it to settle their transactions. As pointed by (Johnson, 1981), even pre-industrial organizations maintained a good account of external transactions and stock without high level techniques, such as, cost accounting. Furthermore (Lebas, 1995) went to the extent of saying that there is no existence of businesses without a performance measurement system, because it is obvious for businesses to collect feedback from employees to manage/improvise the businesses (Sinclair & Zairi, 1995). In late 1980s, after emergence of world economy and globalization of trade, the focus shifted from productivity to quality, time, cost, flexibility and customer satisfaction (Kaplan, 1984; Slack, 1983; Hayes & Abernathy, 1980). This is when the traditional measures were criticized and considered inappropriate for measuring an overall business performance. In fact, (Johnson & Kaplan, 1987) were first to suggest a shift from cost accounting based performance measurement approach to a more integrated performance measurement approach. Performance measurement has become an important area of research and it has gained importance in the business because of the drastic changes in way of doing various operational activities. A performance measurement system plays a significant role in managing a business as it provides the information necessary for making the right decision at right time (Gunasekaran & Kobu, 2007). Both practitioners and research scientists have noted a number of problems regarding measurement activities during the past decade. The problems reported suggest that measurement activities are fragmented both within and across organizations. Good performance measures and metrics will facilitate a more open and transparent communication among the channel partners of supply chain leading to its continuous improvement. As per Kaplan (1990), "No measures, no improvement", the purpose of this study is not to come out with a long list of performance measures and metrics, but to present a list of KPIs categorized in different functional groups with respect to retail supply chain.

Thus the focus is on identifying the relevant performance measures for organized retail environment. "Organized retailing refers to trading activities undertaken by licensed retailers, that is, those who are registered for sales tax, income tax, etc. These include the corporate-backed hypermarkets and retail chains, and also the privately owned large retail businesses" (Parliament of India, Rajya Sabha, 2009).

Retailing has undergone an intense transformation during the last few decades. The Indian retail industry is marked with huge growth potential. However, in spite of the recent developments in the retail sector and its immense contribution to the economy, it continues to be the least evolved industry in India when compared to rest of the world (IJMBS, 2013) . Moreover, Indian retail scenario has been distinguished from developed nations on the following three features:

- Fragmented and multi layered retail distribution market,
- Many retailers of various sizes at many locations vying to serve the final consumer,
- Many buyers for the grower and manufacturer, thus, preventing any retailer from establishing a monopoly and dictating price and credit terms to the growers and manufacturers (Guruswamy, Sharma, & Jos, 2007).

Table 1. India's retail fundamentals

Market size	US$ 350 billion
CAGR	15-20%
Unorganized retail	12 million mom-and pop stores
Organized retail penetration	5-8%
Retail density	6%
Contribution to GDP	14%

Source: (Ernst and Young, 2011)

India is considered as one of the most desirable retail destinations in the world. India's emerging economy is one of the fastest growing across the globe. Since, the Indian economy was liberalized in the 1990s; the average gross domestic product (GDP) of India has been growing at a rate of 8.6% since 20067-07. The India's retail fundamentals are given in Table 1.

A number of performance measurement tools have been identified in where the indicators have been categorized into various groups. As discussed by (Neely, Bourne, & Kennerley, 1995), numerous approaches were developed for performance measurement model, includes: criteria for measurement system design (Globerson, 1985); performance measurement matrix (Keegan, Eiler, & Jones, 1989); performance measurement questionnaires (Dixon, Nanni, & Vollmann, 1990); balanced scorecard (Kaplan & Norton, 1992); Supply chain operations reference model (SCOR) (Supply Chain council, 1996) and, computer aided manufacturing approaches. These approaches have their own limitations with respect to a particular context, and they cannot be generalized in all business scenarios. Some of the reasons identified by (Gunasekran, Patel, & Tirtiroglu, 2001) for the need to study the measures and metrics are lack of balanced approach and lack of a clear distinction between metrics at strategic, tactical and operational levels. ". Thus research in performance measurement holds a promising future and much is required to be done. The chapter is further sub categorized in the following sections; background of the study, main focus of the study covering the research gaps followed by finding and recommendations based on analysis.

BACKGROUND

As supply chains are continuously replacing individual firms by integration for creating value, therefore it is important to understand the relationship between supply chain management [SCM] practices and supply chain performance [SCP]. To determine what performance measures should be evaluated for the SC, the question why supply chain performance is needed should be addressed first. Several factors trigger the business firm's need for the performance measurement, including: Increasing competition which arises from greater customer expectations for cost reductions and value-added products or services (Neely, Bourne, & Kennerley, 1995).

Many companies strive to find specific areas to increase their competitiveness and competency for differentiation. Competition today is no longer between company to company, but rather, supply chain to supply chain. The focus of SCP measurement should go beyond firm focus (Pohlen & Lambert, 2001). Aligning the organizational goals with the supply chain goals is vital to the overall performance success of the supply chain. Moreover, the role of performance measures in the success of Supply Chain cannot be overstated (Gunasekarana, Patelb, & McGaughey, 2004). Performance measurement is an essential element of effective planning and control because it may not only provide necessary feedback information to reveal progress (Chan & Qi, 2003; Fynes & Voss, 2005), but it may also affect strategic, tactical and operational planning and control (Gunasekarana, Patelb, & McGaughey, 2004).

In a SCM context, performance measurement can facilitate integration among the SC members. The authors' emphasized on the importance of the performance measurement system (Gunasekran, Patel, & Tirtiroglu, 2001) in such a way as to enhance the shared destiny principles of partnership and long term relationships. Thus, performance in a SC is defined as the overall efficiency and effectiveness of SCM. To accomplish this, SCM must integrate a number of key business functions, including purchasing, demand management, distribution planning, transportation, quality management, production planning, and materials management throughout the supply chain. Consequently, the output of the processes enabled by the supply chain must then be measured and compared with a set of standards. Control of processes in a supply chain is crucial in improving overall firm performance (Flynn, Huo, & Zhao, 2010; Li, Ragu-Nathanb, Ragu-Nathanb, & Rao, 2006; Tan, Kannan, & Hanfield, 1998).

Performance management in SC is therefore about setting goals within the SC and between functions that will lead to the desired results with balance and without conflict. Every aspect of SC needs to be considered in measuring the SCP. There are many dimensions to measure, and no single measure defines SCP. Thus, there is a need to obtain balance throughout the SC and the preparedness to change for better results. To measure SCP, there are a set of variables that capture the impact of actual working of supply chains on revenues and costs of the whole system (Ramdas & Spekman, 2000). In some of the studies on performance measurement, the predominant method for measuring performance is the use of KPIs that cascaded down from top level business objectives and measures through the organization into a series of functional measures (Storey, Emberson, Godsell, & Harrison, 2006).

Once the SCP measures are developed adequately, managers have to identify the KPIs that need to be improved. Retailers face many challenges in the ever increasing fierce competition. With the increase in product variety, increasing uncertainty in demand and supply, the need to reduce the time to market, shorter and shorter product life cycles with greater efficiency has raised the benchmark for the retailers to deliver value to the customers. However, as a result of the power that comes with control over consumer contacts, retailers today have the opportunity to organize the work in their supply chains in suitable ways. The positive impact of SCM on operational performance can manifest itself in all dimensions. Table 2 shows some of the reasons why organizations measure performance (Neely A. D., 1998). A number of performance measurement tools have been identified in which the indicators have been categorized into various groups.

Companies should start with a small number of performance indicators, which are necessary to examine the meta-level processes (plan, source, make, and delivery), which can be successfully manage and operate (Chae, 2009).

MAIN FOCUS OF THE CHAPTER

It is widely acknowledged that there has been relatively little interest in developing measurement systems and metrics for evaluating supply chain performance (Chen, Paulraj, & Lado, 2004; Gunasekaran, Patel, & Tirtiroglu, 2001; Beamon, 1999) and others (e.g. (Ellinger A. E., 2000; Fynes, Voss, & Búrca, 2005)

Table 2. Why do organizations measure performance?

Why Measure?	Check Position	Communicate Position	Confirm Priorities	Compel Progress
To establish position	✓			
To monitor progress	✓			
Because the organization has to	✓	✓		
Because the organization wants to communicate performance to shareholders or customers		✓		
Because the organization or others want to be able to benchmark performance		✓	✓	
Because measures stimulate interest		✓		✓
Because measures can be used to communicate priorities		✓		✓
Because measures provide a means of motivating people to look for ways of improving performance				✓
Because measures provide a basis for reward				✓
Because measures provide a means of management control	✓			
Because measures provide a mean of cost control	✓			
Because measures provide an insight into what is important for the customer	✓		✓	
Because measures provide an insight into what the business is doing well	✓		✓	
Because measures provide an insight into what the business is not doing well	✓		✓	
Because measures provide an insight into what the business needs to focus on			✓	✓
Because measures provide an insight into where the business should invest			✓	

Source: (Neely A. D., 1998)

found it encouraging that some researchers have developed measures to assess the performance of supply chain relationships, or the performance of a supply chain as a whole A typical firm already has a certain number of KPIs such as return on Investment for assessing its financial performance, but specific supply chain related KPIs have not been widely adopted and businesses are typically uninformed of them (Chae, 2009). Moreover, depending upon the nature and type of supply chain, the set of indicators used for measuring its performance will vary. Traditional BSC and SCOR models generally assume that KPIs are uncoupled (Cai, Liu, Xiao, & Liu, 2009). These approaches could describe business operations well, and serve as a good communication tools, but they are not effective in improving overall performance by accomplishing the critical KPIs (Cai, Liu, Xiao, & Liu, 2009). As pointed out by (Gumbus, 2005; Ho, Chan, W, Wong, & Chan, 2000; Douglas, 1996), categorization must provide the organizations an opportunity to select the performance indicators in which the companies are most interested.

However, (Lambert & Pohlen, 2001) observe that one of the main problems with SC metrics is that, 'they are, in actuality, about internal logistics performance measures', and do not capture how the supply chain as a whole has performed. For example, although measures such as order fill rate are likely to be influenced by activities throughout the entire SC, they ultimately measure performance at the intra, rather than the inter-organizational level. There is a need to incorporate broader relationships, such as, manufacturers and logistics service providers or distributors to collaborate across different levels of supply chain. Some of the concerns that need to be addressed in this direction include the integration issue of SC in varied industries across the countries. Also, there is a requirement to conduct more empirical studies on the effect of management practices on combination of these SCM practices (Gopal & Thakkar, 2012).

To the best of researcher's knowledge, no integrated measurement system exists in Retail supply chain that combines different aspects of performance (e.g. financial and non-financial, qualitative and quantitative) into one measurement system. Therefore, researcher is aiming to identify the performance indicators for measuring Retail Supply Chain Performance. Table 3 provides the list of indicators which were identified through literature review. This list was further used to categorize the KPIs specifically for retail industry.

In the next section the categorization of KPIs is discussed for Retail Supply chain from the identified variables in literature review.

CATEGORIZATION OF KPIS FOR PERFORMANCE MEASUREMENT IN RETAIL

Industries, such as retail sector, are recognizing the significant role of supply chain management (Hill & Scudder, 2002; Mentzer, Foggin, & Golicic, 2000; Ellram, La Londe, & Weber, 1989), and the need to effectively manage the flow of materials, money and information across the supply chain (Gavirneni, R, & S, 1999; Lee & Billington, 1993). The advances in technology (Sahin & Robinson, 2005) have seen a growing trend for organizations to create external linkages based on the sharing of information (e.g. point of sale data (POS), inventory levels, forecasts, etc.) in order to gain increased visibility of their customers and/or suppliers' operations and activities (Fiala, 2005; Shore & Venkatachalam, 2003; Mabert & Venkataramanan, 1998). The purpose of achieving visibility is primarily for improving their own internal decision making and operating performance (Kulp, Lee, & Ofek, 2004; Rungtusanatham, Salvador, Forza, & Choi, 2003). Visibility in information helps to improve supply chain performance (Kulp, Lee, & Ofek, 2004; Rungtusanatham, Salvador, Forza, & Choi, 2003).

According to (Jones & Towill, 1999), information flow at all levels of supply chain is critical, and specifically, the order entry method determines the way and extent to which customer specification are converted into useful information and channelized across the supply chain partners. In previous studies (Aviv, 2007; Chen, Paulraj, & Lado, 2004; Gavirneni, R, & S, 1999; Lee & Billington, 1993; Bourland, Powell, & Pyke, 1996; Cachon & Fisher, 2001), benefits of information sharing have been found mostly from a modeling/simulation perspective. The important characteristics for information sharing are accurate, trusted, timely, useful and in a readily usable format (Whipple, Frankel, & Daugherty, 2002; Closs, Goldsby, & Clinton, 1997; Gustin, Daugherty, & Stank, 1995). Delivery also heavily relies on the quality of information exchanged. For example, once the activities are scheduled continuous monitoring of information derived and supplied takes place (Gunasekran, Patel, & Tirtiroglu, 2001). According to (Bower & Hout, 1988) and (Christopher M., 1992), order cycle time is an important measure for reduction in response time of supply chain and also a source of competitive advantage.

Moreover, it also directly influences the customer satisfaction level (Jones & Towill, 1999) by being more responsive to the customer demand and increasing the delivery reliability and consistency of lead time. Due to fluctuations and uncertainty in the supply chain in handling a large amount of SKUs, a reliable and consistent order lead time reduces the redundancies (Schonberger, 1990). The use of technology and its advances have enforced companies to rearrange the activities of supply chain. The path through which the order travels and spend time in different routes, the non-value adding activities can be identified for elimination.

To necessitate this, it is important to track and trace the products by use of technology (e.g. e-commerce, EDI and internet). The entire planning process of making the final order placement has its impact

Table 3. List of metrics for supply chain performance measurement

Author, Year	Performance Metrics	Context
Akkermans, Bogerd, & Vos, (1999)	Business strategies (functional capabilities), processes (operational efficiencies), stake holders view (risk/return ratio)	SCOR model-(plan, source, make, deliver)
Kim, (2009); Aviv, (2007); Simatupang & Sridharan, (2004)	Order of dominance and decision Sharing	Collaborative planning and production, decision making
Emmet & Crocker, (2006); Dong & Chen, (2005); Lambert & Pohlen, (2001); Beamon, (1999)	Cost, profit, excess inventory, stock-out, resource measure	Collaborative planning and production, decision making
Forslund & Jonsson, (2007); Chang, Fu, Lee, Lin, & Hsueh, (2007); McCarthy & Golicic, (2002); Raghunathan, (2001)	Impact of information quality on Forecasting	Information sharing, Forecasting decision making
Forme, Genoulaz, & Campagne, (2007); Angerhofer & Angelides, (2006); Barratt & Oliveira, (2001)	Reliability, reactivity/flexibility	Information sharing, Forecasting decision making
Aviv, (2007); Simchi-Levi & Zhao, (2005); Chen, Paulraj, & Lado, (2004); Cachon, (2001)	Inventory and stock position, stock out, lead time, internal service rate, cross-functional capability, logistics efficiency	Replenishment, decision making
Chae, (2009)	Cash to cash cycle, Inventory days, planning cycle, supplier fill rate, Automatic PO rate,	Operations strategy
Pyke & Cohen, (1994); Pyke & Morris, (1993); Cohen & Moon, (1990); Lee & Feitzinger, (1995)	Cost	Resource metrics
Arntzen, Gerald, Terry, & Linda, (1995)	Cost and activity time	Resource metrics
Altiok & Raghav, (1995); Christy & Grout, (1994); Davis T., (1993); K, Takahashi, & Muramatsu, (1988); Newhart, Stott, & Vasko, (1993)	Cost and customer responsiveness	Resource metrics
Lee & Billington, (1993)	Customer responsiveness	Output metrics
Cai, Liu, Xiao, & Liu, (2009); Angerhofer & Angelides, (2006); Van der Vorst, (2000)	Information accuracy, Information availability, Information timeliness, Information sharing	Information Metrics
Cai, Liu, Xiao, & Liu, (2009)	Rates of sales in new products, Supply chain stability, Number of new products launched, Process improvement	Innovativeness metrics
Cai, Liu, Xiao, & Liu, (2009)	Manufacturing/production flexibility, New products flexibility, Supply chain responsiveness, Delivery flexibility, Procurement flexibility, Information Systems flexibility, Logistics flexibility	Flexibility
Cai, Liu, Xiao, & Liu, (2009)	Sales (or profit), Percent of on-time deliveries, Rate of stockouts (losing sales), Perfect of order-fulfillment, Fill rate (target fill rate achievement, average item fill rate), Customer satisfaction, Order fulfillment lead time, Rates of customer complaints, Planned process cycle time, Cash to cash cycle time	Output metrics
Cai, Liu, Xiao, & Liu, (2009); Shepherd & Gunter, (2006); Bolstorff,(2003); Chan & Qi, (2003); Beamon, (1999)	Total supply chain management cost, Information management costs, Distribution costs, Value-added employee productivity, Inventory costs, Warranty costs, Manufacturing costs, Return on investment (or ratio of net profits to total assets)	Resource metrics
Voudouris, (1996)	Flexibility	Flexibility metrics
Chia, Goh, & Hum, (2009; Christopher, (1994)	Customer Satisfaction	Output metrics
Nicoll, (1994)	Information flow	Information metrics
Davis T., (1993)	Supplier performance	Output metrics
Johnson & Randolph, (1995)	Risk management	Output metrics

on cost, quality, speed of delivery and delivery reliability and flexibility (Slack, 1983). As the product range has increased, the value added per employee i.e. productivity of human resource is an important parameter to be considered (Gunasekaran, Patel, & Tirtiroglu, 2001). The starting point for any decision of logistics invariably centers on (Drucker, 1962) description of the economy's Dark Continent, which suggested that distribution was one of the last frontiers of business to be discovered. He noted that distribution was viewed as a low status activity by managers, yet major cost savings could be achieved by managing this function more effectively.

By 1970s and 1980s the supply chain was still viewed as series of disparate functions with materials management dealing with the backend of the supply chain, and physical distribution management focusing upon the flow of product from manufacturers to their customers (retailers and wholesalers). For transport efficiency, the distribution mode, the delivery channel, vehicle scheduling and warehouse location play a significant role, and show tremendous opportunities to improve supply chain performance based on lead-time reduction (Gelders, Mannaert, & Maes, 1994). It is determined by on-time delivery/perfect delivery parameter, which ultimately influences the customer service level (Stewart, 1995).

A large number of different types of performance metrics have been used to characterize the various functions of supply chain; production, distribution and inventory systems. Understanding the meaning of a single metric might be easy, but the meaning of the metrics in combination, and their effect on overall company performance, is hard to intuitively understand. One problem is that metrics based on financial accounting systems, and expressed in financial terms, are not easily compared with operational metrics focused on the physical movement of goods and services.

SOLUTIONS AND RECOMMENDATIONS

Through literature review, numerous variables have been identified for supply chain performance measurement. These identified KPIs have been classified into four major categories (KPMG, 2011) which are as follows:

- Transport Optimization,
- Information Technology Optimization
- Inventory Optimization
- Resource Optimization

The importance of these major categories is discussed in the subsequent sections.

1. TRANSPORT OPTIMIZATION

Transport has been identified as a major area of concern in retail sector and its planning needs a trade-off between cost and service. Transport is a key linkage between all activities of supply chain, enabling the physical movement of goods across the multi-echelon of supply chain. Furthermore with the introduction of third party logistics (3PL), this function is being performed in more organized way. It has gained importance with its advancements, like, cold storage, break bulk facilities, which is a value addition to the movement of goods. The inescapable approach of 'Right product at right time and in the right place' across multiple channel touch points have become a challenge for retailers. Moreover the tough cord of fast changing consumer tastes has forced retailers to make their supply chain more effective and efficient.

Logistics is considered as an important activity in the area of operations, marketing and distribution with significant cost implications. According to Thomas & Griffin, (1996), logistics is: 'the single largest cost component of logistics is transportation cost, often comprising half of the total logistics cost'. Bowersox et al., (2005), (Bowersox, Rodrigues, & Calantone, 2005) identified logistics as: *'One of the largest costs involved in international trade'.*

Some of the researchers have argued that SCM has the potential to lower the total logistics costs, while simultaneously improving customer service and satisfaction (Moberg, Whipple, Cutler, & Speh, 2004). With effective and efficient planning, the distribution cost is one component of total supply chain cost which can be optimized. In a physical distribution channel, the total transportation cost can be treated as trucking cost plus local delivery cost. Furthermore (Rushton, Oxley, & Croucher, 2001) shows that, *'trucking cost is always the highest among all costs of total distribution cost'*. To reduce the delivery cost, this total is given high importance (Gunasekaran, Patel, & Tirtiroglu, 2001).

Current Scenario of Logistics in India: Today India is the 4th largest country in terms of purchasing power parity [PPP] (The World Bank, 2013), and constitutes one of the faster growing markets in the world. Ports, airports, roads and railways are all seen as vital for the Indian Economy and have been targeted for investment to keep pace with competition and growing demand. The lead time by road is as high as 9-12 weeks and transporters have a long way to go to achieve milestone- 'A journey towards excellence'. An efficient and inexpensive transport system contributes to greater competition in the market place, greater economies of scale in production and reduced prices of goods. A lot of emphasis is laid on the cost/service balance or revenue growth. Logistics networks are seen as a matter of survival and competitive advantage. The linkage between the production facility, distribution centre and retail outlet involves an extensive planning. That is why; we deal with a lot of routing problems, taking into consideration a number of factors, such as, unit production, transportation and inventory holding cost, distance to be covered, modes of transport available, lead time between the two connecting points, service level, etc.

Table 4 gives the overview of world-wide logistics conditions and solutions, in which it is seen that India is lacking the necessary infrastructural and government support.

In the context of retail logistics, several authors have sought to explain the transformation of logistics practices since 1970s. Also (McKinnon, 1996) identified six trends as follows:

- Retailers increasing their control over secondary distribution/warehouse to shop.
- Restructuring of retailers logistical systems through development of composite distribution and centralization of certain commodities into particular supply chain streams.
- Adoption of quick response techniques to reduce lead time through the implementation of information technology especially electronic data interchange (EDI), EPOS (electronic point of sales) and SBO (sales based ordering).
- Rationalization of primary distribution (factory to warehouse) and attempts to integrate this and secondary distribution into a single network system.
- Introduction of supply chain management and efficient consumer response (ECR).
- Increasing return flow of packaging material and handling equipment for recycle and reuse.

Several authors have discussed the importance of a supply chain focus on the part of transport logistics service providers, as they function to link suppliers, manufacturers, distributors, retailers and customers.

Table 4. World-wide logistics conditions and solutions

Region	Logistics Conditions(s)	Logistics Solution(s)
North America	i. Short term focus on shareholder return and return on capital ii. Excellent infrastructure	i. Extensive logistics finance and performance measures ii. Supply chain integration and logistics information systems to reduce capital assets
Latin America	i. Limited to no logistics infrastructure and / or logistics service providers	i. Leap frog to World class status ii. Import Logistics service providers and education iii. High security designs
Western Europe	i. Transportation heritage ii. Individual rights	i. Transportation heritage makes 3PL providers commonplace ii. Focus on individual rights yields human friendly logistics via excellent logistics ergonomics and green logistics
Japan	i. Lack of land and/ or human resources and high logistics transaction requirements	i. Logistics culture of discipline and order ii. Automated storage and handling systems iii. Multi-story logistics facilities
India	i. Poor Logistics infrastructure, lack of government support	i. Necessary infrastructure and organizational support
China	i. Long delays in property acquisition, varied and complex regulations, difficulty in securing finance	i. Companies have to develop potential strategies for expansion

Source: (Frazelle, 2002b)

They argue that transport logistics service providers must focus on supply chain performance in addition to organizational performance (Lai, Ngai, & Cheng, 2002). The logistics performance construct reflects the organization's performance, as it relates to its ability to deliver goods, and services in the precise quantities and at the precise times required by customers. Authors describe the importance of integrating the Logistics processes of all supply chain partners to better serve the needs of ultimate customers (Stank, Davis, & Fugate, 2005; Lin, 2006). One can observe that Logistics activities are highly integrated in today's retailing sector, and the entire way of doing business has changed due to multiple channels of serving the same customer.

The internal performance measurement mainly focuses on the value chain or logistics supply chain within a single company for its operational functions, like, sourcing, inbound storage/transportation, operations, outbound storage/transportation and consumer distribution (Coyle, Bardi, & Langley, 2003), while the external performance measurement has an emphasis on measuring the performance of the efficient and effective flows of material/products, services, information and financials from the supplier's supplier through various organizations/ companies out to the customer's customer (Coyle, Bardi, & Langley, 2003). Transport is a direct link that connects various activities of supply chain in delivery of goods to the end customers.

Delivery operates in a dynamic ever changing environment, which involves a regular track and trace of flow of goods with accuracy in data. In any typical delivery distribution mode, the delivery channel, vehicle scheduling and warehouse location play an important role in delivery performance. For perfect delivery, certain parameters have to be taken into consideration, like, temperature control during transportation, safety measures so as to avoid the pilferage/damage in transit. Logistic flexibilities can be accomplished by accommodating changes in warehouse locations; transportation networks and the transportation mode have a positive impact on supply chain performance (Green Jr, Whitten, & Inman, 2008; Bowersox, Closs, Stank, & Keller, 2000). Also, some companies own their fleets for their dedicated routes and flexible logistics operations. An increase in delivery performance is possible by selecting the suitable channel, scheduling and location policies.

On time delivery is also a very important aspect of delivery performance, which helps to determines how perfect the delivery has taken place or not, and it acts as a measure of service level (Gunasekaran, Patel, & Tirtiroglu, 2001). Also (Stewart, 1995) has identified the three important delivery performance measures, i.e., delivery-to-request date, delivery-to-commit date and order fill rate. A reduction in fill rate helps to improve the operational efficiency. The vehicle speed, driver reliability, frequency of delivery and the location of depots have been also identified as significant variables for perfect delivery. The customer satisfaction is ultimately the prime concern; hence, the transport flexibility also plays an important role. The number of faultless notes invoiced, which show the date of delivery, time and condition under which goods were received, show the delivery reliability.

Lai et al., (2002), (Lai, Ngai, & Cheng, 2002) distinguished three dimensions of supply chain performance in transport logistics: service effectiveness for shippers; operational efficiency; and service effectiveness for consignees. Perhaps, the most important research concerning logistics is in the area of designing efficient and cost-effective distribution systems. The single largest cost component in logistics is transportation cost (Thomas & Griffin, 1996). The companies have to do a trade-off between the total cost and service; therefore, delivery measures are of high priority in supply chain. The variables used for Transportation Optimization are given below in the Table 5.

2. INFORMATION TECHNOLOGY OPTIMIZATION

Today, India is a global village where MNCs have flourished their businesses and 'Retail' has been identified as the most promising sector of Pan Asia. IT has made the world flat and has integrated the supply chain partners across the world with the help of real time 'Information sharing' which has in turn helped businesses to grow faster and better.

The goal of SCM is to manage the multiple functions of supply chain and information exchange within the supply chain has been cited as a very critical component (Martha, 1997). (Cannon & Perreault

Table 5. Variables for transport optimization

Survey Elements	Representation*
Percent of on-time deliveries is an important indicator for high service level	TO1
Damages due to inefficient delivery (pilferage/ delay/ damage in transit) of the product as % of total sales should be minimal are critical for operational excellence.	TO2
Proper documentation is important for delivery of goods on time	TO3
Temperature control during transportation for perishable commodities is essential for perfect delivery	TO4
Transport connectivity is important for high growth of business	TO5
Usage of GPS/RFID technology for track & trace is essential	TO6
Vehicle optimization is highly significant for logistics operations	TO7
Faster turnaround time of vehicles at loading and unloading time improves efficiency	TO8
Owned vehicles are convenient and cost effective for transportation	TO9
Outsourced vehicles are more efficient for transporting goods	TO10

Representation*: For easy identification of variables, the statements are given a alpha numeric representation

Jr., 1999) With the dynamic changes in the industry, the growing pressures of the industry consolidation, emerging new retail formats, new challenges of managing the fast changing trends, stock-outs and markdowns have resulted for technological support by the retailers. As a key industry in the service sector, retailers accounted for approximately 6.1% or $884.9 billion of the U.S. GDP in 2010 (Bureau of Economic Analysis, 2011).Given the access to data and analytical tools, retailers are finding that information technology has been a great support in managing various disruptions in supply chain, primarily because of the availability of right information at the right time across multiple channels to offer consumers multiple touch points and innovative services (Noble, Haytko, & Phillips, 2009; Wallace, Giese, & Johnson, 2004). Principally, the impact of IT was the main concern in the manufacturing supply chain; recently its role in service industry has also started gaining importance. There is a growing need to understand the connection between the information systems [IS] and the design of service delivery systems in information-intensive service domains (Froehle & Roth, 2007; Spohrer, Maglio, Bailey, & Gruhl, 2007).

All the supply chain partners of upstream and downstream intermediaries are linked with real time data that rely on electronic data interchange [EDI] and the open standard of the internet. Besides, the traditional 'hard' technologies, like, scanners, barcode readers, and wearable computers, the internet-centric software and analytical tools have added value to the large chunk of data to provide specific solutions for strategizing the desired objective in the market (Rao, 2000).

Significance of measuring performance of IT: Prior research has proposed that when IT and business resources are deployed in a complementary manner, performance gains are likely to happen (Barua, Konana, Whinston, & Yin, 2004; Melville, Kraemer, & Gurbaxani, 2004; Ray, Muhanna, & Barney, 2005; Tanriverdi, 2005; Hulland, Wade, & Antia, 2007). Moreover, IT is becoming the most critical resource in service firms (Froehle & Roth, 2007) and it is used to improve the operational and strategic coordination (Sanders, 2008). With the advent of internet, the way of doing business has changed to e-business, and hence, the multiple partners of supply chain across the globe are integrated with real time information.

Technological developments play an important role in retail operations, in terms of streamlining the flow of goods, services and information. Organizational success first depends upon the performance of the supply chains in which the organization functions as a partner (Rosenzweig, Roth, & Dean, 2003). The implementation of an enterprise resource planning [ERP] can be considered one of the most effective ways towards traceability, since one of its major features is integration between modules, data storing/retrieving processes and management, and analysis functionalities, combined with the typical functionalities of standalone applications (Hoffman, 2004; Wortmann, 1998; Davenport, 2000). Moreover, ERP has allowed the business partners to share information under one single roof, and therefore performance track becomes more transparent and real for effective decision making (Barratt & Oke, 2007).

The exchange of information and knowledge is so important that supply chain partners should consider the use of an enterprise planning system to promote the exchange of information and knowledge (Towers & Burnes, 2008). Therefore, it is crucial for multichannel retailers to use IT effectively in integrating their activities across the functional areas, so that the consistency and flow of information regarding customers, orders, and inventory can be ensured (Vickery, Jayaram, Droge, & Calantone, 2003). Information accuracy, Information availability, Information timeliness, Information sharing are some of the measures related to information technology (Cai, Liu, Xiao, & Liu, 2009; Angerhofer & Angelides, 2006; Van der Vorst, 2000). Information and knowledge sharing can help spread the risks, costs, and gains for supply chain partners (Ballou, Gilbert, & Mukherjee, 2000). Generally, information sharing in a retail supply chain context has received a lot of attention (Chae, 2009; Gavirneni, R, & S, 1999; Cachon G., 2001; Sahin & Robinson, 2002), and it has been an important component in improving operational issues (Forrester J. W., 1961; Chen, Drezner, Ryan, & Simchi-Levi, 2000; Lee, Padmanabhan, & Whang, 1997).

Enabling factors of Information Technology: In today's markets, technological and competitive forces are evolving at a very rapid pace. To respond to these forces, radical changes in the organizations have become necessary. To handle these complexities of business, companies have adopted technological advancements in streamlining the flow of various processes of supply chain. Information technology has helped in gaining advantages in terms of cost, speed and flexibility which helps to improve the overall performance of supply chain functions such as warehouse management and network planning. (McCarthy & Golicic, 2002; Forslund & Jonsson, 2007; Rungtusanatham, Salvador, Forza, & Choi, 2003; Chang, Fu, Lee, Lin, & Hsueh, 2007; Chae, 2009) have studied the impact of information quality on Forecasting. Data typically is captured at various interfaces between the supply chain entities, like, customer transactions, shipments to stores, inventory and warehousing activities, interactions pertaining to planning and allocation of stock between retail outlets and corporate head-offices, and the interactions between value chain intermediaries. The data captured at these interfaces is invaluable unless and until it undergoes the proper diagnosis with the help of tools and software for analysis. The analysis of this data has made it possible to make more effective decisions for purchasing, stocking and logistics.

With the help of flexible and adaptable information systems, the reliability and accuracy of data has increased, and it is easily available with the channel partners without any time lag. In addition, Information technologies act as a catalyst for improving the overall performances of the warehouses. Implementation of IT helps to reduce warehouse complexities, which include the number and variety of items to be handled, the nature of product, variety of processes, etc. It helps in tracing items raw material, semi-finished goods and finished goods [RM, SFG and FG] that are stored in a warehouse during material handling operations. Hence, I.T. has made monitoring and managing of inventory surprisingly easier. It provides timely and accurate information about products, which helps to easily track the items, order-picking and sorting of inventory. It is also observed that radio frequency identification [RFID], Warehouse Management Systems [WMS], Electronic Data Interchange [EDI], and Enterprise Resource Planning [ERP], Geographical Information Systems [GIS], etc have become popular tools in automating the logistics operations. Integration and coordination of these information systems are important to supply chain alignment. As Organizational managers are ultimately held accountable for organizational performance (Walters, 2008; Green Jr, Whitten, & In, 2008) it is important to continuously collect feedback and adopt measures to improve the feedback. Information sharing results in benefits for all of the supply chain partners. Here, in this study researchers attempt to identify the KPIs which contribute in improving the overall efficiency of the supply chain operational activities like transport and inventory management. The variables used for Information Technology Optimization are given below in the Table 6.

3. INVENTORY OPTIMIZATION

The complexity in managing the uncertainty of supply chain has increased due to the increase in number of product variants with ever shortening life cycles. Moreover every product has its own supply chain as identified by (Fisher, 1997), where he categorized the supply chains as Efficient and Responsive supply chain depending upon the nature of the product. Managing inventory is one of the biggest challenges in retail chains. With the increase in number of product variants, and with decreasing product life cycles,

Table 6. Variables for information technology optimization

Survey Elements	Representation*
Role of IT in efficient purchasing/inventory management is important	ITO1
EDI helps in faster exchange of data between buyer and seller	ITO2
Quality of the input data *(e.g via POS)* helps in demand forecasting and triggering Re-order point (ROP)	ITO3
Information system should be adaptable and flexible for maximizing benefits.	ITO4
Compliance with latest regulations of information systems is beneficial for overall functioning of organization, hence it is an important indicator for improving SC performance	ITO5
Real time information due to IT usage helps in reducing claims in rupee per month vs monthly turnover	ITO6
IT helps in easy sharing of real time information with channel partners, which increases the accuracy and reliability of the acquired information	ITO7
Investment in IT minimizes the data maintenance and transaction cost	ITO8

Representation*: For easy identification of variables, the statements are given a alpha numeric representation

the supply chain responsiveness increases. Customer responsiveness includes lead time, stock-out probability and fill rate. In order to avoid the stock-outs, companies have to keep the stock to meet such uncertainties in demand and supply Therefore, the coordination of logistics and inventory decisions in a supply chain has a significant effect on the supply chain performance.

One of the major reasons for using information technology, like, EDI is that there should be no discrepancy in the physical flow of goods, and with the accuracy in data, inventory levels should be maintained. Retailers, vendors and suppliers tend to share vital information- such as, demand trend reports, forecasts, and inventory levels order status and transportation plans- in real time. This effective communication and data-sharing helps to identify numerous cost-cutting opportunities that can be beneficial to all parties involved in the supply chain. Various types of costs are associated with inventory like investment value with held inventory, cost associated with obsolete inventory, cost associated with work-in-process inventories and the cost with held with finished goods inventories. Some of the initiatives that retailers have started to optimize regarding their entire supply chain include workforce optimization, inventory planning and revamping of technological infrastructure. Optimization models and algorithms, decision support systems and computerized analysis tools are some of the methods adopted by companies for decisions of different functions (e.g., supply process, distribution, inventory management, production planning, facilities location, etc.) for managing the entire supply chain.

In supply chain, the total cost associated with inventory (Stewart, 1995) (Christopher M., 1992; Slack, 1983; Lee & Billington, 1993; Dobler & Burt, 1996; Levy, 1997) consists of the following:

- Opportunity cost consisting of warehousing, capital and storage;
- Cost associated with inventory as incoming stock level, work in progress;
- Service costs, consisting of cost associated with stock management and insurance;
- Cost held up as finished goods in transit;
- Risk costs, consisting of cost associated with pilferage, deterioration, damage;
- Cost associated with scrap and rework;
- Cost associated with shortage of inventory accounting for lost sales/lost production.

Inventory is where the biggest cost is hidden in most businesses today. (Harrington, 1996)

Significant changes have been seen in global retail sector because the challenge for the companies has been to keep abreast of the best practices in the industry. In terms of cost involved managing inventory has been 20% of the total logistics cost (ELA European Logistics Association / A. T. Kearney Management Consultants, 2004). In a study of an FMCG supply chain, it was found that functional decisions made at each major process are driven by different criteria and this lack of alignment at each link is referred as 'Matrix Ttwist' proposed by (Godsell & Harrison, 1992) (given in Table 7).

Inventory levels in different parts of the supply chain have to be monitored and reported, which can help to reduce the total amount of inventory in the supply chain. Optimizing inventory is an important aspect of supply chain, and it has to be continuously traced so that no disruption or bullwhip effect occurs in the supply chain. (Lee, Padmanabhan, & Whang, 1997) identified four main causes of the bullwhip effect: demand signal processing, batch purchasing, price fluctuations, and shortage gaming. Choice of inventory policies, extent of information sharing, and use of early order commitment are often cited as effective means to achieving better supply chain coordination and alleviating the bullwhip effect. Supply chain coordination is imperative to reducing the inventory and counteracting the demand uncertainty throughout the supply chain. Researchers have focused on missing inventory, inventory record inaccuracy and inventory replenishment, it is reasonable to suspect that, given the high level of problems with inventories (Raman, DeHoratius, & Ton, 2001a; Corsten & Gruen, 2003).

For example, it is not obvious as to how the loading or unloading of goods or the productivity in order picking influence the operational result of a distribution center. Other factors, such as, goods handling damages, delivery quality or inventory accuracy greatly influences the performance of the operations as a whole. Furthermore, the use of financial accounting information promotes a functional perspective within an organization, mainly because; resources are allocated from the top down, whereas goods and services flow horizontally through the firm. Traditionally, retailers make their own inventory replenishment decisions based on their demand forecast and their cost structure (i.e., inventory carrying cost and ordering costs). It was (Ferguson & Ketzenberg, 2006) examining the value of reverse flow of information in which the supplier shares its inventory state with the retailer. In addition, (Zhao, Xie, & Wei, 2002) conducted extensive simulation studies on the effect of Early Order commitment (EOC) on supply chain performance under various operational conditions, including demand pattern, forecast errors, cost structure, number of retailers, and capacity cushion. An inventory model for deteriorating items with time-dependent backlogging rate was studies by (Dye 2007).

Table 7. Supply chain optimization

Supply Chain Process	Supply Chain Decision	Determined By
Source	Which suppliers?	Raw material commodity type
Make	Which manufacturing site?	Product family type
Deliver	Which Manufacturer warehouse?	Historically a function of order size In process of being divided by export paperwork requirements and customer account (arbitrary spilt)
	Which customer regional distribution centre?	Product type and Location of store to serve
	Which products to which store?	Demographics of the store's catchments area which drives layout and range decisions

To achieve the appropriate level of inventory, a focus on correct evaluation, identification, classification and quantification, retrieval and security of goods would provide a clear and accurate view (Chorafas, 1974). The reason for improvement of the warehouse activity is due to the increase in responsiveness and agility of supply chain, which demands higher order accuracy, reduced space requirements, increased volume capacity, control of inventory and increased customer service (Adams, Brown, & Firth, 1996). JIT, MRP and similar methods reflect the need to hold stock or materials for a minimum amount of time. The challenge of holding enough inventories to meet demand, but not incur excess cost, is a perennial supply chain management problem (Forrester J. W., 1961; Lee, Padmanabhan, & Whang, 1997). Warehouses are now redesigned and automated for better efficiency in terms of high throughput and high productivity, to reduce order producing costs. Warehousing activities concern the physical storage and retrieval of materials, and also the processing of information needed about the goods stored. Some of the value adding activities are, production postponement, also conveyor and sortation equipment may be used in a cross-docking facility to direct goods to warehouse areas where such activities as labeling, kitting and hanging may take place, without the goods ever being placed into storage (Marvick & White, 1998) Automation offers flexibility to handle peak throughputs at short notice, particularly in areas where staff availability is a problem, or in operations where the use of additional staffing may result in congestion and productivity issues (Naish & Baker, 2004). The variables used for Inventory Optimization are given below in the Table 8.

Table 8. Variables for inventory optimization

Survey Element	Representation*
Inventory holding cost as % of gross sales shows an impact on overall efficiency	IO1
Accuracy in forecasting sales reduces obsolete inventory	IO2
Stock-outs should be minimum for better profitability	IO3
FIFO is a better method for inventory valuation	IO4
Inventory accuracy *((book inventory – counted inventory)/ book inventory)* gives an insight in your bookkeeping practices and helps to measure stock cover	IO5
Inventory turnover (in rupees per sq. feet) is important to know the average days of inventory	IO6
% of time spent picking orders/back orders impacts the level of operational efficiency	IO7
Inventory replenishment cycle time helps to plan timely orders	IO8
Fill rate is an important indicator in retail operations	IO9
Innovation is a key parameter in Retail Supply Chain *(e.g. automation in warehouse helps to speed up the logistics operations)*	IO10
Optimum number of warehouses are required for maximizing service level	IO11
productivity of MHE (material handling equipment) per square feet of warehouse indicates the level of warehouse efficiency	IO12
Warehouse space/ layout/ future scalability/ use of MHE are critical for warehouse optimization	IO13
Certification of the warehouse-ISO certificates/C-TPAT certification/TAPA certification/Accreditation by WRDA India is essential/desirable for compliance with latest regulations	IO14
Electricity consumption (in Kw-hrs) per sqft of warehouse space reflects the energy efficiency, hence optimizes cost	IO15
Depending on the nature of the goods, the storage facility has to be maintained *(e.g cold storage*	IO16

Representation*: For easy identification of variables, the statements are given a alpha numeric representation

4. RESOURCE OPTIMIZATION

'Resources are generally measured in terms of the minimum requirements (quantity) or a composite efficiency measure' (Beamon, 1999). In fact, resource minimization is one general goal of supply chain analysis, wherein the supply chain is reconfigured to meet the present demand in the market. Companies aim to capitalize maximum on their minimum resources. Efficiency measures the utilization of the resources in the systems that are used to meet the system's objectives. Resource measurement is an important part of the measurement system and resources are directly related to the system's output and flexibility measures.

A supply chain network uses resources of various kinds: manufacturing resources (machines, material handlers, tools, etc.); storage resources (warehouses, automated storage and retrieval systems); logistics resources (trucks, rail transport, air-cargo carriers, etc.); human resources (labor, scientific and technical personnel); and financial (working capital, stocks, etc.). The objective is to utilize these assets or resources efficiently so as to maximize customer service levels, minimize lead times, and optimize inventory levels. In supply chain management, where all the business processes are linked and integrated with all business supply chain members, which makes the structure complex and cumbersome, it is recommended that firms should identify those supply chain members that are critical for successful supply chain performance (Douglas, 1996).

Similarly, the companies have limited resources which have to be optimally used. Resource measures include: inventory levels, personnel requirements, equipment utilization, energy usage and cost. Some of the supply chain resource performance measures include the following cost:

- **Total Cost:** Total cost of resources used.
- **Distribution Cost:** Total cost of distribution, including transportation and handling cost.
- **Manufacturing Cost:** Total cost of manufacturing, including labor, maintenance and re-work costs.
- **Inventory:** Costs associated with held inventory:
    - **Inventory Investment:** Investment value of held inventory.
    - **Inventory Obsolescence:** Costs associated with obsolete inventory; sometimes includes spoilage.
    - **Work-in-Process:** Costs associated with work-in-process inventories.
    - **Finished Goods:** Costs associated with held finished goods inventories.
- **Return on Investment (ROI):** Measures the profitability of an organization. The return on investment is generally given by the ratio of net profit to total assets.

Supply chain assets include, accounts receivable, plant, property and equipment and inventories (Stewart, 1995). It is also important to note here that how the costs associated with each asset, combined with its turnover, affect the 'total cash flow time', which can be measured as the average number of days required to transform the cash invested in assets into the cash collected from a customer. The total cash flow time can be combined with profit with the objective of providing an insight into the rate of return on investment [ROI]. For example, superior customer service leads to improved sales and an increased profit and subsequently a higher ROI. Therefore, ROI is found to be an indicator of financial health of supply chain. Resource based theory [RBT] explains how the rent generating potential of resources and capabilities can lead to sustainable competitive advantage (Wernerfelt, 1984; Barney, 1991; Grant, 1991), and it is also applicable for intangible resources (Conner, 1991; Taylor-Coates and McDermott, 2002). Efficiency describes an input/output relation while effectiveness shows how well supply chain goals have been achieved (see e.g. Bowersox et al., 2010). In this sense, supply chain performance can be seen as a function of the utilization of supply chain resources, or as a function of supply chain results as compared to supply chain targets. The variables used for Resource Optimization are given below in the Table 9.

5. FINANCIAL PERFORMANCE OF THE FIRM

Financial performance is an important indicator for the health of any organization. However, it is challenging to agree on financial measures for performance measurement, as the resources are common, and also because the cost centers are different for trading partners (Papakiriakopoulos & Pramatari, 2010). An early attempt at developing financial measures was made by Du Pont (Walters, 1997). Du Pont is widely acknowledged as being the founder of financial performance measurement, by introducing a pyramid of financial ratios as early as 1903. Thereafter, in the late 1970s and 1980s, numerous authors expressed a general dissatisfaction with traditional backward looking or lag accounting based performance measurement systems (Pitt & Tucker, 2008).

Performance of the firm is mainly measured by financial data, such as the financial ratios. Financial data is useful in providing the measurement of a firm's performance via the market's valuation of the firm's securities. Since the future cash flows of the business entity cannot be estimated with accuracy, measures of financial performance are typically based on accounting data, such as, return on investment [ROI], or, return on assets [ROA]. (Jahera & Lloyd 1992) observed that ROI was a valid performance

Table 9. Variables for resource optimization

Performance Indicator	Representation*
The following cost are important for supply chain efficiency	
Direct labor cost, Direct material cost and Manufacturing cost	RO1a
Cost of goods sold	RO1b
Distribution cost and Inventory cost	RO1c
Information management cost	RO1d
Warranty cost	RO1e
Packaging cost	RO1f
Facility management/ maintenance cost	RO1g
Quality of packaging material used is essential for customer service	RO2
Customer satisfaction is important for the growth of the business/ maximizing profit	RO3
Value added employee productivity helps to measure supply chain efficiency	RO4
Training employees add to their productivity	RO5
Acquiring a new equipment/software/ labour *as and when* business requirement is essential for the supply chain process improvements	RO6
Cargo carried in terms of volumes for fiscal year indicates the benchmark for next year	RO7
Use of renewable/ solar energy/green terminals are the growing need for business efficiency	RO8

Representation*: For easy identification of variables, the statements are given a alpha numeric representation

measure for midsize firms. Moreover, the validity of ROI, as a performance measure, has been challenged (Tobin & Brainard, 1968). A firm's financial leverage can affect its ROI to such an extent that it causes comparisons between firms meaningless (Tan, Kannan, Handfied, & Ghosh, 1999). It is also imperative that firms reporting the highest levels of financial and operational performance emphasize not only on the internal quality initiatives, but also to the management of all elements of their supply chain including customers and suppliers, and the quality of delivered products. Some of the financial indicators identified include: sales, ROA, market share, gross profit (Tan, Kannan, Handfied, & Ghosh, 1999). According to some, financial performance reflects on organization's profitability and return on investment as compared to its competition (Green Jr, Whitten, & Inman, 2008) argue that non-financial performance such as improved quality, innovation and resource planning should actually reduce costs, and thus, have a positive effect on measures of financial performance. The growth of market share and sales growth should impact financial performance through improved revenue numbers (Green Jr, Whitten, & In, 2008). Quality measures help organizations to retain current customers and create greater customer loyalty, which in turn may increase market share and organizational performance (Rust et al., 1994, (Koh, Demirbag, Bayraktar, Tatoglu, & Zaim, 2007).

The supplier-to-customer systems approach taken by P&G, Bridgestone/Firestone (Lampe & Gray, 1998); Pepsi (Bechtel & Jayaram, 1997); Hewlett-Packard (Davis, 1993); and others is based on the 'Seven Rs' of traditional logistics – 'having the Right product, to the Right customer, at the Right place, at the Right time, in the Right condition, in the Right quantity, at the Right cost are essential to market and financial performance' (Tracey, Lim, & Vonderembse, 2005). Anderson et al., (1994) found that marketing performance, as measured by customer satisfaction, positively impacts financial performance, as measured by return on investment.

Any supply chain initiative that results in an improvement of some aspect of supply chain performance must ultimately get translated in to improved business performance. In the final analysis, each firm is primarily interested in improving return on assets (ROA) (Shah, 2009). (LaLonde, 2000), argued that the supply chain management community needed to address an important disconnect between supply chain decisions and financial investment outcomes. Whilst the importance of supply chain management is understood, its influence on organizational financial performance is less explicit (Frohlich & Westbrook, 2001). The financial performance, measured by the retailers, helps the managers to make assumptions about current versus future cash flows. Financial ratios give the retailers an idea about how their assets are being utilized, how the inventory is stocked versus sold. By collecting information on a monthly, quarterly and annual basis, retailers can use the financial data to make decisions, both long term and short term, based on trend analysis for planning the forecast of future sales. The key financial ratios for retailers focus on aspects of income, liquidity and profitability.

From the income point of view, retail managers consider gross margin as an indication of sales remaining after subtracting the cost of purchasing merchandize. It is an indication of converting the existing inventory into future cash. From the liquidity aspect, cash flow is crucial to understand their ability to pay short-term debts. Likewise, quick ratio and inventory turnover ratio provides an insight of business solvency as it uses the most of liquid cash. Cost of goods sold [COGS] gives an idea as to how fast the inventory is being sold at current sales. Yet any other indicator of high importance is, average collection period, which tells managers as to how quickly they are able to collect the outstanding debts.

The profitability ratios, like ROA allows retail managers to identify how productively and effectively they are using the business assets to grow, and it enables them to make necessary decisions about underperforming assets. The variables used for Financial Performance are given below in the Table 10.

Table 10. Variables for financial performance

Survey Element	Representation*
Receivables turnover *(Annual credit sales/Accounts receivables)*	RO9a
Average collection period *(Accounts receivables/ (Annual credit sales/365))*	RO9b
Inventory turnover (COGS/average inventory)	RO9c
Debt Ratio *(Total debt/ Total assets)*	RO9d
Debt-to-equity ratio *(Total debt/ total equity)*	RO9e
Interest coverage *(EBIT/ Interest charges)*	RO9f
Gross profit Margin *((Sales-COGS)/sales))*	RO9g
Return on asset (ROA) is a good measure to study the overall impact of the organization's performance	RO10

Representation*: For easy identification of variables, the statements are given a alpha numeric representation

FUTURE RESEARCH DIRECTIONS

The list of indicators presented above is extensive, covering the major aspects of retail supply chain: transport optimization, information technology optimization, inventory optimization, resource optimization and financial performance. Additionally, the indicators grouped in these categories have been further divided into sub categories, likewise, transport optimization is further segmented into delivery, time, frequency, and capacity related performance indicators; inventory optimization is further segmented into cost, time, quantity, and service performance indicators; information technology optimization is further segmented into level of IT implementation, service, responsiveness, cost related performance indicators; resource optimization is further segmented into cost, service, time and financial related performance indicators. The research can be further carried by empirically measuring the identified indicators for various retail supply chains. Infact the study can be carried out in different context. Since here the emphasis is laid on organized sector, a similar work can be carried out for an unorganized sector and a comparison can be laid out between the operations of organized and unorganized sector. Also it can be studied that what factors differentiate the two broad categories of sectors and what are the essentials applicable to each category.

There is possibility of interrelationship among the various identified indicators and a study can be carried out to develop a model/linkage between the different constructs. Again these linkages will be different depending upon the context of study. It will help to bring the critical factors in limelight and could be very helpful to the managers involved in its operations. Since this study is for retail sector, different sectors can also be opted in a similar way.

As SCM is complex and involves a network of supply chain partners in the effort of producing and delivering the final product, its entire domain cannot be covered in just one study. Further research can be built on using different methodologies for insights in this area of work.

CONCLUSION

This study provides the broad categorization of indicators for measuring retail supply chain performance, taking into consideration the different aspects of supply chain. The variables were extracted from an extensive literature on performance measurement of supply chain in different context and have been categorized specifically for retail supply chain. The challenge identified in measuring the performance is the identification of number and types of indicators for a particular supply chain. It also depends upon the accessibility of information for these indicators. Qualitative data at times, is difficult to be calibrated in to quantitative data. The authors in previous studies believe that there is always a missing factor in development of KPIs, which forces to opt for improved performance indicators. The lack of applicability of these indicators and adopting a holistic approach is one of the main concerns. There have been very limited studies which have focussed on industry specific approach. Further, lack of proper categorization often results in lesser use of performance indicators because the categories often result in no meaning to the industry. Thus an industry specific approach has been adopted such that the interpretation of KPIs on retail supply chain performance can be measured. The managers can prioritize the factors which are important for retail supply chain and can pay more attention to those factors which are more actionable and result oriented in long term.

REFERENCES

Adams, N. D., Brown, T. W., & Firth, R. V. (1996). *Warehouse and Distribution Automation Handbook*. New York: McGraw-Hill.

Angerhofer, B. J., & Angelides, M. C. (2006). A model and a performance measurement system for collaborative supply chains. *Decision Support Systems*, *42*(1), 283–301. doi:10.1016/j.dss.2004.12.005

Aviv, Y. (2007). On the benefits of collaborative forecasting partnerships between retailers and manufacturers. *Management Science*, *53*(5), 777–794. doi:10.1287/mnsc.1060.0654

Ballou, R. H., Gilbert, S. M., & Mukherjee, A. (2000). New managerial challenges from supply chain opportunities. *Industrial Marketing Management*, *29*(1), 7–18. doi:10.1016/S0019-8501(99)00107-8

Barratt, M., & Oke, A. (2007). Antecedents of supply chain visibility in retail supply chains: A resoure-based theory perspective. *Journal of Operations Management*, *25*(6), 1217–1233. doi:10.1016/j.jom.2007.01.003

Barua, A., Konana, P., Whinston, A. B., & Yin, F. (2004). An empirical investigation of net-enabled business value. *Management Information Systems Quarterly*, *28*(4), 585–620.

Beamon, B. M. (1999). Measuring supply chain performance. *International Journal of Operations & Production Management*, *19*(3), 275–292. doi:10.1108/01443579910249714

Bechtel, C., & Jayaram, J. (1997). Supply chain management: a strategic perspective. *The International Journal of Logistics Management*, 15-34.

Bourland, K. E., Powell, S. G., & Pyke, D. K. (1996). Exploiting timely demand information to reduce inventories. *European Journal of Operational Research*, *92*(2), 239–253. doi:10.1016/0377-2217(95)00136-0

Bower, J. L., & Hout, T. M. (1988). Fast cycle capability for competitive power. *Harvard Business Review*, 110–118.

Bowersox, D., Rodrigues, A., & Calantone, R. (2005). Estimation of Global and National Logistics Expenditures: 2002 Data Update. *Journal of Business Logistics*, *26*(2), 1–16. doi:10.1002/j.2158-1592.2005.tb00202.x

Bowersox, D. J., Closs, D. J., Stank, T. P., & Keller, S. B. (2000). How supply chain competency leads to business success. *Supply Chain Management Review*, *4*(4), 70–78.

Cachon, G. (2001). Exact evaluation of batch-ordering policies in two-echelon supply chains with periodic review. *Operations Research*, *49*(1), 79–98. doi:10.1287/opre.49.1.79.11188

Cachon, G. P., & Fisher, M. (2001). Supply chain inventory management and the value of shared information. *Management Science*, *46*(8), 1032–1048. doi:10.1287/mnsc.46.8.1032.12029

Cai, J., Liu, X., Xiao, Z., & Liu, J. (2009). 2009). Improving supply chain performance management: A systematic approach to analyzing iterative KPI accomplishment. *Decision Support Systems*, 612–621.

Cai, J., Liu, X., Xiao, Z., & Liu, J. (2009). Improving supply chain performance management: A systematic approach to analyzing iterative KPI accomplishment. *Decision Support Systems*, *46*(2), 512–521. doi:10.1016/j.dss.2008.09.004

Cannon, J. P., & Perreault, W. D. Jr. (1999). Buyer–seller relationships in business markets. *JMR, Journal of Marketing Research*, *36*(4), 439–460. doi:10.2307/3151999

Chae, B. K. (2009). Developing key performance indicators for supply chain: An industry perspective. *Supply chain Management. International Journal (Toronto, Ont.)*, *14*(6), 422–428.

Chan, F. T., & Qi, H. J. (2003). An innovative performance measurement method for supply chain management. *Supply Chain Management: An International Journal*, *8*(3), 209–223. doi:10.1108/13598540310484618

Chang, T., Fu, H., Lee, W., Lin, Y., & Hsueh, H. (2007). A study of an augmented CPFR model for the 3C retail industry. *Supply Chain Management: An International Journal*, *12*(3), 200–209. doi:10.1108/13598540710742518

Chen, F., Drezner, Z., Ryan, J. K., & Simchi-Levi, D. (2000). Quantifying the bullwhip effect in a simple supply chain:the impact of forecasting, lead times, and information. *Management Science*, *46*(3), 436–443. doi:10.1287/mnsc.46.3.436.12069

Chen, I. J., Paulraj, A., & Lado, A. (2004). Strategic purchasing, supply management and firm performance. *Journal of Operations Management*, *22*(5), 505–523. doi:10.1016/j.jom.2004.06.002

Christopher, M. (1992). *Logistics and Supply Chain Management: Strategies for Reducing Costs and Improving Services.* London: Financial Times/Pitman.

Closs, D. J., Goldsby, T. J., & Clinton, S. R. (1997). Information technology influences on world class logistics capability. *International Journal of Physical Distribution & Logistics Management*, *27*(1), 4–17. doi:10.1108/09600039710162259

Corsten, D., & Gruen, T. (2003). Seeking on-shelf availability—an examination of the extent, the causes and the efforts to address retail out-of-stocks. *International Journal of Retail & Distribution Management, 31*(12), 605–716. doi:10.1108/09590550310507731

Coyle, J. J., Bardi, E. J., & Langley, C. J. (2003). *The Management of Business Logistics: A Supply Chain Perspective* (7th ed.). Mason, OH: South-Western Thomson Learning.

Davenport, T. H. (2000). *Mission critical — realizing the promise of enterprise systems.* Boston, MA: Harvard Business School Press.

Dixon, J. R., Nanni, A. J., & Vollmann, T. E. (1990). *The New Performance Challenge- Measuring Operations for World-class Competition.* Homewood, IL: Dow Jones-Irwin.

Dobler, D. W., & Burt, D. N. (1996). *Purchasing and Supply Management.* New York, NY: The McGraw-Hill Companies.

Douglas, J. (1996). Building performance and its relevance to facilities management. *Facilities, 14*(3/4), 23–32. doi:10.1108/02632779610112508

Drucker, P. F. (1962, April). The economy's dark continent. *Fortune*, 265-270.

ELA European Logistics Association / A. T. Kearney Management Consultants. (2004). *Differentiation for performance excellence in logistics 2004.* Hamburg, Germany: Deutscher Verkehrs-Verlag GmbH.

Ellinger,, A. E. D. P. (2000). The Relationship between Marketing/Logistics Interdepartmental Integration and Performance in U.S. Manufacturing Firms: An Empirical Study. *Journal of Business Logistics, 21*(2), 1–22.

Ellram, L. M., La Londe, B. J., & Weber, M. M. (1989). Retail Logistics. *International Journal of Physical Distribution & Materials Management, 19*(12), 29–39. doi:10.1108/EUM0000000000348

Ernst & Young. (2011). Doing. *Business India.*

Ferguson, M., & Ketzenberg, M. (2006). Sharing Information to Improve Retail Product Freshness of Perishables. *Production and Operations Management, 15*(1), 57–73.

Fiala, P. (2005). Information sharing in supply chains. *Omega, 33*(5), 419–423. doi:10.1016/j.omega.2004.07.006

Fisher, M. L. (1997). What is the Right Supply Chain for your product. *Harvard Business Review.*

Flynn, B. B., Huo, B., & Zhao, X. (2010). The impact of supplychain integration on performance: A contingency and configuration approach. *Journal of Operations Management, 28*(1), 58–71. doi:10.1016/j.jom.2009.06.001

Forrester, J. W. (1961). *Industrial Dynamics.* Cambridge, MA: MIT Press.

Forslund, H., & Jonsson, P. (2007). The impact of forecast information quality on supply chain performance. *International Journal of Operations & Production Management, 27*(1), 90–107. doi:10.1108/01443570710714556

Frazelle, E. H. (2002b). *Supply Chain Strategy: The Logistics of Supply Chain Management*. New York: McGraw-Hill.

Froehle, C., & Roth, A. (2007). A Resource-Process Framework of New Service Development. *Production and Operations Management*, *16*(2), 169–188. doi:10.1111/j.1937-5956.2007.tb00174.x

Frohlich, M. T., & Westbrook, R. (2001). Arcs of integration: An international study of supply chain strategies. *Journal of Operations Management*, *19*(2), 185–200. doi:10.1016/S0272-6963(00)00055-3

Fynes, B., Voss, C., & Búrca, S. (2005). The impact of supply chain relationship dynamics on manufacturing performance. *International Journal of Operations & Production Management*, *25*(1), 6–19. doi:10.1108/01443570510572213

Fynes, B., Voss, C., & de Búrca, S. (2005). The impact of supply chain relationship quality on quality performance. *International Journal of Production Economics*, *96*(3), 339–354. doi:10.1016/j.ijpe.2004.05.008

Gavirneni, S., Kapuscinski, R., & Tayur, S. (1999). Value of Information in Capacitated Supply Chains. *Management Science*, *45*(1), 16–24. doi:10.1287/mnsc.45.1.16

Gelders, L., Mannaert, P., & Maes, J. (1994). Manufacturing strategy performance indicators and improvement programs. *International Journal of Production Research*, *32*(4), 797–805. doi:10.1080/00207549408956971

Globerson, S. (1985). Issues in developing a performance criteria system for an organization. *International Journal of Production Research*, *23*(4), 639–646. doi:10.1080/00207548508904734

Gopal, P. R., & Thakkar, J. (2012). A review on supply chain performance measures and metrics: 2000-2011. *International Journal of Productivity and Performance Management*, *61*(5), 518–547. doi:10.1108/17410401211232957

Green, K. W. Jr, Whitten, D., & In, R. A. (2008). The impact of logistics performance on organizational performance in a supply chain context. *Supply Chain Management: An International Journal*, *13*(4), 317–327. doi:10.1108/13598540810882206

Green, K. W. Jr, Whitten, D., & Inman, R. A. (2008). The impact of logistics performance on organizational performance in a supply chain context. *Supply chain management. International Journal (Toronto, Ont.)*, *13*(4), 317–327.

Gumbus, A. (2005). Introducing the balanced scorecard: Creating metrics to measure performance. *Journal of Management Education*, *29*(4), 617–630. doi:10.1177/1052562905276278

Gunasekaran, A., & Kobu, B. (2007). Performance measures and metrics in logistics and supply chain management: A review of recent literature (1995-2004) for research and applications. *International Journal of Production Research*, *45*(12), 2819–2840. doi:10.1080/00207540600806513

Gunasekaran, A., Patel, C., & Tirtiroglu, E. (2001). Performance measures and metrics in a supply chain environment. *International Journal of Operations & Production Management*, *21*(1), 71–87. doi:10.1108/01443570110358468

Gunasekaran, A., Patel, C., & Tirtiroglu, E. (2001). Performance measures and metrics in a supply chain environment. *International Journal of Operations & Production Management, 21*(1/2), 71–87. doi:10.1108/01443570110358468

Gunasekarana, A., Patelb, C., & McGaughey, R. E. (2004). A framework for supply chain performance measurement. *International Journal of Production Economics, 84*(3), 333–347. doi:10.1016/j.ijpe.2003.08.003

Gunasekran, A., Patel, C., & Tirtiroglu, E. (2001). Performance measures and metrics in a supply chain environment. *International Journal of Operations & Production Management, 21*(1/2), 71–87. doi:10.1108/01443570110358468

Guruswamy, M., Sharma, K., & Jos, M. M. (2007). *FDI in Retail – III Implications of Wal Mart's Backdoor Entry*. Centre for Policy Alternatives.

Gustin, C. M., Daugherty, P. J., & Stank, T. P. (1995). The effects of information availability on logistics integration. *Journal of Business Logistics, 16*(1), 1–21.

Harrington, L. (1996, July 15). Untapped savings abound. *Industry Week*, 53–58.

Hayes, R. H., & Abernathy, W. J. (1980). Managing our way to economic decline. *Harvard Business Review, 62*, 95–101.

Hill, C. A., & Scudder, G. D. (2002). The use of electronic data interchange for supply chain coordination in the food industry. *Journal of Operations Management, 20*(4), 375–387. doi:10.1016/S0272-6963(02)00017-7

Ho, D. C., Chan, W. E., Wong, N., & Chan, M. (2000). Significant metrics for facilities management benchmarking in the Asia Pacific region. *Facilities, 18*(13/14), 545–555. doi:10.1108/02632770010358088

Hoffman, D. (2004). *The hierarchy of supply chain metrics: diagnosing your supply chain health*. Boston, Massachusetts: AMR Research.

Hulland, J., Wade, M., & Antia, K. D. (2007). The impact of capabilities and prior investments on online channel commitment and performance. *Journal of Management Information Systems, 23*(4), 109–142. doi:10.2753/MIS0742-1222230406

IJMBS. (2013, January 24). *India Retail News*. Retrieved January 27, 2013, from Emergence of Retailing Sector in India: Challenges and Opportunities: http://www.indiaretailnews.com/

Johnson, H. T. (1981). Toward a new understanding of nineteenth-century cost accounting. *Accounting Review, 56*(3), 510–518.

Johnson, H. T., & Kaplan, R. S. (1987). *Relevance Lost: The Rise and Fall of Management Accounting*. Harvard Business School Press.

Jones, R. M., & Towill, D. R. (1999). Total cycle time compression and the agile supply chain. *International Journal of Production Economics, 62*(1-2), 61–73. doi:10.1016/S0925-5273(98)00221-7

Kaplan, R. S. (1984). Yesterday's accounting undermines production. *Harvard Business Review, 62*(4), 95–101.

Kaplan, R. S., & Norton, D. P. (1992, January). The balance Scorecard- Measures That Drive performance. *Harvard Business Review*.

Keegan, D. P., Eiler, R. G., & Jones, C. R. (1989, June). Are your performance measures obsolete? *Management Accounting*, 45-50.

Koh, L. S., Demirbag, M., Bayraktar, E., Tatoglu, E., & Zaim, S. (2007). The impact of supply chain management practices on performance of SMEs. *Industrial Management & Data Systems*, *107*(1), 103–124. doi:10.1108/02635570710719089

KPMG. (2011, July). *Issues Monitor: Sharing knowledge on topical issues in Retail Industry*. KPMG.

Kulp, S. C., Lee, H. L., & Ofek, E. (2004). Manufacturer Benefits from Information Integration with Retail Customers. *Management Science*, *50*(4), 431–444. doi:10.1287/mnsc.1030.0182

Lai, K. H., Ngai, E. T., & Cheng, T. E. (2002). Measures for evaluating supply chain performance in transport logistics. *Transp Res Part E*, *38*(6), 439–456. doi:10.1016/S1366-5545(02)00019-4

Lambert, D. M., & Pohlen, T. L. (2001). Supply chain metrics. *The International Journal of Logistics Management*, 1-19.

Lebas, M. J. (1995). Performance measurement and performance management. *International Journal of Production Economics*, *41*(1-3), 23–35. doi:10.1016/0925-5273(95)00081-X

Lee, H. L., & Billington, C. (1993). Material Management in Decentralized Supply Chains. *Operations Research*, *41*(5), 835–847. doi:10.1287/opre.41.5.835

Lee, H. L., Padmanabhan, V., & Whang, S. (1997). Information distortion in a supply chain: The bullwhip effect. *Management Science*, *43*(4), 546–558. doi:10.1287/mnsc.43.4.546

Lee, H. L., Padmanabhan, V., & Whang, S. (1997). The Bullwhip Effect in Supply Chains. *Sloan Management Review*, 93–101.

Levy, D. L. (1997, Winter). Lean production in an international supply chain. *Sloan Management Review*, 94–102.

Li, S., Ragu-Nathanb, B., Ragu-Nathanb, T. S., & Rao, S. S. (2006). The impact of supply chain management practices on competitive advantage and organizational performance. *Omega International Journal of Management Science*, *34*(2), 107–124. doi:10.1016/j.omega.2004.08.002

Lin, C. (2006). Influencing factors on the innovation in logistics technologies for logistics service providers in Taiwan. *The Journal of American Academy of Business*, *9*(2), 257–263.

Mabert, V. A., & Venkataramanan, M. A. (1998). Special research focus on supply chain linkages: Challenges for design and management in the 21st century. *Decision Sciences*, *29*(3), 537–552. doi:10.1111/j.1540-5915.1998.tb01353.x

Marvick, D., & White, J. (1998). Distribution operations: Managing distribution facilities for strategic advantage. In J. G. Gattorna (Ed.), *Strategic Supply Chain Alignment*.

McCarthy, T. M., & Golicic, S. L. (2002). Implementing collaborative forecasting to improve supply chain performance. *International Journal of Physical Distribution & Logistics Management, 32*(6), 431–454. doi:10.1108/09600030210437960

McKinnon, A. C. (1996). *The development of retail logistics in the UK:A position paper, Technology Foresight: Retail and Distribution Panel*. Edinburgh: Heriot-Watt University.

Melville, N., Kraemer, K., & Gurbaxani, V. (2004). Information technology and organizational performance: An integrative model of IT business value. *Management Information Systems Quarterly, 28*(2), 283–322.

Mentzer, J. T., Foggin, J. H., & Golicic, S. L. (2000, September/October). Collaboration: the enablers, impediments, and benefits. *Supply Chain Management Review*.

Moberg, C. R., Whipple, T. W., Cutler, B. D., & Speh, T. W. (2004). Do the Management Components of Supply Chain Management Affect Logistics Performance? *The International Journal of Logistics Management, 15*(2), 15–30. doi:10.1108/09574090410700284

Neely, A., Bourne, M., & Kennerley, M. (1995). Performance measurement system design: Developing and testingaprocess-based approach. *International Journal of Operations & Production Management, 20*(10), 1119–1145. doi:10.1108/01443570010343708

Neely, A. D. (1998). *Measuring Business Performance*. London: Economist Books.

Noble, S. M., Haytko, D. L., & Phillips, J. (2009). What drives college-age Generation Y consumers? *Journal of Business Research, 62*(6), 617–628. doi:10.1016/j.jbusres.2008.01.020

Papakiriakopoulos, D., & Pramatari, K. (2010). Collaborative performance measurement in supply chain. *Industrial Management & Data Systems, 110*(9), 1297–1318. doi:10.1108/02635571011087400

Parliament of India. (2009). *Rajya Sabha*. Ninetienth Report on Foreign and Domestic Investment in Retail Sector.

Pitt, M., & Tucker, M. (2008). Performance measurement in facilities management:driving innovation? *Property Management, 26*(4), 241–254. doi:10.1108/02637470810894885

Pohlen, T. L., & Lambert, D. M. (2001). Supply Chain Metrics. *International Journal of Logistics Management, 12*(1), 1–19. doi:10.1108/09574090110806190

Raman, A., DeHoratius, N., & Ton, Z. (2001a). The Achilles heel of supply chain management. *Harvard Business Review*, 136–152.

Ramdas, K., & Spekman, R. E. (2000). Chain or Shackles: Understanding What Drives Supply-Chain Performance. *Interfaces, 30*(4), 3-21.

Rao, B. P. (2000). Improving retail effectiveness through technology: A survey of analytical tools for physical and on-line retailers. *Technology in Society, 22*, 111–122. doi:10.1016/S0160-791X(99)00032-9

Ray, G., Muhanna, W. A., & Barney, J. B. (2005). Information technology and the performance of the customer service process: A resource-based analysis. *Management Information Systems Quarterly, 29*(4), 625–651.

Rosenzweig, E. D., Roth, A. V., & Dean, J. W. Jr. (2003). The influence of an integration strategy on competitive capabilities and business performance: An exploratory study of consumer products manufacturers. *Journal of Operations Management*, *21*(4), 437–456. doi:10.1016/S0272-6963(03)00037-8

Rungtusanatham, M., Salvador, F., Forza, C., & Choi, T. Y. (2003). Supply-chain linkages and operational performance: A resource-based-view perspective. *International Journal of Operations & Production Management*, *23*(9), 1084–1099. doi:10.1108/01443570310491783

Rushton, A., Oxley, J., & Croucher, P. (2001). The handbook of logistics and distribution management (2nd ed.). London, UK: Academic Press.

Sahin, F., & Robinson, E. P. (2002). Flow coordination and information sharing in supply chains: Review, implications, and directions for future research. *Decision Sciences*, *33*(4), 505–535. doi:10.1111/j.1540-5915.2002.tb01654.x

Sanders, N. R. (2008). Pattern of information technology use: The impact on buyer–suppler coordination and performance. *Journal of Operations Management*, *26*(3), 349–367. doi:10.1016/j.jom.2007.07.003

Schonberger, R. J. (1990). *Building a chain of customers*. London: The Free Press.

Shah, J. (2009). *Supply chain management*. Delhi: Pearson Education.

Shore, B., & Venkatachalam, A. (2003). Evaluating the information sharing capabilities of supply chain partners: A fuzzy logic model. *International Journal of Physical Distribution & Logistics Management*, *33*(9), 804–824. doi:10.1108/09600030310503343

Sinclair, D., & Zairi, M. (1995). Effective process management through performance measurement: Part III - An integrated model of totalquality-based performance measurement. *Business Process Management Journal*, *1*(3), 50–65. doi:10.1108/14637159510103220

Slack, N. (1983). Flexibility as a manufacturing objective. *International Journal of Operations & Production Management*, *3*(3), 4–13. doi:10.1108/eb054696

Spohrer, J., Maglio, P. P., Bailey, J., & Gruhl, D. (2007). Steps Toward a Science of Service Systems. *Computer*, *40*(1), 71–77. doi:10.1109/MC.2007.33

Stank, T. P., Davis, B. R., & Fugate, B. S. (2005). A strategic framework for supply chain oriented logistics. *Journal of Business Logistics*, *26*(2), 27–45. doi:10.1002/j.2158-1592.2005.tb00204.x

Stewart, G. (1995). Supply chain performance benchmarking study reveals keys to supply chain excellence. *Logistics Information Management*, *8*(2), 38–44. doi:10.1108/09576059510085000

Storey, J., Emberson, C., Godsell, J., & Harrison, A. (2006). Supply chain management: Theory, practice and future challenges. *International Journal of Operations & Production Management*, *26*(7), 754–774. doi:10.1108/01443570610672220

Tan, K.-C., Kannan, V. R., Handfied, R. B., & Ghosh, S. (1999). Supply chain management: An empirical study of its impact on performance. *International Journal of Operations & Production Management*, *9*(10), 1034–1052. doi:10.1108/01443579910287064

Tan, K. C., Kannan, V. R., & Hanfield, R. B. (1998). Supply Chain Management: Supplier Performance and Firm Performance. *International Journal of Purchasing and Materials Management, 34*(3), 2–9.

Tanriverdi, H. (2005). Information technology relatedness, knowledge management capability and performance of multibusiness firms. *Management Information Systems Quarterly, 29*(2), 311–334.

The World Bank. (2013). *India Overview*. Retrieved January Saturday, 2013, from http://www.worldbank.org/en/country/india/overview

Thomas, D. J., & Griffin, P. M. (1996). Coordinated supply chain management. *European Journal of Operational Research, 94*(1), 1–15. doi:10.1016/0377-2217(96)00098-7

Towers, N., & Burnes, B. (2008). A composite framework of supply chain management and enterprise planning for small and medium-sized manufacturing enterprises. *Supply Chain Management: An International Journal, 13*(5), 349–355. doi:10.1108/13598540810894933

Tracey, M., Lim, J.-S., & Vonderembse, M. A. (2005). The impact of supply-chain management capabilities on business performance. *Supply Chain Management: An International Journal, 10*(3), 179–191. doi:10.1108/13598540510606232

Van der Vorst, J. A. (2000). *Effective food supply chains Generating, modelling and evaluating supply chain scenarios.* (PhD thesis). Wageningen University.

Vickery, S. K., Jayaram, J., Droge, C., & Calantone, R. (2003). The effects of an integrative supply chain strategy on customer service and financial performance: An analysis of direct versus indirect relationships. *Journal of Operations Management, 21*(5), 523–539. doi:10.1016/j.jom.2003.02.002

Wallace, D. W., Giese, J. L., & Johnson, J. L. (2004). Customer retailer loyalty in the context of multiple channel strategies. *Journal of Retailing, 80*(4), 249–263. doi:10.1016/j.jretai.2004.10.002

Walters, P. G. (2008). Adding value in global B2B supply chains: Strategic directions and the role of the internet as a driver of competitive advantage. *Industrial Marketing Management, 37*(1), 59–68. doi:10.1016/j.indmarman.2007.06.010

Whipple, J. M., Frankel, R., & Daugherty, P. J. (2002). Information support for alliances:performance implications. *Journal of Business Logistics, 23*(2), 67–82. doi:10.1002/j.2158-1592.2002.tb00026.x

Wortmann, J. C. (1998). Evolution of ERP systems. In *Proceedings of International conference of the manufacturing value chain.* Troon: Kluwer Academic Publishing doi:10.1007/978-0-387-35321-0_2

Zhao, X., Xie, J., & Wei, J. C. (2002). The impact of forecast errors on early order commitment in a supply chain. *Decision Sciences, 33*(2), 251–280. doi:10.1111/j.1540-5915.2002.tb01644.x

KEY TERMS AND DEFINITIONS

Average Earliness of Orders: Aggregate earliness divided by the number of orders.

Average Item Fill Rate: Aggregate fill rate divided by the number of items.

Average Lateness of Orders: Aggregate lateness divided by the number of orders.

Backorder/Stockout: Measures item, order, or product availability performance.

Customer Complaints: Number of customer complaints registered.

Customer Response Time: Amount of time between an order and its corresponding delivery.

Distribution Costs: Total cost of distribution, including transportation and handling costs.

Fill Rate: Proportion of orders filled immediately.

Inventory: Costs associated with held inventory include *inventory investment* (Investment value of held inventory), *inventory obsolescence* (Costs associated with obsolete inventory, sometimes includes spoilage), *work-in-process* (Costs associated with work-in-process inventories), and *finished goods* (Costs associated with held finished goods inventories).

Manufacturing Lead Time: Total amount of time required to produce a particular item or batch.

Number of Backorders: Number of items backordered due to stockout.

Number of Stockouts: Number of requested items that are out of stock.

On-Time Deliveries: Measures item, order, or product delivery performance.

Output: Output measures are based on short, finite time horizons, and address issues such as how many did I produce today? Output measures include: customer responsiveness, quality, and the quantity of final product produced. Many output performance measures are easily represented numerically, such as: number of items produced; time required to produce a particular item or set of items; number of on-time deliveries (orders). However, there are also many output performance measures that are much more difficult to express numerically, such as: customer satisfaction; product quality.

Percent On-Time Deliveries: Percent of orders delivered on or before the due date.

Performance Measurement System: It is the overall set of metrics used to quantify both the efficiency and effectiveness of action.

Product Lateness: Delivery date minus due date.

Profit: Total revenue less expenses.

Resource: Resource measures include: inventory levels, personnel requirements, equipment utilization, energy usage, and cost. Resources are generally measured in terms of the minimum requirements (quantity) or a composite efficiency measure. Efficiency measures the utilization of the resources in the systems that are used to meet the system's objectives. Resource measurement is an important part of the measurement system. Too few resources can negatively affect the output and the flexibility of the system, while the deployment of too many resources artificially increases the system's requirements.

Return on Investment (ROI): Measures the profitability of an organization. The return on investment is generally given by the ratio of net profit to total assets.

Sales: Total revenue.

Shipping Errors: Number of incorrect shipments made.

Stockout Probability: Instantaneous probability that a requested item is out of stock.

Supply Chain Management Performance: SCM performance measurement or monitoring is the term for a set of metrics and processes related to assessing and evaluating how accurate the planning is and how well the execution is carried out.

Target Fill Rate Achievement: To what extent a target fill rate has been achieved.

Total Cost: Total cost of resources used.

Chapter 16
The Role of Institutional Pressures on Green Supply Chain Practices in Building the Organizational Image:
An Empirical Study of Indian Hospitals

Manisha Sharma
Gautam Buddha University, India

ABSTRACT

Supply chain management has been used by businesses and the companies have attributed their success to effective supply chain management. But in case of healthcare sector, supply chain management is not that prominent since it deals with finished products. Supply Chain Management is a concept, strategy and approach that is proving its worth in hospital management all over the world. The pressures on hospital supply chains are changing. In response to these pressures some hospitals have initiated green supply chain management (GSCM) practices that provide tremendous opportunities to improve supply chain performance. This paper tries to provide empirical results in order to examine the impact of GSCM practices in building the positive image of the Indian hospitals incorporating two moderating variables namely regulatory and competition. In order to gain unique insights of current levels of awareness/adoption of GSCM and the potential impact GSCM could have in building organizational image, a survey was conducted among 53 hospitals from NCR (National Capital Region) India. Moderated hierarchical regression analysis has been used to arrive at the following results: (1) Indian hospitals have experienced increasing environmental pressure to implement GSCM practices; (2) The implementation of GSCM practices have helped in building positive organizational image (3) The existence of competitive and regulatory pressures influence hospitals to implement green supply chain practices that eventually help in building positive organizational image especially when these pressures cause adoption of green purchasing and green hospital services even though organizational image is most influenced by Internal environment management; (4) Hospitals implement green transportation policies under regulatory and competitive pressures. Implications for operations strategists from these relationships are also discussed.

DOI: 10.4018/978-1-4666-9795-9.ch016

1. INTRODUCTION

The term 'supply chain' describes the network of suppliers, distributors and consumers. It also includes transportation between the supplier and the consumer, as well as the final consumer... the environmental effects of the researching developing, manufacturing, storing, transporting, and using a product, as well as disposing of the product waste, must be considered. (Messelbeck & Whaley, 1999, p. 42). It is the integration of key business processes from end user through original suppliers that provides products, services, and information that add value for customers and other stakeholders" (Lambert et al., 1998, p. 1).

A textbook on supply chain management (Handfield & Nichols, 1999) has described the supply chain as: The supply chain encompasses all activities associated with the flow and transformation of goods from raw materials (extraction), through the end user, as well as associated information flows. Material and information flow both up and down the supply chain. (p. 2).

One of the most accepted definitions of Supply Chain Management is: Supply Chain Management is the systemic, strategic coordination of the traditional business functions and the tactics across these business functions within a particular company and across businesses within the supply chain, for the purposes of improving the long-term performance of the individual companies and the supply chain as a whole (Mentzer *et al.*, 2001).A customer focused definition is given by Gibson et al. (2005, p. 22) through a survey of SC practitioners. They have defined SCM as "encompassing the planning and management of all activities involved in sourcing and procurement, conversion, demand creation and fulfilment and all logistic management activities.

Supply chain management has been used by businesses and the companies have attributed their success to effective supply chain management. But in case of healthcare sector supply chain management is not that prominent since it deals with finished products. Hospitals have to accept what suppliers offer. Companies have been working on building a research base which will help healthcare leverage supply management. Standardization of products may help healthcare sector benefit from supply chain management. The physicians, suppliers and hospitals need to work in coordination so as to enable healthcare industry benefit from efficiencies related to strategic supply chain management. The pressures on hospital supply chains are changing. In the past, a hospital that managed its purchasing costs well could operate efficiently. Today, the cost of materials management can exceed 35% of a hospital's operating budget, with nearly 20-25% attributable to supply costs alone(Stark & Mangione,2004).

As the consumers become more aware of environmental issues and global warming, they do ask more questions about the products they are purchasing. Companies will have to address questions about how green their manufacturing processes and supply chain are, their carbon footprint and how their recycle process. However some companies have realized that this not a bad thing and have been able to convert the consumer's interest in all things green into increased profits. A number of companies have shown the link between improved environmental performance and financial gains. Companies have started addressing the environmental concerns in the way that may produce profits. While attempting to reduce costs in their supply chain, General Motors found that the cost reductions they identified complemented the company's commitment to the environment. GM reduced disposal costs by $12 million with the establishment of a reusable container program with their suppliers.

Companies can actually save cost by reducing the environmental impact of their business processes. By re-evaluating the company's supply chain, savings are often identified as a benefit of implementing green policies.

Despite the consumer's focus on the environment, benefits attributed to reducing a company's environmental impact are not in the priority of supply chain managers' minds. It appears that many executives are still not aware of the fact that improved environmental performance means lower waste-disposal and training costs, fewer environmental-permitting fees, and, often, reduced materials costs.

A number of possible definitions of GSCM have been put forth over the past decade. One definition of green supply chain management (GSCM) is from (Srivastava, 2007). His study collected and classified previous literatures relating to green supply chain management. He defined GSCM as integrating environment thinking into supply chain management, including product design, material sourcing and selection, manufacturing processes, delivery of the final product to the consumers, and end-of-life management of the product after its useful life. "Green supply refers to the way in which innovations in supply chain management and industrial purchasing may be considered in the context of the environment" (Green et al., 1996, p. 188). "Environmental supply chain management consists of the purchasing function's involvement in activities that include reduction, recycling, reuse and the substitution of materials" (Narasimhan & Carter, 1998, p. 6).

2. GREEN SUPPLY CHAIN MANAGEMENT IN HOSPITALS

A green supply chain in a hospital is a traditional supply chain but with an environmental concern and emphasis on wasted material, wasted energy or effort and under-utilized resources.

The conventional supply chain consists mainly of five parts: raw material, manufacturing, distribution, consumer, and waste.

- **Raw Material:** Hospitals use environmentally harmful materials; however, they can compel suppliers to use more environmentally friendly materials and processes. ISO recommends that any organization which is ISO recognized should do business with other ISO recognized organizations or should ask their partners and suppliers to conform to ISO norms.
- **Manufacturing:** Hospitals do not manufacture goods rather produce services, and the processes followed by them provides multiple opportunities to adopt eco-friendly and procedures.
- **Distribution:** Adopting green practices in distribution may include minimizing packaging materials and stress on "reverse distribution."
- **Waste:** Hospitals must abide by the rules and regulations regarding collection and disposal of hazardous materials.

Health care waste is waste from a healthcare establishment, research facilities and laboratories. To ensure proper hospital hygiene and safety of health care workers and communities, proper management of health care waste is necessary. Usually, waste is disposed of in landfills and despite the intense efforts that are directed to the recycling and recovery of solid wastes, landfills still remain and will remain as part of most solid waste management plans. Health care waste requires special attention as it includes different kinds of wastes such as radioactive, chemical, heavy metals and regular municipal wastes. Sterilization, incineration and land filling are to be considered as the final disposal methods. Land filling of such wastes causes soil and ground water pollution resulting in health hazards. The impacts of direct and indirect exposure to health-care wastes include carcinogenic effects, damage to reproductive, respiratory and central nervous systems. Appropriate management can reduce the generation of infectious wastes to a considerable extent. To design an effective waste management supply chain, a detailed study of current practices, types of wastes produced, types of disposal available and quality improvement procedures, is essential.

Health care professionals, who wish to improve their facilities' environmental profiles, must follow the following procedures and develop criteria for the environmental screening of products.

1. **Purchasing:** Various factors like patient safety, quality, standardization, contractual compliance and pricing are considered before selecting products and equipment used for patient care at Hospitals. Hospitals should develop a hospital-wide environmentally focused purchasing program to give additional preferences to environment friendly materials and suppliers providing them. This program will consider the full life cycle of a product from point of origin to disposal, incorporating assessment of its impact at each step in the life cycle.
2. **Resource Conservation:**
    a. **Energy:** Through a variety of strategies, hospitals should maintain awareness of our energy consumption. To reduce use of natural resources, initiatives should include: Upgrading heating, cooling and ventilation systems to more energy efficient systems; Installing energy-efficient motors, equipment, lighting fixtures, installing compact fluorescent lamps (CFL) and light emitting diode lamps (LED); Conducting annual comprehensive energy audits.
    b. **Water:** Initiative encourages water use reduction through the: Using state-of-art washing machines that reduce water consumption and eliminate the need for drying; Closely analysing and improving water treatment process for the cooling/heating system; Planning landscaping for the hospital's expansion that will conserve water and reduce runoff.

ADVANTAGES TO HOSPITALS

1. Green Supply Chain improves operations by employing an environmental solution: by encompassing environmental friendly practices, hospital waste is minimized, hence productivity is improved.
2. **Conformance to Government Norms:** Government norms for environmental protection are followed by adopting green supply chain practices.
3. **Increased Employee Motivation:** Hospital employees have increased morale for being associated with an eco-friendly organization.
4. **Better Corporate Image:** Hospitals following Green supply chain can boast about it and it will add to their corporate image.
5. **Increases Adaptability:** Green supply chain analysis often leads to innovative processes and continuous improvements.
6. **Promotes Alignment:** Green supply chain management involves negotiating policies with suppliers and customers, which results in better alignment of business processes and principles.

Hospitals play a pivotal role in protecting the health of a nation. But they may also have a surprising unhealthy side - inadvertently contributing to illness and pollution by exposing patients and staff to a witch's brew of toxins from building materials, medical waste, hospital supplies and cleaning products (Landro, 2006). Hospitals generate tons of waste per day, and about 80 percent of that is nonhazardous waste that may be infectious, toxic or radioactive. There are ample opportunities to get greener, one such practice is implementation of green supply chain management and some hospitals really are making concerted efforts to do so. Going green can be a big challenge for hospitals that have to rebuild or retrofit older facilities. This paper attempts to explore the green initiatives taken by Indian hospitals in order to meet this challenge as well as mark them as the responsible organizations.

3. REVIEW OF LITERATURE

Previous literatures have seen mixed results on the relationships between environmental and economic performance and environmental practice adoption. Several studies showed that GSCM practices can improve environmental performance (Frosch 1994, Geffen & Rothenberg 2000). Whereas some studies suggest a positive relationship between economic performance and proactive environmental strategies (Sarkis & Cordeiro 2001).On the other hand other researchers have found negative relationships (Cordeiro & Sarkis 1997, Bowen et al. 2001). The reasons for variation in these findings may be due to the heterogeneity of environmental management practices adopted by the firm and industry (e.g. see Elsayed & Paton 2005). This heterogeneity is also represented in the types of pressures that organizations face and can be further explained by Institutional Theory (Gonzalez-Benito & Gonzalez-Benito 2006).

Zhu and Sarkis (2004) concluded that GSCM practice has a positive relationship with environmental performance, positive economic performance, and negative economic performance. Positive economic performance is defined as benefits gained through GSCM while negative economic performance is defined as increased investment and costs.

Zhu and Sarkis (2007) investigated the moderating effects of institutional pressures on emergent green supply chain practices and performance and concluded that there is an increased environmental pressure to implement GSCM practices and also found out that the existence of institutional pressures influence organizations to have improved environmental performance.

This paper tries to examine the moderating effects of pressures for the relationship between GSCM practice adoption and performance. The model posits that Indian hospitals meeting environmental pressure have implemented environmental practice to improve organizational image, which is shown in figure 1.

The research framework, shown in Fig. 1, is developed to investigate the relationships between four GSCM practices (Green purchasing, Transportation, Hospital services and Internal environmental

Figure 1. Research framework for investigation of pressure moderating effects on relationships between GSCM practices and Organizational image in Indian hospitals

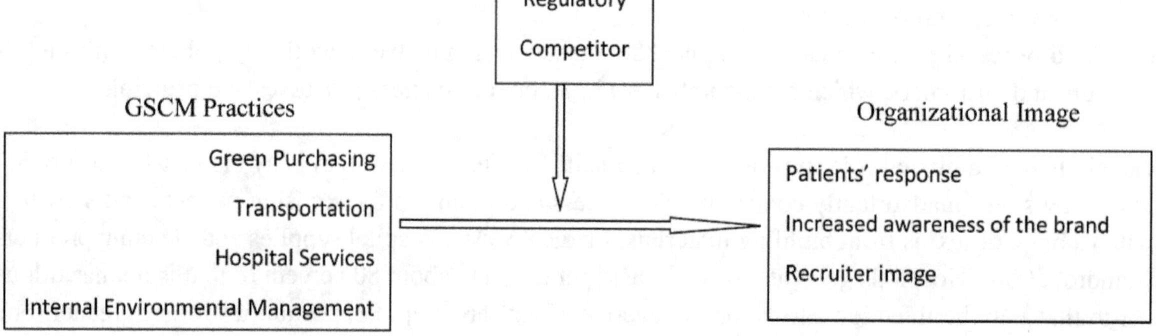

management as supply chain practices in Hospitals. GSCM practices dimensions and items are based on previous literature that addressed various elements of GSCM (e.g. Zhu & Sarkis, 2007.,Carter et al., 1998; Zsidisin & Hendrick, 1998; Walton et al., 1998; Zhu and Cote, 2002) as well as inputs from supply chain managers from various hospitals. Internal environmental management is a key to improving enterprises' performance (Carter et al.,1998). HSCM is a set of approaches to efficiently integrate suppliers or vendors, transport, hospital services (including outpatient, emergency, in-patient, laboratory, radiology, stores and purchase, food, laundry and medicines/equipment) to achieve Total Quality Management (TQM) in health care services by optimum utilization of resources. The HSCM deals with the management of materials/equipments, manpower like doctors, nurses and other supporting staff, to transform sick patient to healthy person at a reasonable cost, in the shortest possible time with utmost satisfaction of the patient (Parray & Kadri, 2007).

Zhu and Sarkis (2007) identified major institutional pressures: market, regulatory, and competitive pressures as three moderators in manufacturing industry. Both non-market and market pressures have moderated relationships between environmental practices and their performance but in varying ways (Hoffman & Ventresca 1999).Jennings and Zandbergen (1995) introduced institutional theory to help explain firms' adoption of environmental management practices. They found out that coercive forces, through regulations and regulatory enforcement, have been the main pressures for environmental management practice adoption, firms within an industry have implemented similar practices. There are also heterogeneous influences of coercive pressures exerted upon different industries, which has led to organizational variations in environmental strategies (Levy & Rothenberg 2002). Related to GSCM issues, Min and Galle (1997) discovered that both state and federal environmental regulations are two key factors affecting a buying firms' green purchasing practice.

Some studies have shown that competitive factors may play a more prevalent role for corporate environmental response (Kagan et al. 2003). For example, Bergh (2002) discovered that firms were most strongly influenced to learn and respond to natural environment issues by paying close attention to other firms such as their competitors and not from their own experiences, legal or regulatory coercive pressures. Competitors have also exerted pressure for manufacturers to implement environmental purchasing (Carter & Carter 1998).

Organizations with long-term international experience, such as those in the electronic industry, have felt higher pressures from competitors to improve their environmental practices, and have more frequently promoted environmental practices with commensurate economic benefits (Hui et al. 2003).

Unfortunately, direct and rapid economic performance gains through environmental practice will probably take longer to envision (Bowen et al. 2001). As a result, Hospitals may suffer short-term economic losses when seeking to improve their environmental performance. However, in the long term, one can argue better environmental practices due to competitive pressure can bring better economic performance (Zhu & Geng 2001).

Additional pressures from competition and support from industrial professional groups, can both lead to adoption of environmental practices and achieve improved organizational image. Resultant economic losses are incurred when organizations face competitive pressures to adopt environmental practices, due to increased investments and shifting resources but positive organizational image will eventually lead to economic gain and benefits that will be relatively strategic implying long-term, rather than short-term gains.

This leads to the construction of the following hypothesis:

Hypothesis: The impact of GSCM practices as moderated by institutional regulation and competition affect organizational image positively than there in no impact of GSCM practices as moderated by institutional regulation and competition on organizational image.

4. METHODOLOGY

4.1. Instrument Development

To test the hypotheses, a survey instrument had been developed with 25 questions for GSCM practice, 3 items for performance, and six items for institutional pressures as moderating variables. The 25 GSCM practice items are based on input from industrial experts and the literature (Walton et al. 1998, Zsidisin and Hendrick 1998, Carter et al. 2000, Parray and Kadri, 2007).

The responses had been taken from 70 hospitals through a structured questionnaire survey across NCR (National Capital region) India. However analysis was done with 53 samples only as 17 hospitals had to be dropped because of the response bias and extreme value effect. NCR had been chosen for the data collection as Delhi is the capital city of India in addition to being a metropolitan. Hospitals for the survey were selected using convenient sampling technique. However, care was taken to ensure coverage of small to big hospitals located at every corner of NCR. Hospital size was determined in terms of number of beds it had that varied from the range of 50 beds up to 650 beds:

GSCM pressure survey items:

- Factors
  - Regulatory:
    1. Central governmental regulation.
    2. State government environmental regulation.
  - Competition:
    1. Competitor's green strategies
    2. Industrial professional group activities

4.2. Control Variable

Since large hospitals typically face higher environmental pressures, and are typically required to implement better environmental practices to improve environmental performance, hospital size has been taken as a control variable. Size is measured as the natural logarithmic transformation of the number of full-time employees (Dean & Snell 1991).

4.3. Analysis

4.3.1. Descriptive Statistics

The descriptive statistics provide the following information: As far as Green Supply chain practices are concerned, hospitals focus most on green purchasing (mean=3.015) followed by green hospital services (mean=2.74). Hospitals abide by regulatory norms (mean=4.035)and do give a lot of importance to competitor's green initiatives while framing their own (mean=3.85). The GSCM practices of hospitals have culminated into very positive response from the patients (mean=3.06). GSCM practices do help in increasing awareness of the hospital brand as well as improve the image of the hospital as a recruiter but the impact has been quite moderate.(mean=2.64 and 2.58 respectively). These statistics show very positive results proving the fact that hospitals have been very forthcoming in their green initiatives (Table 1).

4.3.2. Moderating Hierarchical Regression Analysis

To understand the impact of the moderating variables- regulatory and competitive on green supply chain management practices in order to build the positive image of the hospitals, the moderating hierarchical regression analysis has been used.

Table 2 presents the results for the interaction effects between the two moderating pressures and green purchasing. The results show that the impact of the green purchasing when moderated by institutional regulation in building the organizational image is highly significant (p=0.013 for positive patients' response, p=0.007 for increased awareness of the brand and p =0.007 for improved image as a recruiter). The interaction term of green purchasing and institutional regulation has a significant positive beta for environmental performance which supports the above analysis. The impact of the green purchasing when moderated by competition in building the organizational image is also significant (p=0.049 for positive patients' response, p=0.007 for increased awareness of the brand and p =0.007 for improved image as a

Table 1. Descriptive statistics for GSCM practice, performance and pressure

Descriptive Statistics			
	N	Mean	Std. Deviation
GSCM Practices Green Purchasing	53	3.015	1.227
Transportation	53	2.25	0.806
Hospital Services	53	2.74	0.999
Internal Environment Management	53	2.25	1.1.08
Moderators Regulatory	53	4.035	1.096
Competition	53	3.85	1.221
Organizational image Positive Patients' Response	53	3.06	1.151
Increased awareness of the brand	53	2.64	0.982
Improved image as a recruiter	53	2.58	1.322
Valid N (listwise)	53		

Table 2. Hierarchical moderated regression analysis

	Positive patients' response			Increased awareness of the brand			Improved image as a recruiter		
	b	SEb	β	b	SEb	β	b	SEb	β
Step 1									
Green purchasing	0.85	0.18	0.71	0.93	0.20	0.69	0.69	0.15	0.68
Step 2									
Green purchasing× Regulatory	0.51	0.20	0.43	0.66	0.23	0.49	0.52	0.18	0.51
Step 3									
Green purchasing× competition	0.35	0.17	0.30	0.48	0.22	0.36	0.48	0.17	0.47
Adjusted R^2	0.30	0.39	0.56	0.32	0.34	0.48	0.27	0.28	0.43

recruiter). The interaction term of green purchasing and competition also has a significant positive beta for environmental performance which supports the above analysis. However the moderation impact of competition is less than that of institutional regulation.

Table 3 presents the results for the interaction effects between the two moderating pressures and green transportation services. The results show that the impact of the transportation services when moderated by institutional regulation in building the organizational image is not significant (p=0.062 for positive patients' response, p=0.019 for increased awareness of the brand and p =0.425 for improved image as a recruiter). The impact of the green purchasing when moderated by competition in building the organizational image is also not at all significant (p=0.265 for positive patients' response, p=0.131 for

Table 3.

	Positive patients' response			Increased awareness of the brand			Improved image as a recruiter		
	b	SEb	β	b	SEb	β	b	SEb	β
Step 1									
Transportation	0.66	0.18	0.46	0.84	0.20	0.51	0.17	0.17	0.14
Step 2									
Transportation× Regulatory	0.37	0.19	0.30	0.53	0.22	0.31	-0.16	0.18	-0.12
Step 3									
Transportation× competition	0.23	0.20	0.16	0.35	0.23	0.21	-0.23	0.19	-0.19
Adjusted R^2	0.20	0.30	0.40	0.25	0.33	0.44	0.00	0.16	0.30

Table 4.

		Positive patients' response			Increased awareness of the brand			Improved image as a recruiter		
		b	SEb	β	b	SEb	β	b	SEb	β
Step 1										
	Hospital Services	0.83	0.11	0.95	0.61	0.13	0.61	0.38	0.13	0.51
Step 2										
	Hospital Services× Regulatory	0.82	0.12	0.93	0.58	0.14	0.58	0.37	0.14	0.50
Step 3										
	Hospital Services× competition	0.75	0.13	0.85	0.48	0.15	0.48	0.28	0.16	0.38
	Adjusted R^2	0.72	0.71	0.71	0.66	0.68	0.70	0.46	0.44	0.43

increased awareness of the brand and p =0.226 for improved image as a recruiter). The above analysis is also supported by beta value which is either not a significant positive value or it is negative as in the case of improving image as a recruiter.

Table 4 presents the results for the interaction effects between the two moderating pressures and green hospital services. The interaction term of eco-design and marketing has a significant positive beta for environmental performance. The results show that the impact of the hospital services when moderated by institutional regulation in building the organizational image is highly significant (p=0.000 for positive patients' response, p=0.000 for increased awareness of the brand and p =0.012 for improved image as a recruiter). The impact of the green purchasing when moderated by competition in building the organizational image is also very significant (p=0.000 for positive patients' response, p=0.003for increased awareness of the brand and p =0.078 for improved image as a recruiter). The above analysis is also supported by significant beta values. However the moderation impact of competition is less than that of institutional regulation.

Table 5 presents the results for the interaction effects between the two moderating pressures and Internal environmental management. The results show that the impact of the Internal environment management when moderated by institutional regulation in building the organizational image is highly significant (p=0.000 for positive patients' response, p=0.000 for increased awareness of the brand and p =0.000 for improved image as a recruiter). The impact of the green purchasing when moderated by competition in building the organizational image is also highly significant (p=0.000 for positive patients' response, p=0.000for increased awareness of the brand and p =0.000 for improved image as a recruiter). The significant beta values do support the above analysis. Both the moderating variables play equally important role to manage internal environment and eventually building the organizational image.

The above results support the hypotheses that the positive relationship between GSCM practices and organizational image is stronger in Hospitals facing higher regulatory and higher environmental competitive pressure than in Hospitals facing lower regulatory and lower environmental competitive

Table 5.

	Positive patients' response			Increased awareness of the brand			Improved image as a recruiter		
	b	SEb	β	b	SEb	β	b	SEb	β
Step 1									
Internal Environment Management	0.85	0.08	0.82	0.96	0.09	0.81	0.64	0.09	0.72
Step 2									
Internal Environment Management× Regulatory	0.79	0.10	0.76	0.94	0.12	0.78	0.62	0.10	0.70
Step 3									
Internal Environment Management× competition	0.71	0.10	0.68	0.82	0.11	0.68	0.55	0.11	0.62
Adjusted R^2	0.67	0.66	0.70	0.68	0.69	0.73	0.51	0.51	0.54

pressure respectively. However regulatory factor seems to influence GSCM practices in hospitals little more than the competitive factor.

The results presented below indicate that the pressures: regulatory, and competitive have some moderating effects for certain GSCM practices and in improving organizational image. In the discussion of the results, the focus is on the reasons why certain significant characteristics exist. It is seen that if regulatory pressures may cause hospitals to adopt the green transportation policies, the organizational image improves for the better. Regulatory pressure existence next seems to influence next to green purchasing even though organizational image is most influenced by Internal environment management. This may be because Internal environment management is the least visible among all the four GSCM practices by the regulatory bodies and competitors. It also indicates that Hospitals implement green purchasing as a result of more of regulatory pressures rather than their own initiatives. This result also has policy implications. Making green purchasing as part of regulatory policy, as done by the Indian government by introducing ISO 14001 certification, will improve organizational image, without making a difference in terms of its impact on economic performance in the long run. The hospitals should be encouraged having suppliers possessing the above certification.

5. CONCLUSION AND FURTHER SCOPE

GSCM has become an emerging environmental practice for hospitals to improve environmental image. Hospitals have initiated GSCM practices. Though the adoption level is still relatively immature, the positive relationships to environmental and organizational image are significant. From this study, it can be summarized:

1. Implementation of GSCM practices have helped in building positive organizational image and continue to do so specifically in terms of patients' response to these initiatives.

2. Competitive pressure existence strongly improves the organizational image for a number of GSCM practices. The implication of this result is that having trade association and benchmarking activities (sharing of ideas) will help in implementing a variety of GSCM practices.
3. The existence of regulatory pressures influence hospitals to have better environmental performance, especially when these pressures cause adoption of green purchasing and green hospital services even though organizational image is most influenced by Internal environment management. However, hospitals services in order to be green will require additional investments. The implications are quite clear here; there are situations for certain GSCM practices where it might be prudent for supply chain managers and strategists to champion certain practices down their supply chain to improve internal environmental management, but consequences of doing so, such as additional expenses, will occur for green hospital services and these supply chain members may require some form of economic support or economic encouragement to adopt these practices.
4. Hospitals should face higher regulatory pressures to better implement green transportation practices. Existence of regulatory pressures may cause Indian hospitals to improve their transportation services. The implication of this result is that regulatory policy-makers should push green transportation services, while providing help for hospitals in order to educate them about the long term economic benefit of such practice. Transportation managers in situations with significant regulatory pressure should be made aware of the need to implement GSCM practices.
5. Indian hospitals have experienced increasing environmental pressure to implement GSCM practices however hospitals should be encouraged to share their learning experiences among each other and formation of ‘‘environmental forums’’ can prove quite beneficial in order to have a connect with the citizens that may be their future customers as patients’ response has been very positive towards these initiatives.

This paper examines whether specific institutional pressures can moderate specific GSCM practice and help in the improvement of the organizational image. The theory is advanced by furthering the evidence of the existence of heterogeneity of pressures and their influences on specific GSCM practices. Not all institutional pressures influence in building the organizational image in the same way and that careful consideration of the operational practices will be important for managers in these various situations.

There are some limitations to this study – it has not been fully evaluated why and how many of these isomorphic institutional pressures exist. Also, hospital size may affect most relationships of main effects between GSCM practice and performance as well as the moderating effects of three pressures. That is, different pressures may influence organizations of differing sizes – whereas regulatory pressures may influence larger organizations, smaller organizations may be more influenced by competitive pressures. Future research can group hospitals based on firm size to determine if the moderating effects differ. Investigation for this relationship is warranted but will require additional data acquisition and theoretical justification.

In addition, this paper focuses on organizational image. Future research can focus on environmental and economic performance measures such as reduction of air emission, waste water and solid waste as well as economic performance such as share price, market share, return on assets, etc. It has been identified that some GSCM practices/performance relationships are greatly influenced by the existence of these pressures, while others see less influence. Further investigation of this phenomenon is a direction for potentially insightful contribution to the corporate and academic literature.

REFERENCES

Bergh, J. (2002). *Do social movements matter to organizations? An institutional perspective on corporate responses to the contemporary environmental movement.* (PhD thesis). The Pennsylvania State University, State College, PA.

Bowen, F.E., Cousins, P.D., Lamming, R.C. & Faruk, A.C. (2001). Horse for courses: explaining the gap between the theory and practice of green supply. *Green. Manag. Int.*, 41–60.

Carter, R. C., & Carter, J. R. (1998). Interorganizational determinants of environmental purchasing: Initial evidence from the consumer products industry. *Decision Sciences, 29*(3), 28–38. doi:10.1111/j.1540-5915.1998.tb01358.x

Cordeiro, J., & Sarkis, J. (1997). Environmental proactism and firm performance: Evidence from security analyst forecasts. *Business Strategy and the Environment, 6*(2), 104–114. doi:10.1002/(SICI)1099-0836(199705)6:2<104::AID-BSE102>3.0.CO;2-T

Dean, J. W., & Snell, S. A. (1991). Integrated manufacturing and job design: Moderating effects of organizational inertia. *Academy of Management Journal, 34*(4), 776–804. doi:10.2307/256389

Delmas, M., & Toffel, M. W. (2004). Stakeholders and environmental management practices: An institutional framework. *Business Strategy and the Environment, 13*(4), 209–222. doi:10.1002/bse.409

Elsayed, K., & Paton, D. (2005). The impact of environmental performance on firm performance:static and dynamic panel data evidence. *Structural Change and Economic Dynamics, 16*(3), 395–412. doi:10.1016/j.strueco.2004.04.004

Frosch, R. (1994). Industrial ecology: Minimizing the impact of industrial waste. *Physics Today, 47*(11), 63–68. doi:10.1063/1.881405

Geffen, C., & Rothenberg, S. (2000). Suppliers and environmental innovation: The automotive paint process. *International Journal of Operations & Production Management, 20*(2), 166–186. doi:10.1108/01443570010304242

Gibson, B. J., Mentzer, J. T., & Cook, R. L. (2005). Supply chain management: The pursuit of a consensus definition. *Journal of Business Logistics, 26*(2), 17–25. doi:10.1002/j.2158-1592.2005.tb00203.x

Gonzalez-Benito, J., & Gonzalez-Benito, O. (2006). The role of stakeholder pressure and managerial values in the implementation of environmental logistics practices. *International Journal of Production Research, 44*(7), 1353–1373. doi:10.1080/00207540500435199

Green, K., Morton, B., & New, S. (1998). Case study green purchasing and supply policies: Do they improve companies' environmental performance? *Supply Chain Management, 3*(2), 89–95. doi:10.1108/13598549810215405

Handfield, R. B., & Nichols, E. L. (1999). *Introduction to Supply Chain Management.* Upper Saddle River, NJ: Prentice Hall.

Hoffman, A., & Ventresca, M. (1999). The institutional framing of policy debates. *The American Behavioral Scientist, 42*(8), 1368–1392, 1. doi:10.1177/00027649921954903

Hui, I. K., He, L., & Dang, C. (2003). Hierarchical environmental impact evaluation of a process in printed circuit board manufacturing. *International Journal of Production Research, 41*(6), 1149–1165. doi:10.1080/0020754021000043679

Kagan, R. A., Gunningham, N., & Thornton, D. (2003). Explaining corporate environmental performance: How does regulation matter? *Law & Society Review, 37*(1), 51–90. doi:10.1111/1540-5893.3701002

Lambert, D. M., Cooper, M. C., & Pagh, J. D. (1998). Supply chain management: Implementation issues and research opportunities. *International Journal of Logistic Management, 9*(2), 1–16. doi:10.1108/09574099810805807

Landro, L. (2006, October 4). Hospitals go 'green' to cut toxins. *The Wall Street Journal.*

Levy, D. L. (1995). The Environmental practices and performance of transnational companies. *Trans. Corpor., 4*, 45–67.

Mentzer, J. T., DeWitt, W., Keebler, J. S., Soonhoong, M., Nix, N. W., Smith, C. D., & Zacharia, Z. G. (2001). Defining supply chain management. *Journal of Business Logistics, 22*(2), 1–25. doi:10.1002/j.2158-1592.2001.tb00001.x

Messelbeck, J., & Whaley, M. (1999). Greening the health care supply chain: Triggers of change, models for success. *Corporate Environmental Strategy, 6*(1), 39–45.

Min, H., & Galle, W.P. (1997). Green purchasing strategies: trends and implications. *Int. J. Purch. Mater. Manag.*, 10–17.

Narasimhan, R., & Carter, J. R. (1998). *Environmental Supply Chain Management.* Tempe, AZ: The Center for Advanced Purchasing Studies, Arizona State University.

Parray, S. H., & Kadri, S. M. (2007). Supply chain management in healthcare sector-role of logistics. *Indian Journal for the Practicing Doctor, 4*(1).

Sarkis, J., & Cordeiro, J. J. (2001). An empirical evaluation of environmental efficiencies and firm performance: Pollution prevention versus end-of pipe practice. *European Journal of Operational Research, 135*(1), 102–113. doi:10.1016/S0377-2217(00)00306-4

Srivastava, S. K. (2007). Green supply chain management: A state-of-the-art literature review. *International Journal of Management Reviews, 9*(1), 53–80. doi:10.1111/j.1468-2370.2007.00202.x

Stark, E., & Mangione, T. (2004). Top issues facing the hospital supply chain today. *Healthcare purchasing news.* FindArticles.com.

Walton, S.V., Handfield, R.B., & Melnyk, S.T. (1998). The green supply chain: Integrating suppliers into environmental management process. *Int. J. Purch. Mater. Manag.*, 2–11.

Zhu, Q., & Geng, Y. (2001). Integrating environmental issues into supplier selection and management: a study of Large and Medium-sized State-owned Manufacturers in China. *Green. Manag. Int.*, 27–40.

Zhu, Q., & Sarkis, J. (2004). Relationships between operational practices and performance among early adopters of green supply chain management practices in Chinese manufacturing manufacturers. *Journal of Operations Management, 22*(3), 265–289. doi:10.1016/j.jom.2004.01.005

Zhu, Q., & Sarkis, J. (2007, September 15). The moderating effects of institutional pressures on emergent green supply chain practices and performance. *International Journal of Production Research*, *45*(18–19), 4333–4355. doi:10.1080/00207540701440345

Zhu, Q. H., & Cote, R. (2002). Green supply chain management in China: how and why? *The Fifth International Eco-city Conference*, Shenzhen, China.

Zsidisin, G. A., & Hendrick, T. E. (1998). Purchasing's involvement in environmental issues: a multi-country perspective. *Industrial Management & Data Systems*, *7*(7), 313–320. doi:10.1108/02635579810241773

Chapter 17
Understanding the Port-Centric Logistics Clusters:
Concepts, Characteristics, and Measurements

Amanpreet Singh
RMIT University, Australia

Prem Chhetri
RMIT University, Australia

Rajiv Padhaye
RMIT University, Australia

ABSTRACT

Port-centric logistics clusters are considered as intermodal gateways and/or hubs of international trade, which connect national economies with global production networks. These clusters are spatial agglomeration of logistics related interconnected and interdependent companies. Singapore, Dubai and Rotterdam are world class exemplary of port-centric logistics clusters. The formation of these clusters is a derivative of conducive business environment generated within a geographically defined area. Despite the recognition of port-centric approach to economic development there is insufficient evidence to empirically assess the functional and spatial characteristics of port centric logistics clusters. There is also disagreement on three key questions: how port-centric logistics clusters are defined and identified, what industry types do they constitute and what methods are appropriate to delineate the boundary of port centric logistics cluster. In this paper, a spatial approach is adopted to geographically delineate the spatial congregation of port-centric logistics employment using three major container ports in Australia. Using the Census data from the Australian Bureau of Statistics (ABS 2006), analysis has been conducted on employment data containing information about where people work and what industry they work within the close vicinity of case study seaports. The results show that the spatial extent Australian port centric logistics clusters, which tend to vary both in size and shape. Overall size of employment in port-related industries in Australia has grown substantially. Road freight is dominating industry in the port-centric logistics cluster with a contribution of 23.02% and 37.54% in 2001 and 2006 respectively. The range of port-centric logistics cluster in Melbourne is towards western suburbs and in Botany Bay the port-

DOI: 10.4018/978-1-4666-9795-9.ch017

centric logistics cluster is discerned mainly around eastern suburbs and in Brisbane the spread of port-centred logistics cluster is towards southeast inner Brisbane and Northwest Inner Brisbane suburbs. This shows the impact of land use consolidation by the State Government in their effort to boost transport and warehousing employment closer to Australian container ports. The establishment of port-centric logistics clusters, we argue, could mean the opportunities for organisations to achieve agglomeration economies, increase rivalry among organizations to promote competition, access to greater pool of customers, availability of skilled labour force, closer proximity between customers and supplier, sharing of public infrastructure and resources, increased inter-firm interactions, and knowledge spill-over.

1. INTRODUCTION

Two competing economic forces in the modern world era are 'globalization', which seems to have increased the geographic extent of economic activities, and 'localization' which brings the inter-related industries within a geographically contained area. These forces are also called centripetal and centrifugal forces, which either cluster or disperse economic activities. World trade has grown almost 27 folds, in volume terms, from 1950 to 2006 and the main reason is globalization (WTO 2007). Where globalization supresses the regional distinctiveness, the localization does exactly opposite by attracting businesses to operate within a cluster. Globalization has opened new markets to stimulate trade by removing most of the physical, political, and economical barriers between the nations so industries have the choice to co-locate their production or logistics activities in low-cost locations, which in turn help achieving economies of scale by accessing a larger market. This increased production enhances international trade, outsourcing and offshoring, which create linkages or interconnections among nations to exchange goods and services. The consequence of localized markets has increased the spatial and time distance between manufacturers and consumer. Billions tonnes of freight is traded throughout the world using various modes of transportations in which maritime transport is most preferred mode due to lower transport cost and mass transportation of freight, and better consolidation, though it takes longer transit times (Coyle et al.1996).

Australian shipping and port industry has existed some 200 years ago. Australia, being an island nation needs to consider the maritime competencies to enhance trade, which helps supporting economic growth (Maritime Work Force Development Forum 2013). In Australia almost 99% of international trade is carried by sea which is worth 74% of value (BITRE 2011-12). Investigating the maritime industry behaviour is a very important aspect as most of the trade is transported by sea and almost 1167 million tonnes of cargo moved across the Australian ports in Year 2011-12, which shows the increase of 8.4% from previous financial year with an average growth of 6.5% annually over the last 5 years. The total weight comprises of exports of 973.2 and 94.9 million tonnes of imports (BITRE 2011-12). An Australian government is putting lot of effort to support maritime industry by having more efficient regulation, better business structure and dynamic workforce (Maritime Work Force Development Forum 2013). In year 2010-11 coastal shipping in Australia contributed to around 20% of the total domestic freight which was approximately 101 billion tonnes-kilometres. There are lots of opportunities in Australian shipping not just because of considerable growth in incoming non-bulk containers but also progress in exports which is growing at a rate of 10.7% on an average, by value. This growth in imports and exports will lead the throughput to double in next 10-15 years. It is therefore vital to develop a better understanding of the way transport and logistics value adding services are congregate around key strategic hubs such as ports and airports.

The concept of cluster is not new as it has existed since almost a century ago but it gained greater importance as a planning tool or business strategy over the last two decades. Port-centric logistics cluster is an inter-linkage between firms and organizations by establishing commonalities and complementariness, and to increase the value add in supply chains by co-locating in geographically concentrated area. Firms in the cluster achieve logistics process integration and economic advantage through agglomeration, economies of scale, information exchange, and resource sharing. A cluster-based approach has been regarded as potentially the most effective mechanism to enhance regional competitiveness. Looking at the increased focus on delivered outcomes, of being situated in a cluster, has led us to investigate the application of cluster concept in the Australian port logistics industry. The aim of this paper is therefore to identify the economic activities/industry types that define port centric logistics clusters, in addition to delineating the geographic boundary of port logistics agglomeration.

This paper therefore answers specifically the following two questions:

RQ 1: *What industries are contained within the Australian port centric logistics clusters?*
RQ 2: *What techniques can be employed to delineate the geographic extent of Australian port-centric logistics clusters?*

This paper is organized as follows. The first part describes the cluster concept and concept definition from various viewpoints. The next section presents a review of the prior studies in relation to industrial, business and maritime/port logistics cluster concept and gained benefits of working in geographically concentrated area. Subsequently, research methodology is outlined followed by the survey analysis results. Finally, the conclusion and the aims for future research are presented.

2. THE ROLE OF SEAPORTS IN THE MODERN GLOBAL SUPPLY CHAIN

Seaports play an important role in international trade. They are key strategic intermodal nodes, which drive the efficient functioning of global supply chains both upstream and downstream. Ports are considered to be a critical link in the supply chain as all freight and cargo passes through it. Cluster-based approach is one such strategy, which are of paramount importance for enhancing competency of port services and operations. According to Carbone and Martino (2003), ports are the natural sites for transhipment where goods are moved from one mode to another. Ports mainly provide a connection between maritime and inland transport and plays an important role in managing the freight and flow of information as in both imports and exports the freight has to move through ports (Chhetri et al. 2014). For example, Goss (1990) states that seaports acts as an important connecting node in a national transportation network. As a seaport expedites the exchange of goods in various industry supply chain, and thus, helps enhancing the economic well-being of a nation. Goss also suggested that ports competition which may include competition between ports in different countries, ports within same countries, between operators on the same port, is increasing which assists in enhancing the facilities and infrastructure for value added activities on the port.

Mangan et al (2008) have put forward the concept of port centric logistics where focus of seaports has been changed from traditional view of loading and unloading towards providing more value added services. So, being an important member of supply chain the contemporary role of seaports can be viewed as cluster of organizations/industries where various logistics and transport resources are used collaboratively in order to give value to the final customer (Panayides 2008). Nowadays, industries and industrial activities are more inclined towards spatially concentrated locations and over the last few years the trend supports the revival of localized growth spatially (Gordon and McCann 2000).

One of the strategic recommendations put forward by "the Maritime Workforce Development Forum" in Australia is to establish Australian maritime clusters (Maritime Work Force Development Forum 2013). This strategy with some others important strategies like target on training new entrants, monetary support from government for training, and collaboration with other sectors, is made to secure the future of Australian maritime industry. The breadth of port logistics cluster includes all the firms which are directly (Core Activities) and indirectly related (supporting activities) to the port logistics. To take an initiative in this path *the shipping reform (Tax incentives) Act 2012* has started encouraging companies to operate within a cluster (form and work in a cluster) to be eligible for a range of tax exemptions.

3. THEORETICAL FOUNDATION OF PORT-CENTRIC LOGISTICS CLUSTERS

The concept of port-centric logistics cluster is theoretical founded in the cluster theory. The concept of cluster has been explained by various scholars that include Alfred Marshall (1890), defined cluster as an *industrial districts, externalities and agglomeration*, Paul Krugman (1991) defined cluster as *geographical economies*, Porter (1998) defined the cluster by creating a link between *location and enhanced competition*, Martin and Sunley (1996) explained the cluster concept on *geographical concentration*. The concept of cluster revolves around four theories given by aforementioned scholars which are *agglomeration economics, industrial districts, spatial economics, and economic geography.*

The concept of cluster was originally put forward under the term '*Industrial districts*' by Alfred Marshall in 1890. He explained the phenomenon of cluster as 'the concentration of specialized firms in particular localities' (Marshall 2004). He outlined that the pool of industries achieve economies of scale and get special advantages, such as skilled labour, specialized suppliers and expert knowledge spill over, which normally a big company can attain due to a huge investment.

Later, Michael Porter (1993) developed a competitiveness theory in 1990 and presented in his book, 'Competitive Advantage of Nations' where he defines the cluster on the basis of economic behaviour. Porter is one of most important contributor to the "*cluster theory*". Porter interpretation of *clusters* is based on a deep link between *location and competition*. Porter (1993) comprehended that firms compete and not the nations but presence of competing clusters is a main factor of nation's competitiveness. Porter (1993) suggests that the reason of success of the firms in a particular location /nation is lying under four major components/attributes which are: Factor Conditions, Demand Conditions, Related and Supporting Industries, Firm strategy, structure and rivalry. Though four attributes quoted by Porter plays an important role but "space" should also not be neglected for a successful cluster formation. The main point stated by Porter regards the role played by *industry localization* in contemporary economics. The new affirmative economic agenda should focus in enhancing the role of clusters, as this form of geographic concentration appear to be the way for improving the local economy and hence lead the way for better economic conditions. In a world where the economy is globalised, Porter underlines the role of clusters in creating a positive competition within firms that stimulates productivity and economic growth. (Martin, Sunley (2001) report that Porter use of the term 'cluster' represent a set of industries that are horizontally (common customer, technology and channels) and vertically (buyer and supplier) linked. Krugman (1998) has the merit to draw *"economics and geography"* closer. He proposed the theory where he focusses on importance of internal geography of countries in shaping the performance of international trades. He acknowledges that it is necessary to go beyond the traditional economic theory according to which international trades are driven by differences in resources among countries. In this regard, Krugman focusses his attention on what happens on regional and local level by considering the role of *increasing returns*, and *economies of scale, technological conditions and policies*. Chhetri et al.(2014) in their study on spatial logistics employment cluster specifically taking the case of Melbourne has identified the pattern of spread of employment related to logistics sector. Their research focusses mainly on the identification of industries involved in logistics sector and how are they spatially integrated in few locations, thus translating into a spatial cluster. They conclude that operating business within a locale will help minimising transportation cost, storage cost and impact on last mile problem.

According to Haezendonck (2001) port cluster is a set of interdependent firms which are engaged in port related activities, located at the same part of the region, with common or similar strategies that lead to competitive advantage over external competition. According to Zagkas (2010), clusters are the joining point of three clusters which are maritime cluster, industrial cluster and logistics cluster in the area.

The companies working in a cluster are agglomerated by their similarities and complementariness. The population in a cluster entails business, associations, public and private organizations (Langen 2002). The European networks are considered to be the first in initiating the clusters in maritime industry. European ship owners controls over 40% of world's fleet (European Commision 2008). Enright (Haezendonck 2001) stated that in the modern clusters there is an integration of dissimilar companies as oppose to traditional ideas of clusters having horizontal linkages only. He said the effect of integrating dissimilar companies is to produce new or reengineered product and services which is cross-sectional in nature

There are different types/models of clusters on the basis of their historic background, their characteristics, and relationship with their counterparts. Different countries have different clusters and they are shaped by R&D activities, knowledge base, transactional relation, size, and innovation etc. The clusters can be differentiated on *trade-driven* and *knowledge-driven* clusters. In trade driven clusters companies will pool for better business opportunities and economic gains. In knowledge driven, the companies will share knowledge to innovate new products and services using shared R&D resources. In knowledge driven clusters companies will pool horizontally which compete otherwise but share resources for New product development.

According to Altenburg et al (1999) there are few similarities in a cluster which are stated below

- Upstream and downstream linkage between organizations.
- Information exchange between the integrated firms.
- Common focus and aim of value creation.
- Diversified infrastructure supporting activities of cluster.

There are many renowned clusters in the world such as Silicon valley in California, which is a successful agglomeration of computer related companies (economics, computer and communication industries) emerged in mid to late 1990. Fairchild Semiconductors was the first notable semiconductor company which gained a prominence over its peers. After emergence as a leader many other semiconductor firms began establishing in the valley, which resulted in knowledge spillover, labor pool and vast range of suppliers and customers which helped in achieving the growth and assisted other semiconductor companies to enter into that geographical concentrated area. Hence a cluster was formed. Another example is port cluster in Singapore which is situated in Southern tip of the Malaysian peninsula. Having very low population of around 5 million people, the economy and standard of living is higher than that of the United States. Singapore does not own natural resources but location of makes it important. Singapore prospers on transshipment and a fifth of the world's maritime containers are transshipped from Singapore. It is one of the World's busiest transshipment port with 85% of the containers that come to port never officially enter into the country and more than half of rest of 15% eventually leave Singapore as re-export. The large part of nation's economy depends upon the logistics activities built around seaport and airport. In 2014, Singapore ranks 5th in the world on logistics performance index scale with a score of 4.00 where Germany is on the top with a score of 4.12 (Arvis *et al* 2014). The example of other well established are known clusters may include wine cluster in California and Adelaide, Maritime cluster in Rotterdam, Dubai, Singapore, and Shanghai, financial cluster in London, Fashion cluster in Italy, Electronic cluster in Taiwan, Machine cluster in Germany and many more.

As the concept of cluster has successfully been applied in maritime sector in various countries such as Europe, Dubai, and Hong Kong etc. so we wish to investigate the status/trend of geographical concentration of port logistics related firms around Australian ports areas. The Studies conducted, mainly in Europe, has shown some evidences of benefits to maritime industries after collaborating and collocating their activities at a particular geographical concentrated area. Few of the numerous benefits achieved are availability of specialized labor, R&D institutes, market awareness, knowledge sharing, bigger supplier and customer base and strategic co-operations (European Commision 2008)

According to Zhang and Lam (2013) the cluster concept is widely regarded as a policy making tool to develop maritime industry. Numbers of approaches have been established to map the cluster which may include input/output analysis, geographical density, GIS modelling etc. The drive of this research is to mark/draw the boundary of Australian port logistics cluster geographically and to investigate that what industrial activities outlines Australian port centric logistics cluster. The research is expected to offer a number of contributions for academics, policy making bodies and investing organizations. The European commission has acknowledged maritime sectors such as shipping, ship building, offshore supply, ports and related services and many more which they have included in their studies but in our research we have divided the sectors into two broader categories named core activities and supporting activities.

Cluster can be local (tourism in a city), regional (such as financial cluster in London, industry cluster in Minnesota), interregional (car manufacturing cluster in Germany). The demarcation of the cluster does not happen by itself rather it is shaped by the practitioners. The initial step of demarcation and calling a cluster is to decide a centre where the activities are concentrated and in this research the centre is considered to be the Australian seaports. The centre of cluster does not necessarily have to be near to other economic activities (Langen 2002).

From the detailed research the conceptual definition of port logistics Cluster is '*The outcome of one or more spatial consolidations of cooperating – competing firms and institutions within all sectors, subsectors and economic activities directly or indirectly linked to the shipping industry, maritime transport and generally the utilization of the sea*'.

4. ECOSYSTEM OF PORT CENTRIC LOGISTICS CLUSTERS

Industry specific clusters bring numerous benefits to the companies in that clusters and economy of nation through increased productivity, innovation and increased competition. The firms operating in a cluster achieves agglomeration economies. They cluster to gain the benefit of large labour pool in a geographically centred area which reduces the search cost. . According to Simmie and Sannett (1999) the base of innovation is technical and professional labour. Sanchez (2012) research indicates that within the technology cluster of the Silicon Valley, the technology industry cluster focusses on finding ways of sharing employment, innovative personnel and knowledge to strengthen their organizations. The increased efficiency and performance enhancement created through the incorporation of clusters in turn helps to strengthen the companies and increase industry competitiveness. Industry clusters provide advantages such as access to better inputs and production, increased knowledge sharing, increased availability of skilled labour and evolving competition pressure that pushes for innovation and productivity.

Increased productivity is an impacted outcome of being in cluster through increased access to suitable skilled employees, beneficial supplier networks, information flow, technology developments, infrastructure and related institutions. In large pool of firms it is expected to have high quality labour by recruitment policy and also by information and knowledge sharing in the cluster. Moreover, it is easy for speciality training to be imparted to this labour pool to make them skilled and upgrade their knowledge quality. Apart from this there may be well developed infrastructure and resources to be used by each member in the cluster which otherwise is beyond the scope of an individual firm.

Innovation is another benefit of cluster. Along with improved productivity clusters enhances a company's ability to innovate and increase productivity further. The knowledge and experience of the management in the cluster organizations gives another companies in the cluster a view of the market pace they compete in rather working in isolation. Relationship within the cluster also gives companies access to other knowledge of evolving technology, machinery advancement, market and consumer knowledge. Having majority of the suppliers and customer in the cluster promotes frequent interaction, closer monitoring and reduction in transportation cost due to proximity of location. Hence, large supplier and customer base in the same vicinity makes easier for the companies to compare their performance which in turn becomes a motivator to innovate new techniques (Maskell and Lorenzen 2004). Frequent interactions will also assist in easy knowledge spill over locally and exchange of idea (Tallman et al. 2004). With constant comparison within the cluster comes constant competitive pressure pushing for further innovations within the cluster.

Better economic footprint is one of the many positive impacts/effects of a cluster. The impact can be assessed in terms of employment, productivity gain, added value, knowledge generation, gross domestic product, foreign direct investment etc. The economic impact of maritime cluster in Greece is such an example where shipping sector being a part of maritime cluster contributes to Greek economy. According to Zagkas (2010), only in 2007 $17 billion is the net gain from shipping industry in Greece contributing to 7% of the GDP. It also provides an employment of 76,200 people in the cluster which is around 43.3% of the total maritime employment in Attica, (Attica being concentration of Attica). Another example of economic impact is the Xiqiao textile cluster in Guangdong which provides substantial positive financial impact and economic development in the surrounding region (Zeng 2010).

New business creation is important outcome of cluster and many new businesses choose to operate within the existing clusters rather than independently in remote area. Risk is lowered and the ability to support current and future market opportunities is increased along with the access to available inputs, assets, staff, relationships and infrastructure which is already available within the cluster. These bundles opportunities when teamed with the reorganization of the cluster by financial institutions and investors, lowers the risk in all areas (Brown et al. 2013).

5. DATA SET USED

The primary data set used to analyse the existence of port logistics related employment around the seaport and delineating the boundary of port centric logistics cluster; is the census data captured of year 2001 and year 2006, employment by industry in statistical location areas (SLA) (using the Australian New Zealand Standard Industrial Classification (ANZSIC93)) is used. The SLA is a spatial unit representing each Australian suburb and the second smallest unit defined after CD (collection district) by Australian Standard Geographical Classification (ASGC) in 2006 census. By joining all the SLAs give Australia without gaps and overlaps. In 2001, 1353 SLAs have been identified in Australia. The data addresses the employment types, explicitly port logistics related, and employee number in each geographical area. This refers to the number and types of port related logistics jobs in each area (SLA) and how are they distributed spatially. All major Australian ports state wise have been captured in the data. The selected SLAs for this research are the geographical areas which are near /adjoining to the major ports.

The employment data across industries are divided into core industries (Directly related) and supporting industries (Indirectly related). The purpose of forming the core activities is to create a model where other sub activities will revolve around the core activities, which thus will create a port centric logistics cluster. In this research, the activities which have been defined as core activities in Australian port logistics industry are ship building, water transport, stevedoring, port operators, freight forwarding, custom agencies, warehousing and storage etc. and supporting firms include transport and equipment manufacturers, road and rail freight transport, services to rail, road and pipeline transport industry and postal services etc. There are few logistics activities in manufacturing and service sector which has not been taken into account for this research. The reason for not including manufacturing logistics employment connects to not directly or indirectly relates to port logistics function but defines logistics activities more within the company. The study focusses mainly on logistics activities closely related to maritime sector. The overall revenue of Australian logistics industry is around $94 billion and directly contributes to 3.1% to Australian GDP, which makes this industry very important.

6. RESEARCH METHODOLOGY

A statistical analysis is conducted in 3 stages. Stage 1 identifies the industries and areas that are related to port logistics sector using a four digit ANZSIC code and determining the core and supporting industries. For example, the code range for transport, postal and warehousing is 4600-5399. Where "I" is an alphabet denoted to the industry division, transport in this case. Stage 2 addresses the scale of port centric logistics cluster. From the cumulative data a percentage of each logistics related employment function has been calculated on total port logistics industries and total employment. The data of these selected industries/employment type are then imported in GIS system to map them spatially to analyse the spread and delineating the boundary of maritime logistics cluster, in stage 3 using Geographical information system.

6.1 Stage 1: Identification of Port-Centric Logistics Industries

This stage identifies the industries that define port centric logistics cluster and dividing them into core and supporting industries according to their relationship towards port logistics activities. Statistical analysis was conducted to investigate the employment pattern of diverse firms representing Port logistics sector in Australia. The port logistics sector comprises of the industries directly (core activities) or indirectly related (supporting activities) to port sector to assist in accomplishing port logistics function. The logistics is a broad term which contains many functions such as warehousing, transportation (sea, air, rail and road mainly), and services assisting these function to work smoothly. In this research paper the focus is primarily on logistics firms which assist in port sector (table 1).

'Industry of employment' data comprises of total 633 industries in 2001 and 719 industries types in 2006. Few new industries have been added in 2006 which are not mentioned in 2001 but some industries have been bundled into one category such as transport, postal and warehousing division in 2006 was transport and storage/warehousing as an individual firm in 2001. Among the total industries, 29 industries in 2001 and 24 industries in 2006 have been nominated as the industry types related to port logistics.

6.2 Stage 2: Employment in Port-Centric Logistics Clusters in Australia

Stage 2 quantifies the total number of employees in port centric logistics sector and the analysis of industry types attracting more employment. The total port logistics employment in 2001 is 4.1%, and 3.75% in 2006 of the total employment in Australia, which shows a little declining pattern from 2001 to

Table 1. Type of industries in port logistics sector

Core Activities	Supporting Activities
Shipbuilding	Transport Equipment Manufacturing
Water Transport, undefined	Transport and Storage, undefined
International Sea Transport	Road Freight Transport
Coastal Water Transport	Rail Transport
Inland Water Transport	Other Transport, undefined
Stevedoring	Pipeline Transport
Water Transport Terminals	Road Transport
Port Operators	Services to Transport, undefined
Freight Forwarding (Except Road)	Services to Road Transport, undefined
Customs Agency Services	Services to Road Transport,
Storage, undefined	Services to Water Transport,
Grain Storage	Other Services to Transport, undefined
Storage	Road Freight Forwarding
	Services to Transport,
	Postal and Courier Services, undefined
	Postal Services
	Courier Services

Table 3. Comparative analysis of logistics employment in year 2001 and 2006

	Year 2001	Year 2006
Number of Industry Types	633	719
Port logistics related industries	30	25
% age of port related employment	4.10%	3.75%
Total Employment in selected areas	1493662	1605274
Total employment in all areas	8297561	9104187
Top 50% employment share	Supporting industries	2 supporting and 3 core industries
Dominating industries	Road Freight 23.02%	Road Freight 37.54%

2006 (see table 3). In 2006 though it shows the declining pattern but it is also very important to mention that the number of industries (633 in year 2001 to 719 in year 2006) and employment proportion/ number (1493662 in year 2001 to 1605274 in year 2006) also increased from 2001 to 2006. The result shows that more than 50% of the employment comprises of supporting industries in 2001 (including road freight transport, postal services and rail transport), in contrast, by 2 supporting industries (including road freight transport, postal services) and 3 core industries (including freight forwarding, storage services and water transport services) in 2006.

Please refer the table 2 in text. Results clearly indicate that the road freight transport related employment in all years is dominating with an increasing trend of 23.02% in 2001 to 37.54% in 2006, whereas there has been a downward trend noted in postal services of around 2% from 2001 to 2006 (15.85% in 2001 to 13.14% in 2006). The top 2 employment types in 2001 and 2006 are same, contributing to 38.87% in 2001 and 50.68% in 2006, which are (naming) road freight transport and postal services. It is worth noting that in all these census data's the top 2 dominating industry types are supporting industries. It is also significant to mention that road freight transport has grown almost 14% from year 2001 to 2006. The contribution of rail transport in year 2001 is quite high as compared to year 2006, 14.88% in year 2001 and 4.26% in year 2006. Out of 29 industry types in year 2001 and 24 industry types in year 2006, for data analysis 90% of the total employment is captured by around 11 employment types and majority of the employment is of supporting industry like road and rail freight, stevedoring, warehousing & storage and postal services

It is also evident from the analysis that first 2 industry types, in both 2001 and 2006, which attracts enormous employment, is same but industry types at rank 3, 4 and 5 are different in 2001 and 2006. Analysis represents that the road freight transport is a growing sector because most of the received inbound freight stays locally. The interstate freight distribution is largely conducted using rail and partially using road transportation.

6.3 Stage 3: Mapping the Port-Centred Logistics Employment Clusters

This stage delineates the geographic boundary for port-centred logistics clusters, which will be defined using three criteria. For this research three major Australian ports, which include Melbourne, Botany Bay and Brisbane, have been analysed/mapped to determine their port centric logistics landscape. It is also significant to identify the areas and the degree to which they are spatially disperse /clustered so as to support them with intervention.

Table 2. Industrial composition of port centric logistics sector in Australia

Industries	Number of Port-Centric Logistics Employment (2001)	Number of Port-Centric Logistics Employment (2006)	% of Total Port-Centric Logistics Employment (2001)	% of Total Port-Centric Logistics Employment (2006)
Road Freight Transport	14398	22639	23.02%	37.54%
Postal Services	9912	7919	15.85%	13.14%
Rail Transport	9303	1330	14.88%	2.21%
Courier Services	3633	1968	5.81%	3.27%
Storage, nec	2938	3228	4.70%	5.36%
Shipbuilding	2767	2461	4.42%	4.08%
Freight Forwarding (Except Road)	2537	4919	4.06%	8.16%
Water Transport, undefined	2535	1061	4.05%	1.76%
Stevedoring	1455	2557	2.33%	4.24%
Services to Water Transport, nec	1343	1520	2.15%	2.52%
Water Transport Terminals	1022	2797	1.63%	4.64%
Coastal Water Transport	669	1207	1.07%	2.00%
Inland Water Transport	655	937	1.05%	1.55%
Pipeline Transport	74	104	0.12%	0.17%
Grain Storage	292	542	0.47%	0.90%
Other Services to Transport, undefined	36	6	0.06%	0.01%
Postal and Courier Services, undefined	15	28	0.02%	0.05%
Customs Agency Services	1141	N/A	1.82%	N/A
Port Operators	1084	N/A	1.73%	N/A
Transport and Storage, undefined	4619	N/A	7.39%	N/A
International Sea Transport	592	N/A	0.95%	N/A
Services to Transport, nec	549	N/A	0.88%	N/A
Road Freight Forwarding	338	N/A	0.54%	N/A
Services to Road Transport, nec	309	N/A	0.49%	N/A
Transport, nec	127	N/A	0.20%	N/A
Transport Equipment Manufacturing, nec	75	N/A	0.12%	N/A
Services to Transport, undefined	59	N/A	0.09%	N/A
Storage, undefined	38	N/A	0.06%	N/A
Services to Road Transport, undefined	21	N/A	0.03%	N/A
Rail Freight Transport	N/A	2567	N/A	4.26%
Transport, Postal & Warehousing Division, nfd	N/A	1433	N/A	2.38%
Customs Agency Services	N/A	912	N/A	1.51%
Water Transport Support Services, nfd	N/A	58	N/A	0.10%
Road Transport, nfd	N/A	48	N/A	0.08%
Other Transport, nec	N/A	25	N/A	0.04%
Warehousing and Storage Services, nfd	N/A	16	N/A	0.03%
Port-centric Logistics Employment Cluster	60272	62539		
Total Employment	8297561	9104187		
Total Employment in Port-centric Logistics Cluster	1493662	1605274		

- **Principle of Spatial Adjacency:** Adjacent spatial units are more likely to be related than those which are further apart. Therefore, the areas which are spatially adjacent are more likely to be part of cluster. Adjacency is calculated by creating a binary connectivity matric. Non-adjacent areas are not incorporated in cluster formation.
- **Principle of Concentration:** Concentration is a disproportionate distribution, which refers to a degree of employment density over an area. The proportion of the port logistics related employment should be equal to or greater than 50% to the total employment then only the area would be considered in a cluster.
- **Principle of Distance Decay:** The impact of a phenomenon diminishes with distance. A buffer with a radius of 50 kilometres from the port is created to select the areas, which intersected with the buffer.

In order to be in a cluster, areas are required to fulfil all the above mentioned criteria. Three key container ports in Australia namely Melbourne, Botany Bay and Brisbane are used as case studies. Figure 1 shows the geographic extent of these three ports. Melbourne Port-centred logistics cluster is anchored on Melbourne CBD with a large concentration of logistics employment vis-à-vis industries in the west of Melbourne such as Altona, North Melbourne, Laverton and Footscray. Botany bay Port-centred logistics cluster is discerned mainly around eastern suburbs which include Randwick, Waverly, Woollahra, Hurstville etc. In Brisbane the spread of port-centered logistics cluster is towards southeast inner Brisbane and Northwest Inner Brisbane suburbs such as Wynnum, Lota, Manly, Hamilton, Ascot, Albion. The cluster also outspreads towards Brisbane airport as well

Though Figure 1 shows the concentration of port centric logistics employment in the areas marked but it does not explain the strength of relationship in/of the cluster (i.e. degree of concentration). In this research the new concept has been introduced which is port centric spatial logistics cluster which defines the high concentrated area of logistics employment near to port sector directly or indirectly related in port function; which means the suburbs near to seaport having high concentration of port logistics related employment and adjacent as well. Though it is critical to measure the strength of relationship in the cluster and examining the spill over effect of agglomeration but still it will be an beneficial situation for the company/employment working in a spatial cluster as Tobler first law of Geography depicts that everything is linked to everything but spatially closer functions are more connected to each other and gain advantage.

7. CONCLUSION AND FUTURE RESEARCH

This research examined the existence and spread of port centric logistics cluster taking the case of three major Australian ports such as Melbourne, Botany Bay and Brisbane. The significance entails firstly to identify the industries types which are related closely or indirectly to assist the logistics function at port to happen smoothly. The research concludes that there are 20 and 33 industrial sectors in year 2001 and 2006 respectively which represents the Port centric logistics cluster. It is also important to mention the road freight transportation is a dominating industry with a major contribution of 23.01% and 37% in 2001 and 2006 respectively. The incline in road freight transportation is due to the fact that most of the freight being distributed locally and also reflects clustering of road freight transport companies near to the port vicinity. This might bring down the empty container movement on road and sharing the resources

Figure 1. Port-centric logistics employment spatial cluster spread – Melbourne (a), Botany Bay (b) and Brisbane (c)

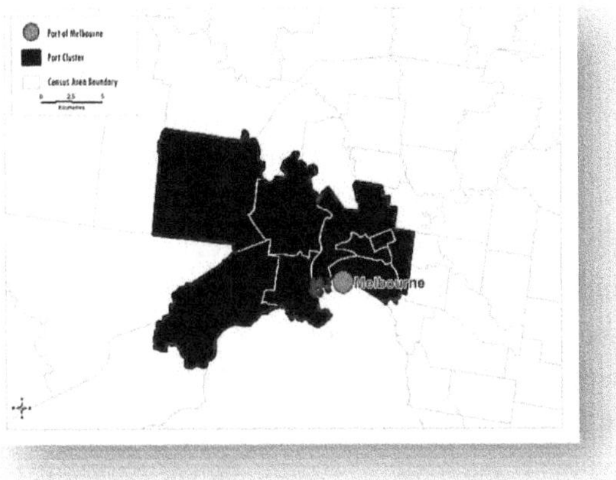

(a) - Melbourne

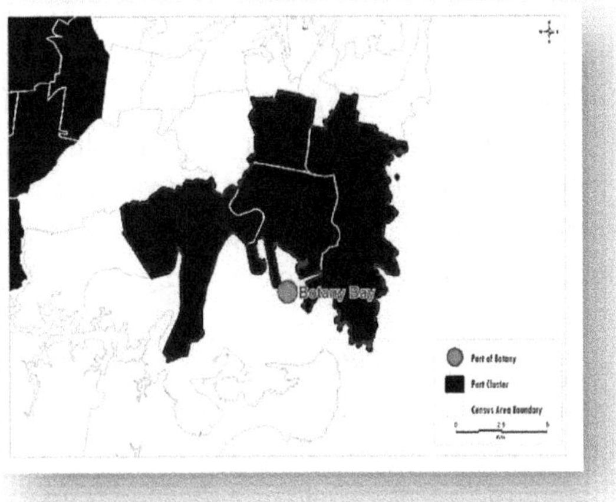

(b) – Botany Bay

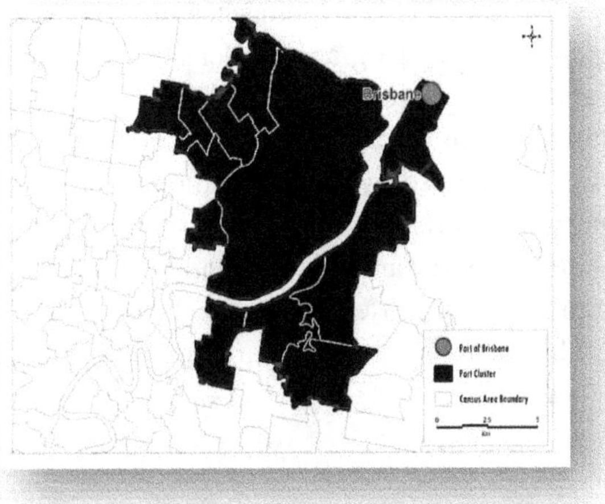

(c) - Brisbane

interchangeably among the companies to gain the competitiveness of being in cluster. This research provided augmented evidence base on cluster-based strategies to promote trade within the cluster and improve efficiency through sharing of infrastructure and technology. Such cluster-based strategy would provide policy makers to devise evidence based policy to attract investment in more established and high performing areas. Thus, it would minimise risk and maximise harnessing business opportunity. It is argued that clustering of firms and employment would improve inter-firm relationships and partnerships and would potentially reduce empty container movements in and out of a port.

There are few areas in this research which forms the basis of future research. Primarily, to investigate quantitatively the impact of spatial clustering on economy of nation/state, production, innovation, new business creation, reduction in empty container movement, achieving economies, and the extent of sharing resources. It will also be interesting to see the trend of employment in the defined cluster specially taking 2011 data and examining the trend over time. Other point to consider will be on how the empirical testing should be done to investigate the inter-firm relationship, nature of collaboration and competition. Finally, to examine on how cluster is governed and what are the prime construct to evaluate the performance of cluster. The performance evaluation would include the value-added generated in the cluster not the performance measurement of individual business working in that cluster. As Langen (2002) has drawn four constructs for the performance measurement of a cluster. Firstly, internal competition in the cluster will add performance to the cluster as competition will bring performance and specialization in the cluster. Secondly, entry and exit barriers influence the performance of cluster though entry barriers may have unfavourable effects but exit barriers will restrict the firms to stay in a cluster and perform to a standard. Thirdly, more diversified firms will have a positive effect on performance as there are more chances of knowledge spill over from various perspectives and also large data base/width of knowledge base. Fourthly, how strongly the firms are interconnected in terms of exchanging resources, infrastructure and utilising skilled labour pool.

The change of ANZSIC codes and amalgamation of the industry types in 2006 may be a limitation in the analysing the situation exactly in comparison to 2001. The 2006 data may be analysed using the same four digit ANZSIC code which has been used in 2001 to investigate the actual impact without combining the industry types. Another limitation of this research is not been able to explore research taking data of 2011 and examine the trend. The influx of employment would illustrate growing cluster base hence performance. Only one approach of mapping has been considered in this research which may not signify an exact representation of cluster, spatially. Other approaches may be used to support the argument which may include input/output approach, and geographical density. As the supply chains are decentralized so the port centric logistics landscape is spreading wider from the port, as a focal point. So, another limitation may be the area selected which is maximum spread of 50 kms taking port as a focal point but there might be another clusters developed over time in some other suburbs away from port which may be virtually linked to the ports and ports functions. Chhetri et al. (2014) in their study suggested that the areas of logistics clusters in Melbourne are predominently Dandenong industrial area in south eastern suburbs and around the airport in north of Melbourne. So, there can be few sub-clusters in the same state representing a big cluster which can be considered in future study.

REFERENCES

ABS. (2006). *Australian and New Zealand Standard Industrial Classification (ANZSIC), Cat 1292.0.* Canberra: Australian Bureau of Statistics, Government of Australia.

Altenburg, T., & Meyer-Stamer, J. (1999). How to promote clusters: Policy experiences from Latin America. *World Development, 27*(9), 1693–1713. doi:10.1016/S0305-750X(99)00081-9

Arvis, J., Saslavsky, D., Ojala, L., Shepherd, B., Busch, C., & Raj, A. (2014). *Connecting to compete 2014 – trade logistics in the global economy: The logistics performance index and its indicators.* Retrieved 8 November 2014 from: www.worldbank.org/lpi

BITRE. (2011-12), *Maritime Australian Sea Freight.* Retrieved 18 November 2013 from: http://www.bitre.gov.au/publications/2013/files/asf_2011_12.pdf

Brown, K., Burgess, J., Festing, M., & Royer, S. (2013). *Resources and Competitive Advantage in Clusters.* Rainer Hampp Verlag.

Carbone, V., & Martino, M. D. (2003). The changing role of ports in supply-chain management: An empirical analysis. *Maritime Policy & Management, 30*(4), 305–320. doi:10.1080/0308883032000145618

Chhetri, P., Butcher, T., & Corbitt, B. (2014). Characterising spatial logistics employment clusters. *International Journal of Physical Distribution & Logistics Management, 44*(3), 221–241. doi:10.1108/IJPDLM-03-2012-0086

Coyle, J. J., Bardi, E. J., & Langley, C. J. (1996). *The management of business logistics.* West Publishing Company New York.

European Commission. (2008). *The role of Maritime Clusters to enhance the strength and development of European maritime sector.* Author.

Gordon, I. R., & McCann, P. (2000). Industrial clusters: Complexes, agglomeration and/or social networks? *Urban Studies (Edinburgh, Scotland), 37*(3), 513–532. doi:10.1080/0042098002096

Goss, R. (1990). Economic policies and seaports: 4. Strategies for port authorities. *Maritime Policy & Management, 17*(4), 273–287. doi:10.1080/03088839000000033

Haezendonck, E. (2001). *Essays on Strategy Analyisis for Seaports.* Garant.

Krugman, P. R. (1991). *Geography and trade.* MIT Press.

Krugman, P. R. (1998). What's new about the new economic geography? *Oxford Review of Economic Policy, 14*(2), 7–17. doi:10.1093/oxrep/14.2.7

Langen, P. D. (2002). Clustering and performance: The case of maritime clustering in The Netherlands. *Maritime Policy & Management, 29*(3), 209–221. doi:10.1080/03088830210132605

Mangan, J., Lalwani, C., & Fynes, B. (2008). Port-centric logistics. *International Journal of logistics Management, 19*(1), 29-41.

Maritime Workforce Development Forum. (2013). *Australian Maritime workforce development strategy*. Retrieved 19 November 2013 from: at http://www.shipping.infrastructure.gov.au/ mwdf/ files/ Australian_Maritime_Workforce_Development_Strategy.pdf

Marshall, A. (1890). *Principles of Economies* (Vol. 1). London: Macmillan.

Marshall, A. (1920). *Principles of Economics*. London: Macmillan.

Martin, R., & Sunley, P. (1996). Paul Krugman's geographical economics and its implications for regional development theyry: A critical assessment. *Economic Geography*, *72*(3), 261–292. doi:10.2307/144401

Martin, R., & Sunley, P. (2001). Deconstructing clusters: Chaotic concept or plicy panacea. *Journal of Economic Geography*, 2–60.

Maskell, P., & Lorenzen, M. (2004). The cluster as market organisation. *Urban Studies (Edinburgh, Scotland)*, *41*(5-6), 991–1009. doi:10.1080/00420980410001675878

Panayides, P. M., & Song, D. W. (2008). Evaluating the integration of seaport container terminals in supply chains. *International Journal of Physical Distribution & Logistics Management*, *38*(7), 562–584. doi:10.1108/09600030810900969

Porter, M. E. (1993). *The competitive advantage of nations*. Harvard Business School Management Programs.

Porter, M. E. (1998). *Clusters and the new economics of competition*. Harvard Business Review Watertown.

Sanchez, T. (2012). The impact of industry clusters on the economy in United States. *Academy of Entrepreneurship*, *18*(1), 85.

Simmie, J., & Sennett, J. (1999). Innovative clusters: Global or local linkages? *National Institute Economic Review*, *170*(1), 87–98. doi:10.1177/002795019917000112

Tallman, S., Jenkins, M., Henry, N., & Pinch, S. (2004). Knowledge, clusters, and competitive advantage. *Academy of Management Review*, *29*(2), 258–271.

Vassilios, K. Z., & Dimitrios, V. L. (2010). *An Analysis of seaport cluster models for the development and competitiveness of maritime sectors: The case of Piraeus*. Academic Press.

Zeng, D. Z. (2010). *Building engines for growth and competitiveness in China*. New York: World Bank. doi:10.1596/978-0-8213-8432-9

Zhang, W., & Lam, J. S. L. (2013). Maritime cluster evolution based on symbiosis theory and Lotka–Volterra model. *Maritime Policy & Management*, *40*(2), 161–176. doi:10.1080/03088839.2012.757375

Compilation of References

Abadias, M., Usall, J., Anguera, M., Solsona, C., & Viñas, I. (2008). Microbiological quality of fresh, minimally-processed fruit and vegetables, and sprouts from retail establishments. *International Journal of Food Microbiology*, *123*(1-2), 121–129. doi:10.1016/j.ijfoodmicro.2007.12.013 PMID:18237811

Abdul Moneim, A. F. (2008). Fuzzy genetic prioritization in multi-criteria decision problems. *Jordan Journal of Mechanical and Industrial Engineering*, *2*, 175–182.

Abele, E., Meyer, T., Naher, U., Strube, G., & Sykes, R. (Eds.). (2008). *Global Production-A handbook for Strategy and Implementation*. Springer-Verlag Berlin Heidelberg. doi:10.1007/978-3-540-71653-2

Abo- Sinna, M., & Abou-El- Enien, T. H. M. (2005). An interactive algorithm for large scale multiple objective programming problems with fuzzy parameters through TOPSIS approach. *Applied Mathematics and Computation*, *177*(2), 515–527.

Abo-Sinna, M.-A., Amer, A.-H., & Ibrahim, A.-S. (2008). Extensions of TOPSIS for large scale multi-objective non-linear programming problems with block angular structure. *Applied Mathematical Modelling*, *32*(3), 292–302. doi:10.1016/j.apm.2006.12.001

ABS. (2006). *Australian and New Zealand Standard Industrial Classification (ANZSIC), Cat 1292.0*. Canberra: Australian Bureau of Statistics, Government of Australia.

Acar, Y., Kadipasaoglu, S., & Schipperijn, P. (2010). A decision support framework for global supply chain modeling: An assessment of the impact of demand, supply and lead-time uncertainties on performance. *International Journal of Production Research*, *48*(11), 3245–3268. doi:10.1080/00207540902791769

Achanga, P., Shehab, E., Roy, R., & Nelder, G. (2006). Critical success factors for lean implementation within SMEs. *Journal of Manufacturing Technology Management*, *17*(4), 460–471. doi:10.1108/17410380610662889

Adams, N. D., Brown, T. W., & Firth, R. V. (1996). *Warehouse and Distribution Automation Handbook*. New York: McGraw-Hill.

Adil, M. (2013a). The relationship between service quality and customer satisfaction in India's rural banking sector: An item analysis and factor-specific approach. *The Lahore Journal of Business*, *1*(2), 43–63.

Adil, M. (2013b). Modelling effect of perceived service quality dimensions on customer satisfaction in Indian bank settings. *International Journal of Services and Operations Management*, *15*(3), 358–373. doi:10.1504/IJSOM.2013.054447

Adil, M., Akhtar, A., & Khan, M. N. (2013). Refinement of internet banking service quality scale: A confirmatory factor analysis approach. *International Journal of Services and Operations Management*, *14*(3), 336–354. doi:10.1504/IJSOM.2013.052094

Adil, M., & Ansari, S. A. (2012). Service quality and customer loyalty at Indian rural banks: an empirical test of an integrative model. In *Proceedings of the National Conference on Contemporary Issues in Business, Management & Finance*. University of Kashmir.

Adil, M., & Rashid, I. (2013). Role Stress as a Barrier to Service Quality in Indian Banks: SEM Approach. *International Journal of Services and Sciences, 15*(1), 40–57. doi:10.1504/IJSSCI.2013.057648

Admin. (2013). Cold Chain in India: Market Opportunities & Challenges. *AFRIS Blog*.

Ajzen, I. (1985). From intentions to actions: A theory of planned behaviour. In J. Kuhl & J. Beckman (Eds.), *Action-Control: From cognition to behaviour* (pp. 11–39). Heidelberg, Germany: Springer. doi:10.1007/978-3-642-69746-3_2

Ajzen, I. (1991). The theory of planned behaviour. *Organizational Behavior and Human Decision Processes, 50*(2), 179–211. doi:10.1016/0749-5978(91)90020-T

Altenburg, T., & Meyer-Stamer, J. (1999). How to promote clusters: Policy experiences from Latin America. *World Development, 27*(9), 1693–1713. doi:10.1016/S0305-750X(99)00081-9

Amiri, M. (2010). Project selection for oil-fields development by using the AHP and fuzzy TOPSIS methods. *Expert Systems with Applications, 37*(9), 6218–6224. doi:10.1016/j.eswa.2010.02.103

Amiri, M., Zandieh, M., Soltani, R., & Vahdani, B. (2009). A hybrid multi-criteria decision-making model for firms competence evaluation. *Expert Systems with Applications, 36*(10), 12314–12322. doi:10.1016/j.eswa.2009.04.045

Anderson, M. G., & Katz, P. B. (1998). Strategic sourcing. *The International Journal of Logistics Management, 9*(1), 1–13. doi:10.1108/09574099810805708

Andrej, O., & Mojca, J. (2009). Maintaining a cold chain from purchase to the home and at home: Consumer opinions. *Food Control, 20*(2), 167–172. doi:10.1016/j.foodcont.2008.03.010

Aneesh, M., & Sinha, A. (2011). *Cold chain: Finally warming up to India?* Infrastructure Today.

Angerhofer, B. J., & Angelides, M. C. (2006). A model and a performance measurement system for collaborative supply chains. *Decision Support Systems, 42*(1), 283–301. doi:10.1016/j.dss.2004.12.005

Anil, G. (2010). *A Survey of Information Technologies in Logistics Management*. University of Texas Research Paper.

Animal and Dairy Tech Asia. (2013). Retrieved on August, 15, 2013, from http://www2.kenes.com/agritech-asia-animal/Pages/Home.aspx

Annual Report. (2011). *Ministry of Food Processing India (MOFPI)*. Retrieved on September 23, 2011 from http://mofpi.nic.in

Anonymous, . (2007). The community summary report on trends and sources of zoonoses, zoonotic agents, antimicrobial resistance and food borne outbreaks in the European Union in 2006. *The EFSA Journal, 130*(5), 1–352.

Archer, N., Wang, S., & Kang, C. (2008). Barriers to the adoption of online supply chain solutions in small and medium enterprises. *Supply Chain Management: An International Journal, 13*(1), 73–82. doi:10.1108/13598540810850337

Arend, R. J., & Wisner, J. D. (2005). Small business and supply chain management: Is there a fit? *Journal of Business Venturing, 20*(3), 403–436. doi:10.1016/j.jbusvent.2003.11.003

Arnold, J. R., Stephen, C., & Lloyd, C. (2008). *Introduction to Materials Management*. Pearson Education International.

Arvanitoyannis, I. S., & Mavropoulos, A. A. (2000). Implementation of the hazard analysis critical control point (HACCP) system to Kaseri/Kefalotyri and Anevato cheese production lines. *Food Control*, *11*(1), 31–40. doi:10.1016/S0956-7135(99)00061-4

Arvanitoyannis, I. S., & Tzouros, N. H. (2006). *ISO 22000 the New Food Quality and Safety Standard*. Athens: Stamoulis S.A.

Arvis, J., Saslavsky, D., Ojala, L., Shepherd, B., Busch, C., & Raj, A. (2014). *Connecting to compete 2014 – trade logistics in the global economy: The logistics performance index and its indicators*. Retrieved 8 November 2014 from: www.worldbank.org/lpi

Asabere, N. Y., Oppong, D., & Kusi-sarpong, S. (2012). A Review of the Roles and Importance of Information and Communication Technologies (ICTs) in Supply Chain Management (SCM) of Organizations and Companies. *International Journal of Computer Science and Network*, *1*(4), 70–78.

Aung, M. M., Chang, Y. S., & Kim, W. R. (2012). Quality Monitoring and Dynamic Pricing in Cold Chain Management. World Academy of Science, Engineering and Technology.

Autry, C. W., & Susan, G. (2010). Evaluating buyer-supplier relationship–performance spirals-A longitudinal study. *Journal of Operations Management*, *24*(2), 87–100. doi:10.1016/j.jom.2009.07.003

Aviv, Y. (2007). On the benefits of collaborative forecasting partnerships between retailers and manufacturers. *Management Science*, *53*(5), 777–794. doi:10.1287/mnsc.1060.0654

Babu, J. (2012). Study on Supply Chain Practices with References to Automobile Industry. *International Journal of Marketing, Financial Services & Management Research, 1*(9).

Badenhorst-Weiss, J., & Nel, J. (2011). *A conceptual framework to analyze supply chain designs. Acta Commercii*. Special.

Bai, R., & Cui, B. (2006). Application in supplier selection using TOPSIS method. *Railway Transport Ecoomies.*, *28*(9), 14–17.

Ballı, S., & Korukoğlu, S. (2009). Operating system selection using fuzzy AHP and TOPSIS methods. *Mathematical and Computational Applications*, *14*(2), 119–130.

Ballou, R. H., Gilbert, S. M., & Mukherjee, A. (2000). New managerial challenges from supply chain opportunities. *Industrial Marketing Management*, *29*(1), 7–18. doi:10.1016/S0019-8501(99)00107-8

Barratt, M., & Oke, A. (2007). Antecedents of supply chain visibility in retail supply chains: A resoure-based theory perspective. *Journal of Operations Management*, *25*(6), 1217–1233. doi:10.1016/j.jom.2007.01.003

Barry, J. (2004). Supply chain risk in an uncertain global supply chain environment. *International Journal of Physical Distribution & Logistics Management*, *34*(9), 695–697. doi:10.1108/09600030410567469

Barua, A., Konana, P., Whinston, A. B., & Yin, F. (2004). An empirical investigation of net-enabled business value. *Management Information Systems Quarterly*, *28*(4), 585–620.

Beamon, B. (1998). Supply Chain Design and Analysis: Models and Methods. *International Journal of Production Economics*, *55*(3), 281–294. doi:10.1016/S0925-5273(98)00079-6

Beamon, B. M. (1999). Designing the green supply chain. *Logistics Information Management*, *12*(4), 332–342. doi:10.1108/09576059910284159

Beamon, B. M. (1999). Measuring supply chain Performance. *International Journal of Operations & Production Management*, *19*(3), 275–292. doi:10.1108/01443579910249714

Bechtel, C., & Jayaram, J. (1997). Supply chain management: a strategic perspective. *The International Journal of Logistics Management*, 15-34.

Becker, T., Billings, R., Eveleth, D., & Gilbert, N. (1996). Foci and Bases of Employee Commitment: Implications for Job Performance. *Academy of Management Journal*, *39*(2), 464–482. doi:10.2307/256788

Behn, R. D. (2003). Why measure performance? Different purposes require different measure. *Public Administration Review*, *63*(5), 586–606. doi:10.1111/1540-6210.00322

Bergh, J. (2002). *Do social movements matter to organizations? An institutional perspective on corporate responses to the contemporary environmental movement.* (PhD thesis). The Pennsylvania State University, State College, PA.

Berrah, L., & Clivillé, V. (2007). Towards an aggregation performance measurement system model in a supply chain context. *Computers in Industry*, *58*(7), 709–719. doi:10.1016/j.compind.2007.05.012

Beynon, M. (2002). DS/AHP method: A mathematical analysis, including an understanding of uncertainty. *European Journal of Operational Research*, *140*(1), 148–164. doi:10.1016/S0377-2217(01)00230-2

Beynon, M. (2005a). A method of aggregation in DS/AHP for group decision-making with the non-equivalent importance of individuals in the group. *Computers & Operations Research*, *32*(7), 1881–1896. doi:10.1016/j.cor.2003.12.004

Beynon, M. (2005b). Understanding local ignorance and non-specificity within the DS/AHP method of multi-criteria decision making. *European Journal of Operational Research*, *163*(2), 403–417. doi:10.1016/j.ejor.2003.11.010

Beynon, M., Cosker, D., & Marshall, D. (2001). An expert system for multi-criteria decision making using Dempster Shafer theory. *Expert Systems with Applications*, *20*(4), 357–367. doi:10.1016/S0957-4174(01)00020-3

Beynon, M., Curry, B., & Morgan, P. (2000). The Dempster–Shafer theory of evidence. An alternative approach to multicriteria decision modeling. *International Journal of Management Sciences*, *28*, 37–50.

Bhagwat, R., & Sharma, M. K. (2007, August). Performance measurement of supply chain management: A balanced scorecard approach. *Computers & Industrial Engineering*, *53*(1), 43–62. doi:10.1016/j.cie.2007.04.001

Bhalla, S. K. (2001). *Quality - A Necessity to face global challenges in dairy industry.* National seminar on technology management for facing global challenge in dairy industry and alumni- convention, Anand, India.

Bhattacherjee, A. (2000). Acceptance of e-commerce services: The case of electronic brokerages. *Systems, Man and Cybernetics, Part A: Systems and Humans, IEEE Transactions on*, *30*(4), 411–420. doi:10.1109/3468.852435

BITRE. (2011-12), *Maritime Australian Sea Freight.* Retrieved 18 November 2013 from: http://www.bitre.gov.au/publications/2013/files/asf_2011_12.pdf

Bolstroff, P. & Rosenbaum, R. (2003). *Supply Chain Excellence.* American Management Association.

Boniface, B. (2012). Producer relationships segmentation in Malaysia's milk supply chains. *British Food Journal*, *114*(10), 1501–1516. doi:10.1108/00070701211263046

Boran, F. E., Genc, S., Kurt, M., & Akay, D. (2009). A multi-criteria intuitionistic fuzzy group decision making for supplier selection with TOPSIS method. *Expert Systems with Applications*, *36*(8), 11363–11368. doi:10.1016/j.eswa.2009.03.039

Borgati, S. (2009). On Social Network Analysis in a Supply Chain Context. *Journal of Supply Chain Management.*

Bortolan, G., & Degani, R. (1985). A review of some methods for ranking fuzzy subsets. *Fuzzy Sets and Systems*, *15*(1), 1–19. doi:10.1016/0165-0114(85)90012-0

Bottani, E., & Rizzi, A. (2006). A fuzzy TOPSIS methodology to support outsourcing of logistics services. *Supply Chain Management: An International Journal*, *11*(4), 294–308. doi:10.1108/13598540610671743

Bourland, K. E., Powell, S. G., & Pyke, D. K. (1996). Exploiting timely demand information to reduce inventories. *European Journal of Operational Research*, *92*(2), 239–253. doi:10.1016/0377-2217(95)00136-0

Bovee, M., Srivastava, R. P., & Mak, B. (2003). A conceptual framework and belieffunction approach to assessing overall information quality. *International Journal of Intelligent Systems*, *18*(1), 51–74. doi:10.1002/int.10074

Bowen, F.E., Cousins, P.D., Lamming, R.C. & Faruk, A.C. (2001). Horse for courses: explaining the gap between the theory and practice of green supply. *Green. Manag. Int.*, 41–60.

Bower, J. L., & Hout, T. M. (1988). Fast cycle capability for competitive power. *Harvard Business Review*, 110–118.

Bowersox, D. J., Closs, D. J., Stank, T. P., & Keller, S. B. (2000). How supply chain competency leads to business success. *Supply Chain Management Review*, *4*(4), 70–78.

Bowersox, D. J., David, J. C., & Theodore, P. S. (1999). *21'' Century Logistics: Making Supply Chain Integration a Reality*. Chicago, IL: Council of Logistics Management.

Bowersox, D., Rodrigues, A., & Calantone, R. (2005). Estimation of Global and National Logistics Expenditures: 2002 Data Update. *Journal of Business Logistics*, *26*(2), 1–16. doi:10.1002/j.2158-1592.2005.tb00202.x

Braglia, M., Frosolini, M., & Montanari, R. (2003). Fuzzy TOPSIS approach for failure mode, effects and criticality analysis. *Quality and Reliability Engineering International*, *19*(5), 425–443. doi:10.1002/qre.528

Brahmaprakash & Dinesh, K. (1997). Infrastructural requirements for the development for the development of Agro-processing industries in rural India. *Agril. Econ. Res. Rev.*, *10*, 325–329.

Brown, K., Burgess, J., Festing, M., & Royer, S. (2013). *Resources and Competitive Advantage in Clusters*. Rainer Hampp Verlag.

Buckley, J. J. (1985). Fuzzy hierarchical analysis. *Fuzzy Sets and Systems*, *17*(3), 233–247. doi:10.1016/0165-0114(85)90090-9

Burki & Khan. (2008). Milk Districts and Efficiency of Smallholder Dairy Producers in Pakistan. Academic Press.

Business Standard. (2013). *Newspaper article*. Retrieved on February 7, 2013 from http://www.business-standard.com/article/companies/cold-chain-industry-to-touch-rs-64-000-cr-by-2017-assocham-113020600684_1.html

Büyüközkan, G., & Çifçi, G. (2012). A novel hybrid MCDM approach based on fuzzy DEMATEL, fuzzy ANP and fuzzy TOPSIS to evaluate green suppliers. *Expert Systems with Applications*, *39*(3), 3000–3011. doi:10.1016/j.eswa.2011.08.162

Cachon, G. (2001). Exact evaluation of batch-ordering policies in two-echelon supply chains with periodic review. *Operations Research*, *49*(1), 79–98. doi:10.1287/opre.49.1.79.11188

Cachon, G. P., & Fisher, M. (2001). Supply chain inventory management and the value of shared information. *Management Science*, *46*(8), 1032–1048. doi:10.1287/mnsc.46.8.1032.12029

Cai, J., Liu, X., Xiao, Z., & Liu, J. (2009). 2009). Improving supply chain performance management: A systematic approach to analyzing iterative KPI accomplishment. *Decision Support Systems*, 612–621.

Cai, J., Liu, X., Xiao, Z., & Liu, J. (2009). Improving supply chain performance management: A systematic approach to analyzing iterative KPI accomplishment. *Decision Support Systems*, *46*(2), 512–521. doi:10.1016/j.dss.2008.09.004

Camerinelli, E. (2009). *Measuring the Value of the Supply Chain*. Ashgate Publishing Group.

Camerinelli, E., & Cantu, A. (2006). Measuring the value of the supply chain: A framework. *Supply Chain Practice*, *8*, 40–59.

Cannon, J. P., & Perreault, W. D. Jr. (1999). Buyer–seller relationships in business markets. *JMR, Journal of Marketing Research*, *36*(4), 439–460. doi:10.2307/3151999

Cano, A., & Cruz, M. P. (2002). Integrated methodology for project risk management. *Journal of Construction Engineering and Management*, *128*(6), 473–485. doi:10.1061/(ASCE)0733-9364(2002)128:6(473)

Carbone, V., & Martino, M. D. (2003). The changing role of ports in supply-chain management: An empirical analysis. *Maritime Policy & Management*, *30*(4), 305–320. doi:10.1080/0308883032000145618

Carlsson, C. (1982). Tackling an MCDM-problem with the help of some results from fuzzy set theory. *European Journal of Operational Research*, *10*(3), 270–281. doi:10.1016/0377-2217(82)90226-0

Carter, R. C., & Carter, J. R. (1998). Interorganizational determinants of environmental purchasing: Initial evidence from the consumer products industry. *Decision Sciences*, *29*(3), 28–38. doi:10.1111/j.1540-5915.1998.tb01358.x

Castonguay, J., McKinley, C., & Kunkel, D. (2013). Health-Related Messages in Food Advertisements Targeting, Children. *Health Education*, *113*(5), 420–432. doi:10.1108/HE-09-2012-0047

Cavalieri, S., Gaiardelli, P., & Ierace, S. (2007). Aligning strategic profiles with operational metrics in after-sales service. *International Journal of Productivity and Performance Management*, *56*(5–6), 436–455.

Cebeci, U. (2009). Fuzzy AHP-based decision support system for selecting ERP systems in textile industry by using balanced scorecard. *Expert Systems with Applications*, *36*(5), 8900–8909. doi:10.1016/j.eswa.2008.11.046

CEC. (1992). Hygienic rules for the production and marketing of raw milk, heat treated milk and products based on milk, Council Directive 92/46/EEC. *Official Journal of the European Communities*, *268*, 1–32.

Celik, M., Kandakoglu, A., & Er, D. (2009). Structuring fuzzy integrated multi-stages evaluation model on academic personnel recruitment in MET institutions. *Expert Systems with Applications*, *36*(3), 6918–6927. doi:10.1016/j.eswa.2008.08.057

Chae, B. K. (2009). Developing key performance indicators for supply chain: An industry perspective. *Supply chain Management. International Journal (Toronto, Ont.)*, *14*(6), 422–428.

Chakravarty, S., Gulen, H., & Mayhew, S. (2005). Informed Trading in Stock and Option Markets. *The Journal of Finance*, *59*(3), 1235–1258. DOI: 10.1111/j.1540-6261.2004.00661.x

Chamodrakas, A., Alexopoulou, N., & Martakos, D. (2009). Customer evaluation for order acceptance using a novel class of fuzzy methods based on TOPSIS. *Expert System Application*, *36*(4), 7409–7415. doi:10.1016/j.eswa.2008.09.050

Chandra, K., & Kumar, S. (2000). Supply chain management in theory and practice: A passing fad or a fundamental change*? Industrial Management & Data Systems*, *100*(3), 100–113. doi:10.1108/02635570010286168

Chan, F. T. S. (2003). Performance measurement in a supply chain. *International Journal of Advanced Manufacturing Technology*, *21*(7), 534–548. doi:10.1007/s001700300063

Chan, F. T. S., & Qi, H. J. (2003). An innovative performance measurement method for supply chain management. *Supply Chain Management: An International Journal*, *8*(3), 209–223. doi:10.1108/13598540310484618

Chan, F. T. S., Tang, N. K. H., Lau, H. C. W., & Ip, R. W. L. (2002). A simulation approach in supply chain management. *Integrated Manufacturing Systems*, *13*(2), 117–122. doi:10.1108/09576060210415455

Chang, C.-W., Wu, C.-R., & Lin, H.-L. (2009). Applying fuzzy hierarchy multiple attributes to construct an expert decision making process. *Expert Systems with Applications*, *36*(4), 7363–7368. doi:10.1016/j.eswa.2008.09.026

Chang, T., Fu, H., Lee, W., Lin, Y., & Hsueh, H. (2007). A study of an augmented CPFR model for the 3C retail industry. *Supply Chain Management: An International Journal*, *12*(3), 200–209. doi:10.1108/13598540710742518

Chen, C. B., & Klein, C. M. (1997). An efficient approach to solving fuzzy MADM problems. *Fuzzy Sets and Systems*, *88*(1), 51–67. doi:10.1016/S0165-0114(96)00048-6

Chen, C. T. (2000). Extensions of the TOPSIS for group decision making under fuzzy environment. *Fuzzy Sets and Systems*, *114*(1), 1–9. doi:10.1016/S0165-0114(97)00377-1

Chen, C. T., Lin, C. T., & Huang, S. F. (2006). A fuzzy approach for supplier evaluation and selection in supply chain management. *International Journal of Production Economics*, *102*(2), 289–301. doi:10.1016/j.ijpe.2005.03.009

Chen, F., Drezner, Z., Ryan, J. K., & Simchi-Levi, D. (2000). Quantifying the bullwhip effect in a simple supply chain:the impact of forecasting, lead times, and information. *Management Science*, *46*(3), 436–443. doi:10.1287/mnsc.46.3.436.12069

Chen, I. J., Paulraj, A., & Lado, A. (2004). Strategic purchasing, supply management and firm performance. *Journal of Operations Management*, *22*(5), 505–523. doi:10.1016/j.jom.2004.06.002

Chen, J. (2008). Inter-organizational communication as a relational competency: Antecedents and performance outcomes in collaborative buyer–supplier relationships. *Journal of Operations Management*, *26*(1), 45–64. doi:10.1016/j.jom.2007.04.001

Chen, K. Y., & Shaw, Y. C. (2011). Applying back propagation network to cold chain temperature monitoring, Advanced Engineering Informatics. *Advanced Engineering Informatics*, *25*(1), 1–22. doi:10.1016/j.aei.2010.05.003

Chen, M. F., & Tzeng, G. H. (2004). Combining grey relation and TOPSIS concepts for selecting an expatriate host country. *Mathematical and Computer Modelling*, *40*(13), 1473–1490. doi:10.1016/j.mcm.2005.01.006

Chen, S. J., & Hwang, C. L. (1992). *Fuzzy Multiple Attribute Decision Making: Methods and Applications*. Berlin: Springer. doi:10.1007/978-3-642-46768-4

Che, Z. H., Wang, H. S., & Chuang, C. L. (2010). A Fuzzy AHP and DEA Approach for Marketing Bank Loan Decisions for Small and Medium Enterprises in Taiwan. *Expert Systems with Applications*, *37*(10), 7189–7199. doi:10.1016/j.eswa.2010.04.010

Chhetri, P., Butcher, T., & Corbitt, B. (2014). Characterising spatial logistics employment clusters. *International Journal of Physical Distribution & Logistics Management*, *44*(3), 221–241. doi:10.1108/IJPDLM-03-2012-0086

Chopra, S., & Meindl, P. (2001). *Supply chain management: Strategy, planning and operation*. Pearson Education Asia.

Chopra, S., Meindl, P., & Kalra, D.V. (2008). Supply Chain Management: Strategy, Planning, and Operations (3rd ed.). Pearson Education.

Chopra, S., Meindl, P., & Kalra, D.V. (2013). *Supply Chain Management Strategy, Planning and Operation*. Dorling Kindersley (India) Pvt. Ltd.

Chopra, S., & Meindl, P. (2001). *Supply Chain Management: Strategy, Planning, and Operation* (2nd ed.). Englewood Cliffs, NJ: Prentice-Hall.

Chountalas, P., Tsarouchas, D., & Lagodimos, A. (2009). Standardized food safety management: The case of industrial yoghurt. *British Food Journal*, *111*(9), 897–914. doi:10.1108/00070700910992835

Chow, G., Heaver, T. D., & Henriksson, L. E. (1994). Logistics performance: Definition and measurement. *International Journal of Physical Distribution & Logistics Management, 24*(1), 117–128. doi:10.1108/09600039410055981

Christiaanse, E., & Kumar, K. (2000). ICT – enabled coordination of dynamic supply webs. *International Journal of Physical Distribution & Logistics, 30*(3/4), 268–285. doi:10.1108/09600030010326019

Christopher, M. (1992). *Logistics and Supply Chain Management: Strategies for Reducing Costs and Improving Services.* London: Financial Times/Pitman.

Christopher, M. (1992). *Logistics & Supply Chain Management.* London: Pitman Publishing.

Christopher, M., & Holweg, M. (2011). Supply Chain 2.0: Managing supply chains in the era of turbulence. *International Journal of Physical Distribution & Logistics Management, 41*(1), 63–82. doi:10.1108/09600031111101439

Chu, T. (2002). Selecting plant location via a fuzzy TOPSIS approach. *International Journal of Advanced Manufacturing Technology, 20*(11), 859–864. doi:10.1007/s001700200227

Chu, T. C. (2002). Facility location selection using fuzzy TOPSIS under group decisions. *International Journal of Uncertainty, Fuzziness and Knowledge-based Systems, 10*(6), 687–701. doi:10.1142/S0218488502001739

Chu, T. C., & Lin, Y. C. (2003). A fuzzy TOPSIS method for robot selection. *International Journal of Advanced Manufacturing Technology, 21*(4), 284–290. doi:10.1007/s001700300033

Chu, T., & Lin, Y. (2002). Improved extensions of the TOPSIS for group decision making under fuzzy environment. *Journal of Information and Optimization Sciences, 23*(2), 273–286. doi:10.1080/02522667.2002.10698996

CII–McKinsey & Co. (1997). *Modernizing the Indian Food Chain, Food & Agriculture Integrated Development Action Plan (FAIDA).* New Delhi: CII and McKinsey & Co.

Closs, D. J., Goldsby, T. J., & Clinton, S. R. (1997). Information technology influences on world class logistics capability. *International Journal of Physical Distribution & Logistics Management, 27*(1), 4–17. doi:10.1108/09600039710162259

Codex. (2003). *Codex Standard for Fermented Milks.* Codex STAN 243-2003. Codex Alimentarius Commission.

Cold Storage Capacity. (2013). Retrieved on June 02, 2013, http://articles.economictimes.indiatimes.com/2013-02-06/news/36949978_1_

Cordeiro, J., & Sarkis, J. (1997). Environmental proactism and firm performance: Evidence from security analyst forecasts. *Business Strategy and the Environment, 6*(2), 104–114. doi:10.1002/(SICI)1099-0836(199705)6:2<104::AID-BSE102>3.0.CO;2-T

Corsten, D., & Gruen, T. (2003). Seeking on-shelf availability—an examination of the extent, the causes and the efforts to address retail out-of-stocks. *International Journal of Retail & Distribution Management, 31*(12), 605–716. doi:10.1108/09590550310507731

Coyle, J. J., Bardi, E. J., & Langley, C. J. (1996). *The management of business logistics.* West Publishing Company New York.

Coyle, J. J., Bardi, E. J., & Langley, C. J. (2003). *The Management of Business Logistics: A Supply Chain Perspective* (7th ed.). Mason, OH: South-Western Thomson Learning.

Coyle, R. G. (1977). *Management System Dynamics.* London: John Wiley and Sons.

CSCMP. (2008). *Supply chain management and logistics management definitions.* Retrieved 19 August from: http://cscmp.org/AboutCSCMP/Definitions/Definitions.asp

D'Avanzo, R., Von Lewinski, H., & Van Wassenhove, L. N. (2003). The link between supply chain and financial performance. *Supply Chain Management Review*, *7*, 40–47.

Dağdeviren, M., Yavuz, S., & Kılınc, N. (2009). Weapon selection using the AHP and TOPSIS methods under fuzzy environment. *Expert Systems with Applications*, *36*(4), 8143–8151. doi:10.1016/j.eswa.2008.10.016

Dalalah, D., Hayajneh, M., & Batieha, F. (2011). A fuzzy multi-criteria decision making model for supplier selection. *Expert Systems with Applications*, *38*(7), 8384–8391. doi:10.1016/j.eswa.2011.01.031

Dalu, R. S., & Deshmukh, S. G. (2001). SWOT analysis of small and medium scale industry: A case study. Productivity, 42(2).

Dangayach, G. S., & Deshmukh, S. G. (2004). Advanced manufacturing technology implementation: Evidence from Indian small and medium enterprises (SMEs). *Journal of Manufacturing Technology Management*, *16*(5), 483–496. doi:10.1108/17410380510600473

Dangayach, G., & Deshmukh, S. (2003). Evidence of manufacturing strategies in Indian industry: A survey. *International Journal of Production Economics*, *83*(3), 279–298. doi:10.1016/S0925-5273(02)00372-9

Dangour, A. D., Dodhia, S. K., Hayter, A., Allen, E., Lock, K., & Uauy, R. (2009). Nutritional quality of organic foods: A systematic review. *The American Journal of Clinical Nutrition*, *90*(3), 680–685. doi:10.3945/ajcn.2009.28041 PMID:19640946

Davenport, T. H. (2000). *Mission critical — realizing the promise of enterprise systems*. Boston, MA: Harvard Business School Press.

Dean, J. W., & Snell, S. A. (1991). Integrated manufacturing and job design: Moderating effects of organizational inertia. *Academy of Management Journal*, *34*(4), 776–804. doi:10.2307/256389

Dehning, B., & Stratopoulos, T. (2002). DuPont analysis of an IT-enabled competitive advantage. *International Journal of Accounting Information*, *3*(3), 165–176. doi:10.1016/S1467-0895(02)00032-5

Delmas, M., & Toffel, M. W. (2004). Stakeholders and environmental management practices: An institutional framework. *Business Strategy and the Environment*, *13*(4), 209–222. doi:10.1002/bse.409

Deloitte Touche Tohmatsu India Private Limited. (2009). *Bringing value to the changing Indian Landscape*. Author.

Deloitte. (2003). Mastering Complexity in Global Manufacturing: Powering Profits and Growth through Value Chain Synchronization. Retrieved from https://www.entrepreneur.com/tradejournals/article/ 113306163.html

Deloitte. (2007). A Report on the Summit on Indian Manufacturing Competitiveness. *Deloitte Research: Competing in Global Manufacturing and Service Networks*. Retrieved from http://www.deloitte.com/research

Deng, H., Yeh, C. H., & Willis, R. J. (2000). Inter-company comparison using modified TOPSIS with objective weights. *Computers & Operations Research*, *27*(10), 963–973. doi:10.1016/S0305-0548(99)00069-6

Deng, Y. (2006). Plant location selection based on fuzzy TOPSIS. *International Journal of Advanced Manufacturing Technology*, *28*(7-8), 839–844. doi:10.1007/s00170-004-2436-5

Department of Statistics (DOS). (2013). *Labour Force Survey Report Malaysia 2012*. Author.

Deveshwar, A., & Rathee, R. (2010, July). Challenges For Supply Chain Management In Today's Global Competitive Environment. *International Review of Business Research Papers*, *6*(2), 194–203.

Dixon, J. R., Nanni, A. J., & Vollmann, T. E. (1990). *The New Performance Challenge- Measuring Operations for World-class Competition*. Homewood, IL: Dow Jones-Irwin.

Dobler, D. W., & Burt, D. N. (1996). *Purchasing and Supply Management*. New York, NY: The McGraw-Hill Companies.

Dorian, L. (2011). *Understanding Risk Mitigation, Industry Insights*. Institute of Chartered Accountants.

Douglas, J. (1996). Building performance and its relevance to facilities management. *Facilities*, *14*(3/4), 23–32. doi:10.1108/02632779610112508

Douligeris, C., & Pereira, I. J. (1994). A Telecommunications Quality Study Using the Analytic Hierarchy Process. *IEEE Journal on Selected Areas in Communications*, *12*(2), 241–250. doi:10.1109/49.272873

Drucker, P. F. (1962, April). The economy's dark continent. *Fortune*, 265-270.

Drucker, P. (1995). The information executives truly need. *Harvard Business Review*, *73*, 54–62.

Duschek, S. (2004). Inter-Firm Resources and Sustained Competitive Advantage. *Management Review*.

Economics Times. (2013). *Newspaper article*. Retrieved on February 2013, available at http://articles.economictimes.indiatimes.com/keyword/cold-chain

Egels-Zandén, N. (2007). Suppliers' Compliance with MNCs' Codes of Conduct: Behind the Scenes at Chinese Toy Suppliers. *Journal of Business Ethics, 75*(1), 45-62.

ELA European Logistics Association / A. T. Kearney Management Consultants. (2004). *Differentiation for performance excellence in logistics 2004*. Hamburg, Germany: Deutscher Verkehrs-Verlag GmbH.

Eldukair, Z. A., & Ayyub, B. M. (1992). Multi-attribute fuzzy decisions in construction strategies. *Fuzzy Sets and Systems*, *46*(2), 155–165. doi:10.1016/0165-0114(92)90128-Q

Elgazzar, S., Tipi, N. S., Hubbard, N. J., & Leach, D. Z. (2011a). Linking supply chain operations' performance to the company's financial strategy: a case study of an Egyptian natural bottled water company. In *Proceedings from: The 16th International Symposium on Logistics*.

Elgazzar, S., Tipi, N. S., Hubbard, N. J., & Leach, D. Z. 2011b. A SWapplication system for measuring supply chain operations' performance using SCOR FAHP technique. In *Proceedings from: The 2011 International Conference on Business and Economics Research*. Retrieved November 2011 from: http://www.ipedr.com/vol16.htm

Ellinger,, A. E. D. P. (2000). The Relationship between Marketing/Logistics Interdepartmental Integration and Performance in U.S. Manufacturing Firms: An Empirical Study. *Journal of Business Logistics*, *21*(2), 1–22.

Ellram, L. M., La Londe, B. J., & Weber, M. M. (1989). Retail Logistics. *International Journal of Physical Distribution & Materials Management*, *19*(12), 29–39. doi:10.1108/EUM0000000000348

Ellram, L. M., Tate, W. L., & Billington, C. (2004). Understanding and managing the services supply chain. *Journal of Supply Chain Management*, *40*(4), 17–32. doi:10.1111/j.1745-493X.2004.tb00176.x

Elovitz & Kenneth, M. (2010). Green design: Risk or reward? *Consulting-Specifying Engineer*, 47.

Elsayed, K., & Paton, D. (2005). The impact of environmental performance on firm performance:static and dynamic panel data evidence. *Structural Change and Economic Dynamics*, *16*(3), 395–412. doi:10.1016/j.strueco.2004.04.004

Emond, J. P. (2008). The cold chain. In S.B. Miles, S.E. Sarma, & J.R. Williams (Eds.), RFID Technology and Applications. Cambridge University Press.

Eric, L., & Abhijeet, K. (2003). *Pharmaceutical Distribution in India*. Bio Pharm International.

Erkan, T. E., & Bac, U. (2011). Supply chain performance measurement: A case study about applicability of SCOR model in a manufacturing industry firm. *International Journal of Business and Management Studies, 3*(1).

Ernst & Young. (2011). Doing. *Business India*.

Ertuğrul, İ., & Karakaşoğlu, N. (2008). Comparison of fuzzy AHP and fuzzy TOPSIS methods for facility location selection. *International Journal of Advanced Manufacturing Technology, 39*(7-8), 783–795. doi:10.1007/s00170-007-1249-8

European Agency for Safety and Health at Work (EU-OSHA). (2000). *Occupational safety and health in marketing and procurement*. Luxembourg, Belgium: Office for Official Publications of the European Communities.

European Commission. (2008). *The role of Maritime Clusters to enhance the strength and development of European maritime sector.* Author.

European Commission. (2011). *Rapid Alert System for Food and Feed*. Retrieved on August 15, 2013 from http://ec.europa.eu/food/food/rapidalert/archive

Ezekiel, R., Singh, N., Sharma, S., & Kaur, A. (2013). Beneficial phytochemicals in potato a review. *Food Research International, 50*(1), 487–496. doi:10.1016/j.foodres.2011.04.025

Fatehpuria, S. (2013). Indian cold supply chain: A case study. *ASM's International E-Journal of Ongoing Research in Management and IT*, 1–10.

Fathi, T. (2011). *Automotive Supply Chain Trends – Innovation and Solutions*. DHL.

Fawcett, S. E., Ogden, J. A., Magnan, G. M., & Cooper, M. B. (2006). Organisational commitment and governance for supply chain success. *International Journal of Physical Distribution & Logistics Management, 36*(1), 22–35. doi:10.1108/09600030610642913

Feng, B., Fan, Z., & Li, Y. (2011). A decision method for supplier selection in multi-service outsourcing. *International Journal of Production Economics, 132*(2), 240–250. doi:10.1016/j.ijpe.2011.04.014

Feng, S., & Xu, L. D. (1999). Decision support for fuzzy comprehensive evaluation of urban development. *Fuzzy Sets and Systems, 105*(1), 1–12. doi:10.1016/S0165-0114(97)00229-7

Ferguson, M., & Ketzenberg, M. (2006). Sharing Information to Improve Retail Product Freshness of Perishables. *Production and Operations Management, 15*(1), 57–73.

Fiala, P. (2005). Information sharing in supply chains. *Omega, 33*(5), 419–423. doi:10.1016/j.omega.2004.07.006

Field, A. (2000). *Discovering statistics using SPSS*. London: Sage Publications.

Fisher, M. L. (1997). What is the Right Supply Chain for your product. *Harvard Business Review*.

Fleischmann, B., Ferber, S., & Henrich, P. (2006). Strategic Planning of BMW´s Global Production Network. *Interfaces*.

Flint, J. A., Van Duynhoven, Y. T., Angulo, F. J., DeLong, S. M., Braun, P., Kirk, M., & Walke, H. B. P. et al. (2005). Estimating the burden of acute gastroenteritis, foodborne disease, and pathogens commonly transmitted by food, an international review. *Clinical Infectious Diseases, 41*(5), 698–704. doi:10.1086/432064 PMID:16080093

Flynn, B. B., Huo, B., & Zhao, X. (2010). The impact of supplychain integration on performance: A contingency and configuration approach. *Journal of Operations Management, 28*(1), 58–71. doi:10.1016/j.jom.2009.06.001

Food and Agriculture Organization (FAO) of the United Nations. (2004). Retrieved on September,18, 2012, from http://www.fao.org/docrep/006/Y5160E/Y5160E00.htm

Forecast & Opportunities. (2013). *Cold Chain Market in India*. ADS Reports.

Forman, E., & Gass, S. (2001). The Analytic Hierarchy Process – An Exposition. *Operations Research*, *49*(4), 469–486. doi:10.1287/opre.49.4.469.11231

Forrester, J. W. (1961). *Industrial Dynamics*. Cambridge, MA: MIT Press.

Forrester, J. W. (1961). *Industrial, Dynamics*. Cambridge, MA: MIT Press.

Forrester, J. W. (1968). *Principals of Systems*. Cambridge, MA: Wright Allen Press.

Forslund, H. (2007). The impact of performance management on customers' expected logistics performance. *International Journal of Operations & Production Management*, *27*(8), 901–918. doi:10.1108/01443570710763822

Forslund, H., & Jonsson, P. (2007). The impact of forecast information quality on supply chain performance. *International Journal of Operations & Production Management*, *27*(1), 90–107. doi:10.1108/01443570710714556

Frazelle. (2001). *Supply Chain Strategy*. Academic Press.

Frazelle, E. H. (2002b). *Supply Chain Strategy: The Logistics of Supply Chain Management*. New York: McGraw-Hill.

Froehle, C., & Roth, A. (2007). A Resource-Process Framework of New Service Development. *Production and Operations Management*, *16*(2), 169–188. doi:10.1111/j.1937-5956.2007.tb00174.x

Frohlich, M. T., & Westbrook, R. (2001). Arcs of integration: An international study of supply chain strategies. *Journal of Operations Management*, *19*(2), 185–200. doi:10.1016/S0272-6963(00)00055-3

Frosch, R. (1994). Industrial ecology: Minimizing the impact of industrial waste. *Physics Today*, *47*(11), 63–68. doi:10.1063/1.881405

Fullerton, D., & Kinnaman, T. C. (1995). Garbage, Recycling, and Illicit burning or dumping. *Journal of Environmental Economics and Management*, *29*(1), 78–91. doi:10.1006/jeem.1995.1032

Fullerton, D., & Wu, W. (1998). Policies for green design. *Journal of Environmental Economics and Management*, *36*(2), 131–148. doi:10.1006/jeem.1998.1044

Fynes, B., Voss, C., & Búrca, S. (2005). The impact of supply chain relationship dynamics on manufacturing performance. *International Journal of Operations & Production Management*, *25*(1), 6–19. doi:10.1108/01443570510572213

Fynes, B., Voss, C., & de Búrca, S. (2005). The impact of supply chain relationship quality on quality performance. *International Journal of Production Economics*, *96*(3), 339–354. doi:10.1016/j.ijpe.2004.05.008

Gabriel, Bitran, Suri, Gurumurthio, & Shiou. (2007). *Emerging Trends in Supply Chain Governance*. MIT Sloan School of Management.

Gardner, D. (2004). *Supply Chain Vector Methods for Linking the Execution of Global Business Models with Financial Performance*. J. Ross Publishing.

Gavirneni, S., Kapuscinski, R., & Tayur, S. (1999). Value of Information in Capacitated Supply Chains. *Management Science*, *45*(1), 16–24. doi:10.1287/mnsc.45.1.16

Geffen, C., & Rothenberg, S. (2000). Suppliers and environmental innovation: The automotive paint process. *International Journal of Operations & Production Management*, *20*(2), 166–186. doi:10.1108/01443570010304242

Gelders, L., Mannaert, P., & Maes, J. (1994). Manufacturing strategy performance indicators and improvement programs. *International Journal of Production Research*, *32*(4), 797–805. doi:10.1080/00207549408956971

Ghobadian, A., & Gallear, D. N. (1996). Total quality management in SMEs. *Omega*, *24*(1), 83–106. doi:10.1016/0305-0483(95)00055-0

Ghose, A. K. (2001). SMEs and environment protection. Productivity, 42(2).

Gibson, B. J., Mentzer, J. T., & Cook, R. L. (2005). Supply chain management: The pursuit of a consensus definition. *Journal of Business Logistics*, *26*(2), 17–25. doi:10.1002/j.2158-1592.2005.tb00203.x

Global Automotive Logistics. (2013). *Report*. Research and Markets.

Globerson, S. (1985). Issues in developing a performance criteria system for an organization. *International Journal of Production Research*, *23*(4), 639–646. doi:10.1080/00207548508904734

Goetschalckx, M., Huangb, E., & Mital, P. (2012). Robust global supply network design. *Information, Knowledge, Systems Management*, *11*, 119–130.

Golden, B., Wasil, E., & Harker, P. (1989). *The Analytic Hierarchy Process: Applications and Studies*. Heidelberg, Germany: Springer-Verlag. doi:10.1007/978-3-642-50244-6

Gonzalez-Benito, J., & Gonzalez-Benito, O. (2006). The role of stakeholder pressure and managerial values in the implementation of environmental logistics practices. *International Journal of Production Research*, *44*(7), 1353–1373. doi:10.1080/00207540500435199

Gopal, P. R., & Thakkar, J. (2012). A review on supply chain performance measures and metrics: 2000-2011. *International Journal of Productivity and Performance Management*, *61*(5), 518–547. doi:10.1108/17410401211232957

Gordon, I. R., & McCann, P. (2000). Industrial clusters: Complexes, agglomeration and/or social networks? *Urban Studies (Edinburgh, Scotland)*, *37*(3), 513–532. doi:10.1080/0042098002096

Goss, D., & Jone, R. (1997). Organization structure and SME training provision. *International Small Business Journal*, *10*(4), 13–25. doi:10.1177/026624269201000401

Goss, R. (1990). Economic policies and seaports: 4. Strategies for port authorities. *Maritime Policy & Management*, *17*(4), 273–287. doi:10.1080/03088839000000033

Grant, R. M. (2005). *Contemporary Strategy Analysis*. Oxford, UK: Blackwell Publishing.

Green, K. W. Jr, Whitten, D., & Inman, R. A. (2008). The impact of logistics performance on organizational performance in a supply chain context. *Supply chain management. International Journal (Toronto, Ont.)*, *13*(4), 317–327.

Green, K. W. Jr, Whitten, D., & In, R. A. (2008). The impact of logistics performance on organizational performance in a supply chain context. *Supply Chain Management: An International Journal*, *13*(4), 317–327. doi:10.1108/13598540810882206

Green, K., Morton, B., & New, S. (1998). Case study green purchasing and supply policies: Do they improve companies' environmental performance? *Supply Chain Management*, *3*(2), 89–95. doi:10.1108/13598549810215405

Grigg, N. P., & McAlinden, C. (2001). A new role for ISO 9000 in the food industry? Indicative data from the UK and mainland Europe. *British Food Journal*, *103*(9), 644–656. doi:10.1108/00070700110407003

Grönroos, C. (1984). A service quality model and its marketing implications. *European Journal of Marketing*, *18*(4), 36–44. doi:10.1108/EUM0000000004784

Gumbus, A. (2005). Introducing the balanced scorecard: Creating metrics to measure performance. *Journal of Management Education, 29*(4), 617–630. doi:10.1177/1052562905276278

Gumus, A.-T. (2009). Evaluation of hazardous waste transportation firms by using a two step fuzzy-AHP and TOPSIS methodology. *Expert Systems with Applications, 36*(2), 4067–4074. doi:10.1016/j.eswa.2008.03.013

Gunasekaran, A. (2004). Supply Chain management: Theory and applications. *European Journal of Operational Research, 159*(2), 265–268. doi:10.1016/j.ejor.2003.08.015

Gunasekaran, A., & Kobu, B. (2007). Performance measures and metrics in logistics and supply chain management: A review of recent literature (1995–2004) for research and applications. *International Journal of Production Research, 45*(12), 2819–2840. doi:10.1080/00207540600806513

Gunasekaran, A., Okko, P., Martikainen, T., & Yli-Olli, P. (1996). Improving productivity and quality in small and medium enterprises: Cases and analysis. *International Small Business Journal, 15*(1), 59–72. doi:10.1177/0266242696151004

Gunasekaran, A., Patel, C., & McGaughey, R. E. (2004). A framework for supply chain performance measurement. *International Journal of Production Economics, 87*(3), 333–347. doi:10.1016/j.ijpe.2003.08.003

Gunasekaran, A., Patel, C., & Tirtiroglu, E. (2001). Performance measures and metrics in a supply chain environment. *International Journal of Operations & Production Management, 21*(1/2), 71–87. doi:10.1108/01443570110358468

Gupta, T. K., & Vikram, S. (2012). Service Quality in Supply Chain: A Review. *IACSIT International Journal of Engineering and Technology, 2*(8), 1395–1398.

Guruswamy, M., Sharma, K., & Jos, M. M. (2007). *FDI in Retail – III Implications of Wal Mart's Backdoor Entry.* Centre for Policy Alternatives.

Gustin, C. M., Daugherty, P. J., & Stank, T. P. (1995). The effects of information availability on logistics integration. *Journal of Business Logistics, 16*(1), 1–21.

Gyan Prakash. (2011). *Poor's Supply Chain: Indian Public Distribution System Revisited.* Submission of Paper for the Biennial Supply Chain Management Conference, organized by IIM Bangalore.

Haezendonck, E. (2001). *Essays on Strategy Analyisis for Seaports.* Garant.

Hafliðason, T., Ólafsdóttir, G., Bogason, S., & Stefánsson, G. (2012). Criteria for temperature alerts in cod supply chains. *International Journal of Physical Distribution & Logistics Management, 42*(4), 355–371. doi:10.1108/09600031211231335

Haghighi, M., Divandari, A., & Keimasi, M. (2010). The impact of 3D e-readiness on e-banking development in Iran: A fuzzy AHP analysis. *Expert Systems with Applications, 37*(6), 4084–4093. doi:10.1016/j.eswa.2009.11.024

Hair, J., Andeson, R., Black, B., Barry, B., & Tatham, R. (2007). *Multivariate Data Analysis* (5th ed.). New York, NY: Macmillan Publishing Company.

Handfield, R. B., & Nichols, E. L. (1999). *Introduction to Supply Chain Management.* Upper Saddle River, NJ: Prentice Hall.

Harrington, L. (1996, July15). Untapped savings abound. *Industry Week*, 53–58.

Harrison, A., & New, C. (2002). The role of coherent supply chain strategy and performance management in achieving competitive advantage: An international survey. *The Journal of the Operational Research Society, 53*(3), 263–271. doi:10.1057/palgrave.jors.2601193

Harrison, A., & van Hoek, R. (2008). *Logistics Management and Strategy: Competing Through The Supply Chain*. London: Prentice-Hall.

Haydock, P. M. (2003). Supply Chain Intelligence. *ASCET*, *5*, 15–21.

Hayes, R. H., & Abernathy, W. J. (1980). Managing our way to economic decline. *Harvard Business Review*, *62*, 95–101.

Hazen, B. T., & Byrd, T. A. (2012). Toward creating competitive advantage with logistics information technology. *International Journal of Physical Distribution & Logistics Management*, *42*(1), 8–35. doi:10.1108/09600031211202454

Hertz, S., & Alfredsson, M. (2003). Strategic development of third party logistics providers. *Industrial Marketing Management*, *32*(2), 139–149. doi:10.1016/S0019-8501(02)00228-6

Hervani, A. A., Helms, M. M., & Sarkis, J. (2005). Performance measurement for green supply chain management. *Benchmarking: An International Journal*, *12*(4), 330–353. doi:10.1108/14635770510609015

Higginson, J. K., & Alam, A. (1997). Supply Chain Management Techniques in Medium-to-Small Manufacturing Firms. *The International Journal of Logistics Management*, *8*(2), 19–32. doi:10.1108/09574099710805646

Hill, C. A., & Scudder, G. D. (2002). The use of electronic data interchange for supply chain coordination in the food industry. *Journal of Operations Management*, *20*(4), 375–387. doi:10.1016/S0272-6963(02)00017-7

History of HACCP. (1959). Retrieved on August, 15, 2013, from http://foodsafety.unl.edu/haccp/start/gettingstarted.html

Ho, D. C., Chan, W. E., Wong, N., & Chan, M. (2000). Significant metrics for facilities management benchmarking in the Asia Pacific region. *Facilities*, *18*(13/14), 545–555. doi:10.1108/02632770010358088

Hoffman, A., & Ventresca, M. (1999). The institutional framing of policy debates. *The American Behavioral Scientist*, *42*(8), 1368–1392, 1. doi:10.1177/00027649921954903

Hoffman, D. (2004). *The hierarchy of supply chain metrics: diagnosing your supply chain health*. Boston, Massachusetts: AMR Research.

Holweg, M. (2008). The Evolution of Competition in the Automotive Industry. In G. Parry & A. Graves (Eds.), *Build To Order*. London: Springer. doi:10.1007/978-1-84800-225-8_2

Hong, P., & Jeong, J. (2006). Supply chain management practices of SMEs: From a business growth perspective. *Journal of Enterprise Information Management*, *19*(3), 292–302. doi:10.1108/17410390610658478

Ho, W. (2008). Integrated analytic hierarchy process and its applications - A literature review. *European Journal of Operational Research*, *186*(1), 211–228. doi:10.1016/j.ejor.2007.01.004

Hsu, H. M., & Chen, C. T. (1996). Aggregation of fuzzy opinions under group decision making. *Fuzzy Sets and Systems*, *79*(3), 279–285. doi:10.1016/0165-0114(95)00185-9

Hsu, S. H., Kao, C.-H., & Wu, M.-C. (2009). Design facial appearance for roles in video games. *Expert Systems with Applications*, *36*(3), 4929–4934. doi:10.1016/j.eswa.2008.05.049

Hsu, Y.-L., Lee, C.-H., & Kreng, V. B. (2010). The application of Fuzzy Delphi Method and Fuzzy AHP in lubricant regenerative technology selection. *Expert Systems with Applications*, *37*(1), 419–425. doi:10.1016/j.eswa.2009.05.068

Huang, S. H., Sheoran, S. K., & Keskar, H. (2005). Computer-assisted supply chain configuration based on supply chain operations reference (SCOR) model. *Computers & Industrial Engineering*, *48*(2), 377–394. doi:10.1016/j.cie.2005.01.001

Huan, S. H., Sheoran, S. K., & Wang, G. (2004). A review and analysis of supply chain operations reference (SCOR) model. *Supply Chain Management*, *9*(1), 23–29. doi:10.1108/13598540410517557

Hui, I. K., He, L., & Dang, C. (2003). Hierarchical environmental impact evaluation of a process in printed circuit board manufacturing. *International Journal of Production Research*, *41*(6), 1149–1165. doi:10.1080/0020754021000043679

Hulland, J., Wade, M., & Antia, K. D. (2007). The impact of capabilities and prior investments on online channel commitment and performance. *Journal of Management Information Systems*, *23*(4), 109–142. doi:10.2753/MIS0742-1222230406

Humphrey, P. K., Li, W. L., & Chan, L. Y. (2004). The impact of supplier development on buyer–supplier performance. *International Journal of Management Sciences*, *32*(2), 131–143.

Hum, S. H. (2000). A Hayes-Wheelwright framework approach for strategic management of third party logistics service. *Integrated Manufacturing Systems*, *11*(2), 132–137. doi:10.1108/09576060010314125

Hutchison, P.D., Farris, M.T., & Gary, M.F. (2009). Supply chain cash-to-cash: a strategy for the 21st century. *Strategic Finance*, 41–48.

Hwang, C. L., & Yoon, K. (1981). *Multiple Attribute Decision Making: Methods and Applications*. Berlin: Springer. doi:10.1007/978-3-642-48318-9

IBM. (2012). *The Smarter Supply Chain of the future*. IBM.

Ic, Y. (2012). An experimental design approach using TOPSIS method for the selection of computer-integrated manufacturing technologies. *Robotics and Computer-integrated Manufacturing*, *28*(2), 245–256. doi:10.1016/j.rcim.2011.09.005

Iç, Y. T., & Yurdakul, M. (2009). Development of a decision support system for machining center selection. *Expert Systems with Applications*, *36*(2), 3505–3513. doi:10.1016/j.eswa.2008.02.022

IFST. (1991). *Food and Drink-Good Manufacturing Practice: A Guide to Its Responsible Management, 3*. Institute of Food Science and Technology.

IJMBS. (2013, January 24). *India Retail News*. Retrieved January 27, 2013, from Emergence of Retailing Sector in India: Challenges and Opportunities: http://www.indiaretailnews.com/

International Labour Office (ILO). (2001). *Self-regulation in the workplace: Codes of conduct, social labelling and socially responsible investment*. MCC Working Paper No. 1, Series on Management Systems and Corporate Citizenship, Geneva, Management and Corporate Citizenship Program, Job Creation and Enterprise Development Department.

Interview articles of logistics executives from various companies

Iyad, K. (2013). *20/20 Vision - Logistics at the end of 2020*. Paper presented at the 13th Annual Logistics CIO and Supply Chain Technology Forum.

Jackson, L. (2006). Towards a South African model for food safety auditor registration. *British Food Journal*, *108*(1), 16–26. doi:10.1108/00070700610637607

Jedermann, R., Garcia, L.R., & Lang, W. (2009). Spatial temperature profiling by semi-passive RFID loggers for perishable food transportation. *Computers and Electronics in Agriculture*, *65*(2), 145–154.

Johnson, H. T. (1981). Toward a new understanding of nineteenth-century cost accounting. *Accounting Review*, *56*(3), 510–518.

Johnson, H. T., & Kaplan, R. S. (1987). *Relevance Lost: The Rise and Fall of Management Accounting*. Harvard Business School Press.

Jones, R. M., & Towill, D. R. (1999). Total cycle time compression and the agile supply chain. *International Journal of Production Economics*, *62*(1-2), 61–73. doi:10.1016/S0925-5273(98)00221-7

Kafetzopoulos, D. P., Evangelos, L., Psomas, E. L., & Kafetzopoulos, P. D. (2013). Measuring the effectiveness of the HACCP Food Safety Management System. *Food Control*, *33*(2), 505–513.

Kagan, R. A., Gunningham, N., & Thornton, D. (2003). Explaining corporate environmental performance: How does regulation matter? *Law & Society Review*, *37*(1), 51–90. doi:10.1111/1540-5893.3701002

Kahraman, C., Ates, N. Y., Çevik, S., Gülbay, M., & Erdogan, S. A. (2007). Hierarchical fuzzy TOPSIS model for selection among logistics information technologies. *Journal of Enterprise Information Management*, *20*(2), 143–168. doi:10.1108/17410390710725742

Kahraman, C., & Kaya, I. (2010In Press). *A fuzzy multicriteria methodology for selection among energy alternatives. Expert Systems with Applications.* Corrected Proof; doi:10.1016/j.eswa.2010.1002.1095

Kannan, G., Pokharel, S., & Kumar, S. P. (2009). A hybrid approach using ISM and fuzzy TOPSIS for the selection of reverse logistics provider. *Resources, Conservation and Recycling*, *54*(1), 28–36. doi:10.1016/j.resconrec.2009.06.004

Kaplan, R. S. (1984). Yesterday's accounting undermines production. *Harvard Business Review*, *62*(4), 95–101.

Kaplan, R. S., & Norton, D. P. (1992, January). The balance Scorecard- Measures That Drive performance. *Harvard Business Review*.

Kara, S. S. (2011). Supplier selection with an integrated methodology in unknown environment. *Expert Systems with Applications*, *38*(3), 2133–2139. doi:10.1016/j.eswa.2010.07.154

Kasi, V. (2005). Systemic Assessment of SCOR for Modeling Supply Chains. *Proceedings of the 38th Hawaii International Conference on System Sciences*. doi:10.1109/HICSS.2005.574

Kathikeyan, M. & Teshome, T., (2010). *Dairy supply chain management: A case of Ada'a Dairy Cooperative in Ethiopia.* LAP Lambert Academic Publishing GmbH & Co. KG.

Kaya, T., & Kahraman, C. (2011). Multicriteria decision making in energy planning using a modified fuzzy TOPSIS methodology. *Expert Systems with Applications*, *38*(6), 6577–6585. doi:10.1016/j.eswa.2010.11.081

Kearney, A. T. (1994). Achieving customer satisfaction through logistics excellence. *Managing Service Quality*, *4*(2), 47–50. doi:10.1108/09604529410796224

Keebler, J. S., Manradt, K. B., Durtsche, D. A., & Ledyard, D. M. (n.d.). *Keeping score: Measuring the business value of logistics in the supply chain.* Accession No-00770539.

Keegan, D. P., Eiler, R. G., & Jones, C. R. (1989, June). Are your performance measures obsolete? *Management Accounting*, 45-50.

Khosla, R., Goonesekera, T., & Chu, M.-T. (2009). Separating the wheat from the chaff: An intelligent sales recruitment and benchmarking system. *Expert Systems with Applications*, *36*(2), 3017–3027. doi:10.1016/j.eswa.2008.01.090

Khurrum,, S., Bhutta, M., Arif, I. R., & Asad, U. (2007). SCM practices and the health of SMEs in Pakistan. *Supply Chain Management: An International Journal*, *12*(6), 412–422. doi:10.1108/13598540710826344

Kim, B., & H. Oh, (2005). The impact of decision making sharing between supplier and manufacturer on their collaboration performance. *Supply Chain Management: An International Journal*, 223-36.

Kim, S., Lee, K., Cho, J. K., & Kim, C. O. (2011). Agent-based diffusion model for an automobile market with fuzzy TOPSIS-based product adoption process. *Expert Systems with Applications*, *38*(6), 7270–7276. doi:10.1016/j.eswa.2010.12.024

Kim, Y., Chung, E., Jun, S.-M., & Kim, S. U. (2013). Prioritizing the best sites for treated waste water in stream use in an urban watershed using fuzzy TOPSIS. *Resources, Conservation and Recycling*, *73*, 23–32. doi:10.1016/j.resconrec.2012.12.009

Kittipanya-ngam, P., Shi, Y., & Gregory, M. J. (2011). Exploring geographical dispersion in Thailand-based food supply chain (FSC), Benchmarking. *An International Journal*, *8*(6), 802–833.

Klug. (2006). *Logistics implications of global production networks in car manufacturing*. University of Applied Sciences Munich.

Koh, S. C. L., & Maguire, S. (2004). Identifying the adoption of e-business and knowledge management within SMEs. *Journal of Small Business and Enterprise Development*, *11*(3), 338–348. doi:10.1108/14626000410551591

Konecka-Matyjek, E., Turlejska, H., Pelzner, U., & Szponar, L. (2005). Actual situation in the area of implementing quality assurance systems GMP, GHP and HACCP in Polish food production and processing plants. *Food Control*, *16*(1), 1–9. doi:10.1016/j.foodcont.2003.10.007

Koo, K. E., Nurulazam, M. D. A., Rohaida, M. Z. S., Teo, T. G., & Salleh, Z. (2014). Examining The Potential of Safety Knowledge As Extension Construct For Theory of Planned Behaviour: Explaining Safety Practices of Young Adults At Engineering Laboratories And Workshops. *Procedia: Social and Behavioral Sciences*, *116*, 1513–1518. doi:10.1016/j.sbspro.2014.01.426

Kopicki, R., Berg, M. B., & Legg, L. (1993). *Reuse and Recycling – Reverse Logistics Opportunities*. Oak Brook, IL: Council of Logistics Management.

KPMG. (2011, July). *Issues Monitor: Sharing knowledge on topical issues in Retail Industry*. KPMG.

Kremers, L. (2010). The link between supply chain and finance. *Supply Chain Asia*, 22–25.

Krugman, P. R. (1991). *Geography and trade*. MIT Press.

Krugman, P. R. (1998). What's new about the new economic geography? *Oxford Review of Economic Policy*, *14*(2), 7–17. doi:10.1093/oxrep/14.2.7

Kruse, G. (2004). *Managing The Extended Supply Chain, Opportunities and Solutions For Manufacturing Migration*. Retrieved April 6, 2014 from http://www.littoralis.info/iom/secure/assets/iom20040901.775512_417681c8304c.pdf

Kukalis, S. (1989). The relationship among firm characteristics and design of strategic planning systems in large organizations. *Journal of Management*, *15*(4), 565–579. doi:10.1177/014920638901500406

Kulkarni, M. B. (1979). Collection of Milk: Some Lacunae and Remedies. *Dairy Guide*, *1*, 31–34.

Kulkarni, S. (1997). A case for the review of present food laws relating to dairy products in India. *Indian Journal of Dairy and Biosciences*, *8*, 45–47.

Kulp, S. C., Lee, H. L., & Ofek, E. (2004). Manufacturer Benefits from Information Integration with Retail Customers. *Management Science*, *50*(4), 431–444. doi:10.1287/mnsc.1030.0182

Kumar, S., & Vaidya, O. (2006). Analytic hierarchy process: An overview of applications. *European Journal of Operational Research*, *169*(1), 1–29. doi:10.1016/j.ejor.2004.04.028

Kunadhamraks, P., & Hanaoka, S. (2008). Evaluating the logistics performance of intermodal transportation in Thailand. *Asia Pacific Journal of Marketing and Logistics*, *20*(3), 323–342. doi:10.1108/13555850810890084

Labib, A., & Shah, J. (2001). Management decisions for a continuous improvement process in industry using the Analytical Hierarchy Process. *Journal of Work Study*, *50*(5), 189–193. doi:10.1108/EUM0000000005678

Lai, K. H., Ngai, E. T., & Cheng, T. E. (2002). Measures for evaluating supply chain performance in transport logistics. *Transp Res Part E*, *38*(6), 439–456. doi:10.1016/S1366-5545(02)00019-4

Lai, W.-H., & Tsai, C.-T. (2009). Fuzzy rule-based analysis of firm's technology transfer in Taiwan's machinery industry. *Expert Systems with Applications*, *36*(10), 12012–12022. doi:10.1016/j.eswa.2009.03.054

Lambert, D. M., & Pohlen, T. L. (2001). Supply chain metrics. *The International Journal of Logistics Management*, 1-19.

Lambert, D. M., & Cooper, M. C. (2000). Issues in supply chain management. *Industrial Marketing Management*, *29*(1), 65–83. doi:10.1016/S0019-8501(99)00113-3

Lambert, D. M., Cooper, M. C., & Pagh, J. D. (1998). Supply chain management: Implementation issues and research opportunities. *International Journal of Logistic Management*, *9*(2), 1–16. doi:10.1108/09574099810805807

Lambert, D. M., García-Dastugue, S. J., & Croxton, K. L. (2005). An evaluation of processoriented supply chain management frameworks. *Journal of Business Logistics*, *26*(1), 25–51. doi:10.1002/j.2158-1592.2005.tb00193.x

Lamming, R., Johnsen, T., Zheng, J., & Harland, C. (2000). An initial classification of supply networks. *International Journal of Operations & Production Management*, *20*(6), 675–691. doi:10.1108/01443570010321667

Landro, L. (2006, October 4). Hospitals go 'green' to cut toxins. *The Wall Street Journal*.

Langen, P. D. (2002). Clustering and performance: The case of maritime clustering in The Netherlands. *Maritime Policy & Management*, *29*(3), 209–221. doi:10.1080/03088830210132605

Lassar, W., Haar, J., Montalvo, R., & Hulser, L. (2012). Determinants of Strategic Risk Management in Emerging Markets Supply Chains: The Case of Mexico. *Journal of Economics, Finance and Administrative Science*.

Leavitt, H. (1972). *The Volatile Organization: Everything Triggers Everything Else*. Chicago: University of Chicago Press.

Lebas, M. J. (1995). Performance measurement and performance management. *International Journal of Production Economics*, *41*(1-3), 23–35. doi:10.1016/0925-5273(95)00081-X

Lee, C. W., Kwon, I. G., & Severance, D. (2007). Relationship between supply chain performance and degree of linkage among supplier, internal integration, and customer. *Supply Chain Management: An International Journal*, *12*(6), 444–452. doi:10.1108/13598540710826371

Lee, E., Park, Y., & Shin, J. G. (2009). Large engineering project risk management using a Bayesian belief network. *Expert Systems with Applications*, *36*(3), 5880–5887. doi:10.1016/j.eswa.2008.07.057

Lee, H. L., & Billington, C. (1993). Material Management in Decentralized Supply Chains. *Operations Research*, *41*(5), 835–847. doi:10.1287/opre.41.5.835

Lee, H. L., Padmanabhan, V., & Whang, S. (1997). Information distortion in a supply chain: The bullwhip effect. *Management Science*, *43*(4), 546–558. doi:10.1287/mnsc.43.4.546

Lee, H. L., Padmanabhan, V., & Whang, S. (1997). The Bullwhip Effect in Supply Chains. *Sloan Management Review*, 93–101.

Lee, S.-H. (2010In Press). *Using fuzzy AHP to develop intellectual capital evaluation model for assessing their performance contribution in a university. Expert Systems with Applications*. Corrected Proof; doi:10.1016/j.eswa.2009.1012.1020

Lenny, S. C. (2007). The impact of supply chain management practices on performance of SMEs. *Industrial Management & Data Systems, 107*(1), 103–124. doi:10.1108/02635570710719089

Levy, D. L. (1995). The Environmental practices and performance of transnational companies. *Trans.Corpor., 4*, 45–67.

Levy, D. L. (1997, Winter). Lean production in an international supply chain. *Sloan Management Review*, 94–102.

Levy, M., & Powell, P. (1998). SME flexibility and the role of information systems. *Small Business Economics, 11*(2), 183–196. doi:10.1023/A:1007912714741

Liang, G. S. (1999). Fuzzy MCDM based on ideal and anti-ideal concepts. *European Journal of Operational Research, 112*(3), 682–691. doi:10.1016/S0377-2217(97)00410-4

Liang, L., Yang, F., Cook, W. D., & Zhu, J. (2006). DEA models for supply chain efficiency evaluation. *Annals of Operations Research, 145*(1), 35–49. doi:10.1007/s10479-006-0026-7

Liberatore, M., & Nydick, R. (2008). The analytic hierarchy process in medical and health care decision making: A literature review. *European Journal of Operational Research, 189*(1), 194–207. doi:10.1016/j.ejor.2007.05.001

Li, D. F. (2005). Multi attribute decision making models and methods using intuitionistic fuzzy sets. *Journal of Computer and System Sciences, 70*(1), 73–85. doi:10.1016/j.jcss.2004.06.002

Lim, J., Gan, B., & Chang, T.-T. (2002). A Survey on NSS Adoption Intention. *Proceedings of the 35th Hawaii International Conference on System Sciences (HICSS)*. doi:10.1109/HICSS.2002.993888

Lim, R. Y. G., Baines, T., Tjahjono, B., & Chandraprakaikul, W. (2006). Integrated strategic supply chain positioning for SMEs: An empirical study. *The International Journal of Logistics Management, 17*(2), 260–276. doi:10.1108/09574090610689989

Lin, C. (2006). Influencing factors on the innovation in logistics technologies for logistics service providers in Taiwan. *The Journal of American Academy of Business, 9*(2), 257–263.

Lin, C. T., & Chen, Y. T. (2004). Bid/no-bid decision making – a fuzzy linguistic approach. *International Journal of Project Management, 22*(7), 585–593. doi:10.1016/j.ijproman.2004.01.005

Lin, C.-L., Chen, C.-W., & Tzeng, G.-H. (2010In Press). *Planning the development strategy for the mobile communication package based on consumers' choice preferences. Expert Systems with Applications*. Corrected Proof. doi:10.1016/j.eswa.2009.1011.1009

Lin, F., & Shaw, M. J. (1998). Reengineering the order fulfillment process in supply chain networks. *International Journal of Flexible Manufacturing Systems, 10*(3), 197–229. doi:10.1023/A:1008069816606

Lin, H.-F., & Lee, G.-G. (2004). Perceptions of senior managers toward knowledge-sharing behaviour. *Management Decision, 42*(1), 108–125. doi:10.1108/00251740410510181

Lin, M.-C., Wang, C.-C., Chen, M.-S., & Chang, C.-A. (2008). Using AHP and TOPSIS approaches in customer-driven product design process. *Computers in Industry, 59*(1), 17–31. doi:10.1016/j.compind.2007.05.013

Li, R. J. (1999). Fuzzy method in group decision making. *Computers & Mathematics with Applications (Oxford, England), 38*(1), 91–101. doi:10.1016/S0898-1221(99)00172-8

Li, S., & Li, J. Z. (2009). Hybridising human judgment, AHP, simulation and a fuzzy expert system for strategy formulation under uncertainty. *Expert Systems with Applications, 36*(3), 5557–5564. doi:10.1016/j.eswa.2008.06.095

Li, S., Ragu-Nathanb, B., Ragu-Nathanb, T. S., & Rao, S. S. (2006). The impact of supply chain management practices on competitive advantage and organizational performance. *Omega International Journal of Management Science*, *34*(2), 107–124. doi:10.1016/j.omega.2004.08.002

Li, T.-S., & Huang, H.-H. (2009). Applying TRIZ and Fuzzy AHP to develop innovative design for automated manufacturing systems. *Expert Systems with Applications*, *36*(4), 8302–8312. doi:10.1016/j.eswa.2008.10.025

Little, A. D. (2007). *Innovation Excellence in Logistics*. European Logistics Association.

Liu, C.-C., & Chen, S.-Y. (2009). Prioritization of digital capital measures in recruiting website for the national armed forces. *Expert Systems with Applications*, *36*(5), 9415–9421. doi:10.1016/j.eswa.2008.12.051

Li, Y. W., Nie, X. T., & Chen, S. Y. (2007). Fuzzy approach to prequalifying construction contractors. *Journal of Construction Engineering and Management*, *133*(1), 40–49. doi:10.1061/(ASCE)0733-9364(2007)133:1(40)

Li, Y., Tang, J., & Luo, X. (2010In Press). *An ECI-based methodology for determining the final importance ratings of customer requirements in MP product improvement. Expert Systems with Applications*. Uncorrected Proof. doi:10.1016/j.eswa.2010.1002.1100

Llore'ns-Montes, F. J., & Molina, L. M. (2006). Six Sigma and management theory: Processes, content and effectiveness. *Total Quality Management and Business Excellence*, *17*(4), 485–506. doi:10.1080/14783360500528270

Lockamy, A. III, & McCormack, K. (2004). Linking SCOR planning practices to supply chain performance. *International Journal of Operations & Production Management*, *24*(12), 1192–1218. doi:10.1108/01443570410569010

Lummus, R. R., & Vokurka, R. J. (1999). Defining supply chain management: A historical perspective and practical guidelines. *Industrial Management & Data Systems*, *99*(1), 11–17. doi:10.1108/02635579910243851

Mabert, V. A., & Venkataramanan, M. A. (1998). Special research focus on supply chain linkages: Challenges for design and management in the 21st century. *Decision Sciences*, *29*(3), 537–552. doi:10.1111/j.1540-5915.1998.tb01353.x

Macpherson, A., & Wilson, A. (2003). Enhancing SMEs' capability: Opportunities in supply chain relationships? *Journal of Small Business and Enterprise Development*, *10*(2), 167–179. doi:10.1108/14626000310473193

Maguire, S., & Magrys, A. (2001). The potential use of IT to gain competitive advantage for small and medium-sized enterprises. In *Proceedings of the 9th Annual High Technology Small Firms Conference*. Manchester Business School.

Mahajan, R., Garg, S., & Sharma, P. B. (2011). Indian frozen peas market: A case study on FPIL. *International Journal of Globalisation and Small Business*, *4*(2), 154–169. doi:10.1504/IJGSB.2011.042250

Mangan, J., Lalwani, C., & Fynes, B. (2008). Port-centric logistics. *International Journal of logistics Management*, *19*(1), 29-41.

Mangan, J., & Christopher, M. (2005). Management development and supply chain manager of the future. *The International Journal of Logistics Management*, *16*(2), 178–191. doi:10.1108/09574090510634494

Manjunath, M S. (2013). *India centric challenges*. Ingersoll Rand International (India) Ltd.

Manufacturing Technology Committee-Risk Management Working Group. (n.d.). *Risk Management Training Guide: Risk Ranking and Filtering*. Retrieved June 14, 2013 from http://www.pqri.org/pdfs/MTC/Risk_Rank_Filter_Training_Guide.pdf

Maritime Workforce Development Forum. (2013). *Australian Maritime workforce development strategy*. Retrieved 19 November 2013 from: at http://www.shipping.infrastructure.gov.au/ mwdf/ files/ Australian_Maritime_Workforce_Development_Strategy.pdf

Marquez, A. C. (2010). *Dynamic Modelling for Supply Chain Management: Dealing with Front-End, Back-End and Integration Issues*. London: Springer. doi:10.1007/978-1-84882-681-6

Marshall, A. (1890). *Principles of Economies* (Vol. 1). London: Macmillan.

Marshall, A. (1920). *Principles of Economics*. London: Macmillan.

Martin, R., & Sunley, P. (1996). Paul Krugman's geographical economics and its implications for regional development theyry: A critical assessment. *Economic Geography*, *72*(3), 261–292. doi:10.2307/144401

Martin, R., & Sunley, P. (2001). Deconstructing clusters: Chaotic concept or plicy panacea. *Journal of Economic Geography*, 2–60.

Marvick, D., & White, J. (1998). Distribution operations: Managing distribution facilities for strategic advantage. In J. G. Gattorna (Ed.), *Strategic Supply Chain Alignment*.

Marwah, A. K., Thakar, G., & Gupta, R. C. (2013). An Empirical Study of Supplier-Buyer Relations as a Key Determinant of Supply Chain Performance of Indian Manufacturing Organizations. In *Proceedings of the International Conference on Managing Supply Chain for Global Competitiveness*. IIIE.

Marwah, A. K., Thakar, G., & Gupta, R. C. (2014). A Confirmatory Study of Supply Chain Performance and Competitiveness of Indian Manufacturing Organizations. *International Journal of Quality Research*, *8*(1), 23–38.

Maskell, P., & Lorenzen, M. (2004). The cluster as market organisation. *Urban Studies (Edinburgh, Scotland)*, *41*(5-6), 991–1009. doi:10.1080/0042098041000l675878

Massilia, R. M. R., Melgar, J. M., Fortuny, R. S., & Belloso, O. M. (2009). Control of Pathogenic and Spoilage Microorganisms in Fresh-cut Fruits and Fruit Juices by Traditional and Alternative Natural Antimicrobials. *Comprehensive Reviews in Food Science and Food Safety*, *8*(3), 157–180. doi:10.1111/j.1541-4337.2009.00076.x

Matapurkar, A., & Sinha, A. (2011). Cold chain: finally warming up to India. *Infrastructure Today*, 22-25.

McAdam, R. (2002). Large scale innovation – reengineering methodology in SMEs: Positivistic and phenomenological approaches. *International Small Business Journal*, *20*(1), 33–50. doi:10.1177/0266242602201004

McCartan-Quinn, D., & Carson, D. (2003). Issues which impact upon marketing in the small firm. *Small Business Economics*, *21*(2), 201–213. doi:10.1023/A:1025070107609

McCarthya, M., Cluzel, E., Dressel, K., & Newton, R. (2013). Food and health research in Europe: Structures, gaps and futures. *Food Policy*, *39*, 64–71. doi:10.1016/j.foodpol.2012.12.005

McCarthy, T. M., & Golicic, S. L. (2002). Implementing collaborative forecasting to improve supply chain performance. *International Journal of Physical Distribution & Logistics Management*, *32*(6), 431–454. doi:10.1108/09600030210437960

McKinnon, A. C. (1996). *The development of retail logistics in the UK:A position paper, Technology Foresight: Retail and Distribution Panel*. Edinburgh: Heriot-Watt University.

McMeekin, T. A., Smale, N., Jenson, I., Ross, T., & Tanner, D. (2006b). Combining microbial growth models with near real-time temperature monitoring technologies to estimate the shelf life and safety of foods during processing and distribution. In *Proceedings of the 2nd International Workshop Cold Chain-Management*. University of Bonn.

McMeekin, T., Baranyi, J., Bowman, J., Dalgaard, P., Kirk, M., Ross, T., & Zwietering, M. H. et al. (2006). Information systems in food safety management. *International Journal of Food Microbiology*, *112*(3), 181–194. doi:10.1016/j.ijfoodmicro.2006.04.048 PMID:16934895

Meera, K., & Rajan, H. (2007). *Machine to Machine: A fresh approach to profits from perishables*. Infosys Whitepaper.

Mejza, M., Barnard, R., Corsi, T. M., & Keane, T. (2003). Driver management practices of motor carriers with high compliance and safety performance. *Transportation Journal*, *42*, 16–29.

Mello, J. E., & Stank, T. P. (2005). Linking firm culture and orientation to supply chain success. *International Journal of Physical Distribution & Logistics Management*, *35*(8), 542–554. doi:10.1108/09600030510623320

Melnyk, S. A., Theodore, P. S., & David, J. C. (2000). Supply chain management at Michigan state university: The journey and the lessons learned. *Production and Inventory Management Journal*, *41*, 13–18.

Melnyk, S., Narasimhan, R., & DeCampos, H. (2014). Supply chain design: Issues, challenges, frameworks and solutions. *International Journal of Production Research*, *52*(7), 1887–1896. doi:10.1080/00207543.2013.787175

Melville, N., Kraemer, K., & Gurbaxani, V. (2004). Information technology and organizational performance: An integrative model of IT business value. *Management Information Systems Quarterly*, *28*(2), 283–322.

Mentzer, J. T., Foggin, J. H., & Golicic, S. L. (2000, September/October). Collaboration: the enablers, impediments, and benefits. *Supply Chain Management Review*.

Mentzer, J. T. (Ed.). (2001). *Supply Chain Management*. Thousand Oaks, CA: Sage Publications.

Mentzer, J. T., DeWitt, W., Keebler, J. S., Min, S., Nix, N. W., Smith, C. D., & Zacharia, Z. G. (2001). Defining supply chain management. *Journal of Business Logistics*, *22*(2), 1–25. doi:10.1002/j.2158-1592.2001.tb00001.x

Mentzer, J. T., Flint, D. J., & Hult, G. T. (2001). Logistics service quality as a segment-customized process. *Journal of Marketing*, *65*(4), 82–104. doi:10.1509/jmkg.65.4.82.18390

Mentzer, J. T., Flint, D. J., & Kent, J. L. (1999). Developing a logistics service quality scale. *Journal of Business Logistics*, *20*(1), 9–32.

Messelbeck, J., & Whaley, M. (1999).Greening the health care supply chain: Triggers of change, models for success. *Corporate Environmental Strategy*, *6*(1), 39–45.

Mezgar, I., Kovacs, G., & Paganelli, P. (2000). Co-operative production planning for small- and medium-sized enterprises. *International Journal of Production Economics*, *64*(1-3), 37–48. doi:10.1016/S0925-5273(99)00032-8

Mikhailov, L. (2004). A fuzzy approach to deriving priorities from interval pair-wise comparison judgments. *European Journal of Operational Research*, *159*(3), 687–704. doi:10.1016/S0377-2217(03)00432-6

Miller, J. (1966). *The assessment of Worth: a systematic procedure and its experimental validation*. MIT.

Miller, J. (1969). Assessing alternative transportation systems. In *Memorandum RM-5865-DOR*. The RAND Corporation.

Miller, J. (1970). *Professional decision-making: a procedure for evaluation complex alternatives*. New York: Praeger Publishers.

Min, H., & Galle, W.P. (1997). Green purchasing strategies: trends and implications. *Int. J. Purch. Mater. Manag.*, 10–17.

Ministry of Food Processing Industries. (2012). Retrieved on July 10, 2012, from http://mofpi.nic.in

Moberg, C. R., Whipple, T. W., Cutler, B. D., & Speh, T. W. (2004). Do the Management Components of Supply Chain Management Affect Logistics Performance? *The International Journal of Logistics Management*, *15*(2), 15–30. doi:10.1108/09574090410700284

Mohapatra, B. B. (2013). *Indian Vaccine Market and Ensuring Quality from Production to Distribution*. Paper presented at the 2nd Annual Cold Chain India Conference.

Monczka, R.M. & Morgan, J. (1997, January). What's wrong with supply chain management?. *Purchasing*, 69-72.

Morgan, C. (2004). Structure, speed and salience: Performance measurement in the supply chain. *Business Process Management Journal*, *10*(5), 522–536. doi:10.1108/14637150410559207

Morgan, C. (2007). Supply network performance measurement: Future challenges. *The International Journal of Logistics*, *18*(2), 255–273. doi:10.1108/09574090710816968

Motawa, I. A., Anumba, C. J., & El-Hamalawi, A. (2006). A fuzzy system for evaluating the risk of change in construction projects. *Advances in Engineering Software*, *37*(9), 583–591. doi:10.1016/j.advengsoft.2006.01.006

Moureh, J., & Flick, D. (2004). Airflow pattern and temperature distribution in a typical refrigerated Truck configuration loaded with pallets. *International Journal of Refrigeration*, *27*(5), 464–474. doi:10.1016/j.ijrefrig.2004.03.003

Mueller, R., Buergelt, D., & Seidel-Lass, L. (2007, February). *Supply Chains And Social Network Analysis*. Paper presented at 1st International European Forum on Innovation and System Dynamics in Food Networks. Innsbruck-Igls, Austria.

Muralidhar, P., Ravindranath, K., & Srihari, V. (2013). The influence of GRA and TOPSIS for assortment of green supply chain management strategies in Cement industry. *International Journal of Supply Chain Management*, *2*(1), 49–55.

Muriati, Norleyza, Salha, & Zuraidah. (2009). A Framework for Analyzing e-Supply Chains. *European Journal of Scientific Research*, *25*(4).

Naesens, K., Gelders, L., & Pintelon, L. (2007). A swift response tool for measuring the strategic fit for resource pooling: A case study. *Management Decision*, *45*(3), 434–449. doi:10.1108/00251740710745061

Naesens, K., Gelders, L., & Pintelon, L. (2009). A swift response framework for measuring the strategic fit for a horizontal collaborative initiative. *International Journal of Production Economics*, *121*(2), 550–561. doi:10.1016/j.ijpe.2007.04.004

Naghadehi, M. Z., Mikaeil, R., & Ataei, M. (2009). The application of fuzzy analytic hierarchy process (FAHP) approach to selection of optimum underground mining method for Jajarm Bauxite Mine, Iran. *Expert Systems with Applications*, *36*(4), 8218–8226. doi:10.1016/j.eswa.2008.10.006

Narasimhan, R., & Carter, J. R. (1998). *Environmental Supply Chain Management*. Tempe, AZ: The Center for Advanced Purchasing Studies, Arizona State University.

Naslund, D., & Williamson, S. (2010). What is management in supply chain management? – A critical review of definitions, frameworks and terminology. *Journal of Management Policy and Practice*, *11*, 11–28.

National Centre for Cold Chain Development. (2012). Retrieved on May 02, 2012, from http://nhm.dacnet.nic.in/NCCD/AboutUs.html

NBACFS. (2005). Good manufacturing practices for milk collection center. *Royal Gazette*, *122*, 1-20.

Neeliah, S. A., Neeliah, H., & Goburdhun, D. (2013). Assessing the relevance of EU SPS measures to the food export sector: Evidence from a developing agro-food exporting country. *Food Policy*, *41*(1), 53–62. doi:10.1016/j.foodpol.2013.04.002

Neely, A. (2003). *Business Performance Measurement: Theory and Practice*. Cambridge: Cambridge University Press.

Neely, A. (2005). The evolution of performance measurement research developments in the last decade and a research agenda for the next. *International Journal of Operations & Production Management*, *25*(12), 1264–1277. doi:10.1108/01443570510633648

Neely, A. D. (1998). *Measuring Business Performance*. London: Economist Books.

Neely, A., Bourne, M., & Kennerley, M. (1995). Performance measurement system design: Developing and testin-gaprocess-based approach. *International Journal of Operations & Production Management*, *20*(10), 1119–1145. doi:10.1108/01443570010343708

Neely, A., Gregory, M., & Platts, K. (2005). Performance measurement system design: A literature review and research agenda. *International Journal of Operations & Production Management*, *25*(12), 1228–1263. doi:10.1108/01443570510633639

Niaraki, A. S., & Kim, K. (2009). Ontology based personalized route planning system using a multi-criteria decision making approach. *Expert Systems with Applications*, *36*(2), 2250–2259. doi:10.1016/j.eswa.2007.12.053

Nissim, D., & Penman, S. H. (2001). Ratio analysis and equity valuation: From research to practice. *Review of Accounting Studies*, *6*(1), 109–154. doi:10.1023/A:1011338221623

Nix, N. (2001). Customer service in supply chain management context. In Supply Chain Management. Sage.

Noble, S. M., Haytko, D. L., & Phillips, J. (2009). What drives college-age Generation Y consumers? *Journal of Business Research*, *62*(6), 617–628. doi:10.1016/j.jbusres.2008.01.020

Ölçer, A. I., & Odabasi, A. Y. (2005). A new fuzzy multiple attributive group decision making methodology and its application to propulsion/maneuvering system selection problem. *European Journal of Operational Research*, *166*(1), 93–114. doi:10.1016/j.ejor.2004.02.010

Omta, S. W. F., Trienekens, J. H., & Beers, G. (2001). Chain and network science: A research framework. *Journal on Chain and Network Science*, *1*(1), 1–6. doi:10.3920/JCNS2001.x001

Önüt, S., Efendigil, T., & Soner Kara, S. (2009). A combined fuzzy MCDM approach for selecting shopping center site: An example from Istanbul, Turkey. *Expert Systems with Applications.* doi:.10.1016/j.eswa.2009.1006.1080

Onut, S., & Soner, S. (2008). Transshipment site selection using the AHP and TOPSIS approaches under fuzzy environment. *Waste Management (New York, N.Y.)*, *28*(9), 1552–1559. doi:10.1016/j.wasman.2007.05.019 PMID:17768038

Orton, J. D., & Weick, K. E. (1990). Loosely coupled systems: A reconceptualization. *Academy of Management Review*.

Paksoy, T., Pehlivan, N. Y., & Kahraman, C. (2012). Organizational strategy development in distribution channel management using fuzzy AHP and hierarchical fuzzy TOPSIS. *Expert Systems with Applications*, *39*(3), 2822–2841. doi:10.1016/j.eswa.2011.08.142

Panayides, P. M., & Song, D. W. (2008). Evaluating the integration of seaport container terminals in supply chains. *International Journal of Physical Distribution & Logistics Management*, *38*(7), 562–584. doi:10.1108/09600030810900969

Pan, N. (2009). Selecting an appropriate excavation construction method based on qualitative assessments. *Expert Systems with Applications*, *36*(3), 5481–5490. doi:10.1016/j.eswa.2008.06.097

Panta, M., & Vrysagotis, V. (2011). A new managerial approach to safety at workplace in warehousing systems: A case study. *Advances in Management & Applied Economics*, *1*(3), 115–126.

Papakiriakopoulos, D., & Pramatari, K. (2010). Collaborative performance measurement in supply chain. *Industrial Management & Data Systems*, *110*(9), 1297–1318. doi:10.1108/02635571011087400

Parasuraman, A., Zeithaml, V. A., & Berry, L. L. (1985). A conceptual model of service quality and its implications for future research. *Journal of Marketing*, *49*(3), 41–50. doi:10.2307/1251430

Parliament of India. (2009). *Rajya Sabha*. Ninetienth Report on Foreign and Domestic Investment in Retail Sector.

Parray, S. H., & Kadri, S. M. (2007). Supply chain management in healthcare sector-role of logistics. *Indian Journal for the Practicing Doctor*, *4*(1).

Perspectives, A. (2007). *The Sun Rises on the Cold Chain Industry: The Impact of Organized Retail*. Retrieved from consultavalon.com

Pery, M., & Sohal, A. (1999). Improving service quality within the supply chain: An Australian study. *Total Quality Management*, *10*(4/5), 673–679. doi:10.1080/0954412997677

Pitt, M., & Tucker, M. (2008). Performance measurement in facilities management:driving innovation? *Property Management*, *26*(4), 241–254. doi:10.1108/02637470810894885

Pohlen, T. L., & Lambert, D. M. (2001). Supply Chain Metrics. *International Journal of Logistics Management*, *12*(1), 1–19. doi:10.1108/09574090110806190

Porter, M. E. (1993). *The competitive advantage of nations*. Harvard Business School Management Programs.

Porter, M. E. (1998). *Clusters and the new economics of competition*. Harvard Business Review Watertown.

Prakash, J. (2013). The challenges for global harmonisation of food safety norms and regulations: Issues for India. *Journal of the Science of Food and Agriculture*. doi:10.1002/jsfa.6147 PMID:23553336

Presutti, W. D. Jr, & Mawhinney, J. R. (2007). The supply chain-finance link. *Supply Chain Management Review*, *11*, 32–38.

PRTM Management Consultants. (2008). *Global Supply Chain Trends 2008 – 2010*. Author.

Punt, H., & Huysamer, M. (2005). Supply chain technology and assessment—temperature variances in a 12 m integral reefer container carrying plums under a dual temperature shipping regime. *Acta Horticulturae*, *687*(687), 289–296. doi:10.17660/ActaHortic.2005.687.35

PWC. (2012). *Next-generation supply chains Efficient, fast and tailored*. PWC.

Qrunfleh, S., Tarafdar, M., & Ragu-Nathan, T. S. (2012). Examining alignment between supplier management practices and information systems strategy. *Benchmarking: An International Journal*, *19*(4/5), 604–617. doi:10.1108/14635771211258034

Quayle, M. (2003). A study of supply chain management practice in UK industrial SMEs. *Supply Chain Management: An International Journal*, *8*(1), 79–86. doi:10.1108/13598540310463387

Rae-Smith, J. B., & Ellinger, A. E. (2002). Insights from the introduction of an online logistics service system, Supply Chain Management. *International Journal (Toronto, Ont.)*, *7*(1), 5–11.

Rahman, S. (2001). A comparative study of TQM practice and organizational performance of SMEs with and without ISO 9000 certification. *International Journal of Quality & Reliability Management*, *18*(1), 35–49. doi:10.1108/02656710110364486

Raisinghani, M., & Meade, L. L. (2005). Strategic decisions in supply-chain intelligence using knowledge management: An analytic-network-process framework. *Supply Chain Management: An International Journal*, *10*(2), 114–121. doi:10.1108/13598540510589188

Ramaa, A., Rangaswamy, T. M., & Subramanya, K. N. 2009. A review of literature on performance measurement of supply chain network. In *Proceedings from:Second International Conference on Emerging Trends in Engineering and Technology*. doi:10.1109/ICETET.2009.18

Raman, A., DeHoratius, N., & Ton, Z. (2001a). The Achilles heel of supply chain management. *Harvard Business Review*, 136–152.

Ramanathan, R., & Ramanathan, U. (2013). Adoptionof RFID technologies in UK logistics: Moderating roles of size, barcode experience and government support. *Expert Systems with Applications*, *39*(12), 11115–11124.

Ramdas, K., & Spekman, R. E. (2000). Chain or Shackles: Understanding What Drives Supply-Chain Performance. *Interfaces, 30*(4), 3-21.

Ran, T. (2009). *Internal Logistics as a Part of Supply Chain, Case: Nokia-China, Dongguang Branch.* (Bachelor's Thesis). International Business Program, Lahti University of Applied Sciences, Faculty of Business Studies.

Rangasamy, N., & Dhaka, J. P. (2007). Marketing procurement cost for cooperative and private dairy plants in Tamil Nadu: A comparison. *Indian Journal of Agricultural Economics Research Review*, *62*(4), 679–69.

Rao, B. P. (2000). Improving retail effectiveness through technology: A survey of analytical tools for physical and on-line retailers. *Technology in Society*, *22*, 111–122. doi:10.1016/S0160-791X(99)00032-9

Ray, G., Muhanna, W. A., & Barney, J. B. (2005). Information technology and the performance of the customer service process: A resource-based analysis. *Management Information Systems Quarterly*, *29*(4), 625–651.

Reference. (n.d.). Retrieved from https://supply-chain.org/f/SCOR%2090%20Overview%20Booklet.pdf

Rennie, S. (2011). *Country Report: India's cold chain industry*. Department of Commerce.

Report of the Symposium on Supply Chain Management for Small and Medium Enterprises, Taipei, Republic of China. (2001). *Asian Cases on Supply Chain Management in SMEs.*

Report on Cold Storage of India. (2013). *Agmark.* Retrieved on August 29, 2013, from http://agmarknet.nic.in/coldstorage.htm

Robinson, C. J., & Malhotra, M. K. (2005). Defining the concept of supply chain quality management and its relevance to academic and industrial practice. *International Journal of Production Economics*, *96*(3), 315–337. doi:10.1016/j.ijpe.2004.06.055

Rodríguez-Bermejo, J., Barreiro, P., Robla, J. I., & Ruiz-Garcia, L. (2007). Thermal study of a transport container. *Journal of Food Engineering*, *80*(2), 517–527. doi:10.1016/j.jfoodeng.2006.06.010

Rohit, Joshi, Kumar, Banwet, & Shankar. (2009). Indian cold chain: modeling the inhibitors. *British Food Journal, 111*.

Rosenzweig, E. D., Roth, A. V., & Dean, J. W. Jr. (2003). The influence of an integration strategy on competitive capabilities and business performance: An exploratory study of consumer products manufacturers. *Journal of Operations Management*, *21*(4), 437–456. doi:10.1016/S0272-6963(03)00037-8

Roy, B. C. (1997). "Growth and prospects of fruit and vegetables processing industry in India," *The Bihar J. Agril. Mkting*, *5*, 356–365.

Roy, D., & Thorat, A. (2008). Success in High Value Horticultural Export Markets for the Small Farmers: The Case of Mahagrapes in India. *World Development*, *36*(10), 1874–1890. doi:10.1016/j.worlddev.2007.09.009

Rungtusanatham, M., Salvador, F., Forza, C., & Choi, T. Y. (2003). Supply-chain linkages and operational performance: A resource-based-view perspective. *International Journal of Operations & Production Management*, *23*(9), 1084–1099. doi:10.1108/01443570310491783

Rushton, A., Oxley, J., & Croucher, P. (2001). The handbook of logistics and distribution management (2nd ed.). London, UK: Academic Press.

Saarijarvi, H., Karjaluoto, H., & Kuusela, H. (2013). Extending customer relationship management : From empowering firms to empowering customers. *Journal of Systems and Information Technology, 15*(2), 140–158. doi:10.1108/13287261311328877

Saaty, T. (1972). *An eigenvalue allocation model for prioritization and planning*. Working Paper. Energy Management and Policy Center, University of Pennsylvania.

Saaty, T. (1977). A scaling method for priorities in hierarchical structures. *Journal of Mathematical Psychology, 15*(3), 234–281. doi:10.1016/0022-2496(77)90033-5

Saaty, T. (1994). Highlights and critical points in the theory and application of the Analytic Hierarchy Process. *European Journal of Operational Research, 74*(3), 426–447. doi:10.1016/0377-2217(94)90222-4

Saaty, T. L. (1980). *The Analytic Hierarchy Process*. New York: McGraw-Hill.

Saaty, T. L. (1983). Priority Setting in Complex Problems. *IEEE Transactions on Engineering Management, 30*(3), 140–155. doi:10.1109/TEM.1983.6448606

Saaty, T., & Forman, E. (1992). *The Hierarchon: A Dictionary of Hierarchies* (Vol. 5). Pittsburgh, PA: RWS Publications.

Saeed, M., Anjum, F. M., Khan, M. R., Khan, M. I., & Nadeem, M. (2013). Isolation, characterization and utilization of starter cultures for the development of wheyghurt drink. *British Food Journal, 115*(8), 1169–1186. doi:10.1108/BFJ-10-2011-0274

Sahin, F., & Robinson, E. P. (2002). Flow coordination and information sharing in supply chains: Review, implications, and directions for future research. *Decision Sciences, 33*(4), 505–535. doi:10.1111/j.1540-5915.2002.tb01654.x

Salin & Nayga, Jr. (2003). A cold chain network for food exports to developing countries. *International Journal of Physical Distribution & Logistics Management, 33*.

Salin, V., & Nayga, R. M. Jr. (2003). A cold chain network for food exports to developing countries. *International Journal of Physical Distribution & Logistics Management, 33*(10), 918–933. doi:10.1108/09600030310508717

Sanchez, T. (2012). The impact of industry clusters on the economy in United States. *Academy of Entrepreneurship, 18*(1), 85.

Sanders, N. R. (2008). Pattern of information technology use: The impact on buyer–suppler coordination and performance. *Journal of Operations Management, 26*(3), 349–367. doi:10.1016/j.jom.2007.07.003

Saremi, M. S., Mousavi, F., & Sanayei, A. (2008). TQM consultant selection in SME with TOPSIS under fuzzy environment. *Expert Systems with Applications*, 36.

Sarkis, J. (2003). A strategic decision framework for green supply chain management. *Journal of Cleaner Production, 11*(4), 397–409. doi:10.1016/S0959-6526(02)00062-8

Sarkis, J., & Cordeiro, J. (2001). An Empirical Evaluation of Environmental Efficiencies and Firm Performance: Pollution Prevention Versus End-of-Pipe Practice. *European Journal of Operational Research, 135*(1), 102–113. doi:10.1016/S0377-2217(00)00306-4

Schonberger, R. J. (1990). *Building a chain of customers*. London: The Free Press.

Secme, N. Y., Bayrakdarog˘lu, A., & Kahraman, C. (2009). Fuzzy performance evaluation in Turkish Banking Sector using Analytic Hierarchy Process and TOPSIS. *Expert Systems with Applications, 36*(9), 11699–11709. doi:10.1016/j.eswa.2009.03.013

Sekaran, U. (2003). *Research methods for business: A skill-building approach*. John Wiley & Sons.

Sen, C. G., & Çınar, G. (2010). Evaluation and pre-allocation of operators with multiple skills: A combined fuzzy AHP and max-min approach. *Expert Systems with Applications*, *37*(3), 2043–2053. doi:10.1016/j.eswa.2009.06.075

Senge, P. M. (1990). *The fifth discipline: The art & Practice of the Learning organization*. Double Day Currency.

Seo, S., Aramaki, T., Hwang, Y. W., & Hanaki, K. (2004). Fuzzy decision-making tool for environmental sustainable buildings. *Journal of Construction Engineering and Management*, *130*(3), 415–423. doi:10.1061/(ASCE)0733-9364(2004)130:3(415)

Shafer, G. (2008). Belief functions. In *Classic Works of the Dempster–Shafer Theory of Belief Functions*. Retrieved February 2010, from http://www.glennshafer.com/assets/downloads/rur_chapter7.pdf>

Shah, J. (2009). *Supply chain management*. Delhi: Pearson Education.

Sharma, M. K., & Bhagwat, R. (2006). Practice of information systems: Evidence from select Indian SMEs. *Journal of Manufacturing Technology Management*, *17*(2), 199–223. doi:10.1108/17410380610642278

Shaverdi, M., Akbari, M., & Tafti, S. F. (2011). Combining Fuzzy MCDM with BSC Approach in Performance Evaluation of Iranian Private Banking Sector. *Advances in Fuzzy Systems*, *2011*, 1–12. doi:10.1155/2011/148712

Shepherd, C., & Gunter, H. (2006). Measuring supply chain performance: Current research and future directions. *International Journal of Productivity and Performance Management*, *55*(3/4), 242–258. doi:10.1108/17410400610653219

Shih, H. S., Syur, H. J., & Lee, E. S. (2007). An extension of TOPSIS for group decision making. *Mathematical and Computer Modelling*, *45*(7-8), 801–813. doi:10.1016/j.mcm.2006.03.023

Shim, J. (1989). Bibliography research on the analytic hierarchy process (AHP). *Socio-Economic Planning Sciences*, *23*, 161–167. doi:10.1016/0038-0121(89)90013-X

Shore, B., & Venkatachalam, A. (2003). Evaluating the information sharing capabilities of supply chain partners: A fuzzy logic model. *International Journal of Physical Distribution & Logistics Management*, *33*(9), 804–824. doi:10.1108/09600030310503343

Simchi-Levi, D., Kaminsky, P. & Simchi-Levi, E. (2000). *Designing and Managing the Supply Chain: Concepts, Strategies, and Case Studies*. Irwin McGraw-Hill.

Simmie, J., & Sennett, J. (1999). Innovative clusters: Global or local linkages? *National Institute Economic Review*, *170*(1), 87–98. doi:10.1177/002795019917000112

Sinclair, D., & Zairi, M. (1995). Effective process management through performance measurement: Part III - An integrated model of totalquality-based performance measurement. *Business Process Management Journal*, *1*(3), 50–65. doi:10.1108/14637159510103220

Singh, A., Mishra, P., Jain, R., & Khurana, M. (2012). Design of global supply chain network with operational risks. *International Journal of Advanced Manufacturing Technology*, *60*(1-4), 273–290. doi:10.1007/s00170-011-3615-9

Singh, P. J., & Power, D. (2009). The nature and effectiveness of collaboration between firms, their customers and suppliers : A supply chain perspective. *Supply Chain Management: An International Journal*, *14*(3), 189–200. doi:10.1108/13598540910954539

Singh, P., Wani, A. L. I. A., Karim, A. A., & Langowski, H. (2012). The use of carbon dioxide in the processing and packaging of milk and dairy products. *International Journal of Dairy Technology*, *65*(2), 161–177. doi:10.1111/j.1471-0307.2011.00744.x

Singh, R. K., & Garg, S. K. (2008). Strategy development by SMEs for competitiveness: A review", Bechmarking. *International Journal (Toronto, Ont.), 15*(5), 525–547.

Singh, R. K., Garg, S. K., & Deshmukh, S. G. (2008). Competency and Performance analysis of Indian SMEs and large organizations: An exploratory study. *Competitiveness Review: An International Business Journal, 18*(4), 308–321. doi:10.1108/10595420810920798

Singh, R. K., Garg, S., & Deshmukh, S. G. (2004). Competitiveness of small and medium enterprises, case of an Indian auto component manufacturing organization. *IIMB Management Review*, (December), 94–102.

Sipahi, S., & Timor, M. (2010). The analytic hierarchy process and analytic network process: An overview of applications. *Management Decision, 48*(5), 775–808. doi:10.1108/00251741011043920

Sivakumar & Patel. (2013). *Cold chain industry-India.* International Exhibition & Conference on Cold Chain Industry.

Slack, N. (1983). Flexibility as a manufacturing objective. *International Journal of Operations & Production Management, 3*(3), 4–13. doi:10.1108/eb054696

Smarandache, F. (2003). Neutrosophic logic – A generalization of the intuitionistic fuzzy logic. In *Proceedings from: The Third Conference of the European Society for Fuzzy Logic and Technology.* University of Applied Sciences. Retrieved April 2010 from: http://www.arxiv.org/abs/math/0303009

Smith, G.B. (1995). *Good manufacturing practices for dairy processing plant.* The Dairy Practice Council.

Snider, B., da Silveira, G. J. C., & Balakrishnan, J. (2009). ERP implementation at SMEs: Analysis of five Canandian Cases. *International Journal of Operations & Production Management, 29*(1), 4–29. doi:10.1108/01443570910925343

Social Security Organisation (SOCSO). (2012). *Annual Report 2011.* Author.

Soffer, P., & Wand, Y. (2007). Goal-driven multi-process analysis. *Journal of the Association for Information Systems, 8*(3), 175–204.

Soliman, M. T. (2007). *The Use of DuPont Analysis by Market Participants.* Social Science Research Network. Retrieved June 2010 from: http://www.papers.ssrn.com/sol3/papers.cfm?abstract_id=1101981

Soon, J. M., Bharu, K., & Manning, L. (2012). Fresh produce-associated outbreaks: A call for HACCP on farms? *British Food Journal, 114*(4), 553–597. doi:10.1108/00070701211219568

Spekman, R. E., Kamauff, J. W. Jr, & Myhr, N. (1998). An empirical investigation into supply chain management: A perspective on partnerships. *Supply Chain Management: An International Journal, 3*(2), 53–67. doi:10.1108/13598549810215379

Spohrer, J., Maglio, P. P., Bailey, J., & Gruhl, D. (2007). Steps Toward a Science of Service Systems. *Computer, 40*(1), 71–77. doi:10.1109/MC.2007.33

Srinivas, M.R., Balamuralidhara, V., & Kumar, T.M.P., & khattri,S. (2013). Cold Chain Management in Vaccines Logistics - A Review. *Journal for Clinical Studies, 5*(3), 50–56.

Srivastava, R. K., Tasadduq, A. S., & Liam, F. (1999). Marketing, business processes, and shareholder value: An organisationally embedded view of marketing activities and the discipline of marketing. *Journal of Marketing, 63*, 168–179. doi:10.2307/1252110

Srivastava, S. K. (2007). Green Supply-Chain Management: A State-of-the-Art Literature Review. *International Journal of Management Reviews, 9*(1), 53–80. doi:10.1111/j.1468-2370.2007.00202.x

Stank, T. P., Davis, B. R., & Fugate, B. S. (2005). A strategic framework for supply chain oriented logistics. *Journal of Business Logistics*, *26*(2), 27–45. doi:10.1002/j.2158-1592.2005.tb00204.x

Stanley, L. L., & Wisner, J. D. (2002). "The determinants of service quality: Issues for purchasing", *European. Journal of Purchasing and Supply Management*, *8*(2), 97–109. doi:10.1016/S0969-7012(01)00009-0

Stark, E., & Mangione, T. (2004). Top issues facing the hospital supply chain today. *Healthcare purchasing news*. FindArticles.com.

Stevenson, G. (1998). "Integrating the Supply Chain", International Journal of Physical Distribution and Materials Management "Supply chain performance measurement: A literature review". *International Journal of Production Research*, *48*(17), 2010.

Stevens, S. (1957). On the psychophysical law. *Psychological Review*, *64*(3), 153–181. doi:10.1037/h0046162 PMID:13441853

Stewart, G. (1995). Supply chain performance benchmarking study reveals keys to supply chain excellence. *Logistics Information Management*, *8*(2), 38–44. doi:10.1108/09576059510085000

Stock, J. R., & Boyer, S. L. (2009). Developing a consensus definition of supply chain management; a qualitative study. *International Journal of Physical Distribution & Logistics Management*, *39*(8), 690–711. doi:10.1108/09600030910996323

Stonebraker & Afifi. (2004). Toward a contingency theory of supply chains. *Management Decision*, *42*(5), 1131–1144.

Storey, J., Emberson, C., Godsell, J., & Harrison, A. (2006). Supply chain management: Theory, practice and future challenges. *International Journal of Operations & Production Management*, *26*(7), 754–774. doi:10.1108/01443570610672220

Subbaiah, K. V., Rao, K. N., & Nookesh, K. (2009). *Supply Chain Management in a Dairy Industry – A Case Study*. Academic Press.

Subrahmanya. (2007). Development strategies for Indian SMEs: Promoting linkages with global transnational corporations. *Management Research News*, *30*(10), 762-774.

Sun, C. C. (2010). A performance evaluation model by integrating fuzzy AHP and fuzzy TOPSIS methods. *Expert Systems with Applications*, *37*(12), 7745–7754. doi:10.1016/j.eswa.2010.04.066

Sun, C. C., & Lin, G. T. R. (2009). Using fuzzy TOPSIS method for evaluating the competitive advantages of shopping websites. *Expert Systems with Applications*, *36*(9), 11764–11771. doi:10.1016/j.eswa.2009.04.017

Supply Chain Value. (n.d.). UPS Whitepaper.

Supply-Chain Council. (2008). *Supply-chain Operations Reference-Model: SCOR Model – Version 9*. Retrieved May 2011 from: http://www.supply-chain.org

Swami & Shah. (2011). *Channel Coordination in Green Supply Chain Management: The Case of Package Size and Shelf-Space Allocation*. IIMB, Supply Chain Management Centre, Working Paper No: 348

Sweeney, D.W., & Stulen, R.H. (1999). Quantum Electronics. *IEEE Journal of Sandia Nat. Labs.*, *35*(5), 694 – 699.

Tah, J. H. M., & Carr, V. (2000). A proposal for construction project risk assessment using fuzzy logic. *Construction Management and Economics*, *18*(4), 491–500. doi:10.1080/014461900050024905

Tallman, S., Jenkins, M., Henry, N., & Pinch, S. (2004). Knowledge, clusters, and competitive advantage. *Academy of Management Review*, *29*(2), 258–271.

Tamime, A. Y., & Robinson, R. K. (2004). Yoghurt Science and Technology. Woodhead Publishing.

Tan, C., & Chen, X. (2011). Induced intuitionistic fuzzy choquet integral operator for multicriteria. *International Journal of Intelligent Systems*, *26*(7), 659–686. doi:10.1002/int.20489

Tan, K. C., Kannan, V. R., Handfield, R. B., & Ghosh, S. (1999). Supply chain management: An empirical study of its impact on performance. *International Journal of Operations & Production Management*, *19*(10), 1034–1052. doi:10.1108/01443579910287064

Tan, K. C., Kannan, V. R., & Hanfield, R. B. (1998). Supply Chain Management: Supplier Performance and Firm Performance. *International Journal of Purchasing and Materials Management*, *34*(3), 2–9.

Tanner, D., & Amos, N. D. (2003b). Modelling Product Quality Changes as a Result of Temperature Variability in Shipping Systems. *International Congress of Refrigeration*, Washington, DC.

Tanriverdi, H. (2005). Information technology relatedness, knowledge management capability and performance of multibusiness firms. *Management Information Systems Quarterly*, *29*(2), 311–334.

Taticchi, P., Tonelli, F., & Cagnazzo, L. (2010). Performance measurement and management: A literature review and a research agenda. *Measuring Business Excellence*, *14*(1), 4–18. doi:10.1108/13683041011027418

Technopak. (2010). *Public Private Partnership in India Dairy Industry*. Retrieved from: http://www.technopak.com/resouces/food/PPP%20indian%20dairy%20industrytechnopak_CII_background%20papermay08,2010.pdf

Tejas, F., & Srikanth, K. (2007). Supply chain metrics that measure up. *Supply and Demand Chain Executive*, *8*, 50–53.

Terpstra, M. J., Steenbekkers, L. P. A., de Maertelaere, N. C. M., & Nijhuis, S. (2005). Food storage and disposal:consumer practices and Knowledge. *British Food Journal*, *107*(7), 526–533. doi:10.1108/00070700510606918

Thakkar, J., Kanda, A., & Deshmukh, S. G. (2008). A conceptual role interaction model for supply chain management in SMEs. *Journal of Small Business and Enterprise Development*, *15*(1), 74–95. doi:10.1108/14626000810850856

Thakkar, J., Kanda, A., & Deshmukh, S. G. (2008). Supply chain management in SMEs: Development of constructs and propositions. *Asia Pacific Journal of Marketing and Logistics*, *20*(1), 97–13. doi:10.1108/13555850810844896

Thakkar, J., Kanda, A., & Deshmukh, S. G. (2012). Supply chain issues in Indian manufacturing SMEs: Insights from six case studies. *Journal of Manufacturing Technology Management*, *23*(5), 634–664. doi:10.1108/17410381211234444

The smarter supply chain of the future. (2009). IBM.

The World Bank. (2013). *India Overview*. Retrieved January Saturday, 2013, from http://www.worldbank.org/en/country/india/overview

Thomas, D. J., & Griffin, P. M. (1996). Coordinated supply chain management. *European Journal of Operational Research*, *94*(1), 1–15. doi:10.1016/0377-2217(96)00098-7

Thurstone, L. (1927). A law of comparative judgments. *Psychological Review*, *34*(4), 273–286. doi:10.1037/h0070288

Todd, P. R., & Javalgi, R. G. (2007). Internalization of SMEs in India: Fostering entrepreneurship by leveraging information technology. *International Journal of Emerging Markets*, *2*(2), 166–180. doi:10.1108/17468800710739234

Tony, W., & Kevin, H. (2007). The human factors in managing China supply chain. In *Proceedings of the 92nd Annual International Supply Management Conference* (pp. 1-6). Nokia (China) Investment Co. Ltd. Shanghai Branch

Torlak, G., Sevkli, M., Sanal, M., & Zaim, S. (2010). Analyzing business competition by using fuzzy TOPSIS method: An example of Turkish domestic airline industry. *Expert Systems with Applications*, *38*(4), 3396–3406. doi:10.1016/j.eswa.2010.08.125

Towers, N., & Burnes, B. (2008). A composite framework of supply chain management and enterprise planning for small and medium-sized manufacturing enterprises. *Supply Chain Management: An International Journal, 13*(5), 349–355. doi:10.1108/13598540810894933

Tracey, M., Lim, J.-S., & Vonderembse, M. A. (2005). The impact of supply-chain management capabilities on business performance. *Supply Chain Management: An International Journal, 10*(3), 179–191. doi:10.1108/13598540510606232

Triantaphyllou, E., & Lin, C. T. (1996). Development and evaluation of five fuzzy multi attribute decision-making methods. *International Journal of Approximate Reasoning, 14*(4), 281–310. doi:10.1016/0888-613X(95)00119-2

Tsaur, S. H., Chang, T. Y., & Yen, C. H. (2002). The evaluation of airline service quality by fuzzy MCDM. *Tourism Management, 23*(2), 107–115. doi:10.1016/S0261-5177(01)00050-4

Tseng & Yue. (2005). *The role of transportation in logistics chain.* Eastern Asia Society for Transportation Studies.

Tseng, Y.-F., & Lee, T.-Z. (2009). Comparing appropriate decision support of human resource practices on organizational performance with DEA/AHP model. *Expert Systems with Applications, 36*(3), 6548–6558. doi:10.1016/j.eswa.2008.07.066

Turner, S. (2002). *Tools for Success: A Manager Guide.* Retrieved June 11, 2013 from http://www.google.co.in/url?sa=t&rct=j&q=&esrc=s&source=web&cd=5&ved=0CD8QFjAE&url=http%3A%2F%2Fwww2.warwick.ac.uk%2Ffac%2Fsci%2Fwmg%2Feducation%2Ftranscap%2Fsupportmaterials%2Fhandbooks%2Flvc%2Fgeneral%2Fputtick_grid.ppt&ei=Ltc7U-2iMcWWrgfS-4GwAQ&usg=AFQjCNE2USzvMYgyWc4gw8sGI89zGsXy0g

Umali-Deininger, D., & Sur, M. (2006). *Food safety in a globalizing world: Opportunities and challenges for India.* Paper presented at the International Association of Agricultural Economists Conference, Gold Coast Australia.

USAID. (2010). *Agency Sustainability Plan.* U.S. Agency for International Development Earth Day Statement.

Valand, T. I., & Heides, M. (2007). Can the SME survive the supply chain challenge? *Supply Chain Management: An International Journal, 12*(1), 20–31. doi:10.1108/13598540710724374

Valli, C., & Traill, W. B. (2005). Culture and food: A model of yoghurt consumption in the EU. *Food Quality and Preference, 16*(4), 291–304. doi:10.1016/j.foodqual.2004.05.006

Van der Velde, L. N. J., & Meijer, B. R. (2003). *A System Approach to supply chain design with a multinational for colorant and coatings.* Retrieved September 20, 2006, from http://www.ifm.eng.cam.ac.uk/mcn/pdf_files/part6_5.pdf

Van der Vorst, J. A. (2000). *Effective food supply chains Generating, modelling and evaluating supply chain scenarios.* (PhD thesis). Wageningen University.

Van Hoek, R.I. (1998). Measuring the unmeasurable - Measuring and improving performance in the supply chain. *Supply Chain Management: An International Journal, 3*(40), 187-192.

van Hoek, R. I. (1999). From reversed logistics to green supply chains", Supply Chain Management. *International Journal (Toronto, Ont.), 4*(3), 129–135.

Vargas, L. (1990). An overview of the analytic hierarchy process and its applications. *European Journal of Operational Research, 48*(1), 2–8. doi:10.1016/0377-2217(90)90056-H

Varzakas, T. H., & Arvanitoyannis, I. S. (2008). Application of ISO22000 and comparison to HACCP for processing of ready to eat vegetables: Part I. *International Journal of Food Science & Technology, 43*(10), 1729–1741. doi:10.1111/j.1365-2621.2007.01675.x

Vassilios, K. Z., & Dimitrios, V. L. (2010). *An Analysis of seaport cluster models for the development and competitiveness of maritime sectors: The case of Piraeus.* Academic Press.

Vasudeva, M. (2009). *Evolution of Bulk Handling from Conventional Bagged Food-grains handling and its socio-economic impact on India.* Retrieved on November 15, 2013, http://www.geaps.com/knowledge/proceedings/article.cfm

Vickery, S. K., Jayaram, J., Droge, C., & Calantone, R. (2003). The effects of an integrative supply chain strategy on customer service and financial performance: An analysis of direct versus indirect relationships. *Journal of Operations Management, 21*(5), 523–539. doi:10.1016/j.jom.2003.02.002

Vidal, L.-A., Sahin, E., Martelli, N., Berhoune, M., & Bonan, B. (2010). Applying AHP to select drugs to be produced by anticipation in a chemotherapy compounding unit. *Expert Systems with Applications, 37*(2), 1528–1534. doi:10.1016/j.eswa.2009.06.067

Vilella, A., & Neves, F. (2008). Strategic management in the Argentina dairy supply cha*in. 8th International Conference on Management in AgriFood Chains and Networks Ede*, The Netherlands.

Wagner, B. A., Fillis, I., & Johansson, U. (2003). E-business and e-supply in small and medium sized businesses. *Supply Chain Management: An International Journal, 8*(4), 343–354. doi:10.1108/13598540310490107

Wallace, D. W., Giese, J. L., & Johnson, J. L. (2004). Customer retailer loyalty in the context of multiple channel strategies. *Journal of Retailing, 80*(4), 249–263. doi:10.1016/j.jretai.2004.10.002

Walters, D., & Lancaster, G. (2000). Implementing value strategy through the value chain. *Management Decision, 38*(3), 160–178. doi:10.1108/EUM0000000005344

Walters, P. G. (2008). Adding value in global B2B supply chains: Strategic directions and the role of the internet as a driver of competitive advantage. *Industrial Marketing Management, 37*(1), 59–68. doi:10.1016/j.indmarman.2007.06.010

Walton, S.V., Handfield, R.B., & Melnyk, S.T. (1998). The green supply chain: Integrating suppliers into environmental management process. *Int. J. Purch. Mater. Manag.*, 2–11.

Wang, H. S., Che, Z. H., & Wu, C. (2010). Using analytic hierarchy process and particle swarm optimization algorithm for evaluating product plans. *Expert Systems with Applications, 37*(2), 1023–1034. doi:10.1016/j.eswa.2009.05.073

Wang, T. C., & Chang, T. H. (2007). Application of TOPSIS in evaluating initial training aircraft under a fuzzy environment. *Expert Systems with Applications, 33*(4), 870–880. doi:10.1016/j.eswa.2006.07.003

Wang, T.-Y., & Yang, Y.-H. (2009). A fuzzy model for supplier selection in quantity discount environments. *Expert Systems with Applications, 36*(10), 12179–12187. doi:10.1016/j.eswa.2009.03.018

Wang, Y., & Feng, H. (2012). Customer relationship management capabilities Measurement, antecedents and consequences. *Management Decision, 50*(1), 115–129. doi:10.1108/00251741211194903

Wang, Y.-M., & Elhag, T.-M.-S. (2006). Fuzzy TOPSIS method based on alpha level sets with an application to bridge risk assessment. *Expert Systems with Applications, 31*(2), 309–319. doi:10.1016/j.eswa.2005.09.040

Waring, P., & Lever, J. (2004). The impact of Socially Responsible Investment on Human Resource Management: A conceptual Framework. *Journal of Business Ethics, 52*(1), 99–108. doi:10.1023/B:BUSI.0000033110.21900.da

Weibert, M., & Plunkett, C. (2006). Motivational Safety System: One Organisation's Experience Moving toward World-Class Performance. *Professional Safety*, (February), 34–39.

Westhead, P., & Storey, D. (1996, July01). Management training and small firm performance: Why is the link so weak? *International Small Business Journal, 14*(4), 13–24. doi:10.1177/0266242696144001

Whipple, J. M., Frankel, R., & Daugherty, P. J. (2002). Information support for alliances:performance implications. *Journal of Business Logistics, 23*(2), 67–82. doi:10.1002/j.2158-1592.2002.tb00026.x

Wikipedia. (2008). *The Free Encyclopedia.* Wikimedia Foundation, Inc. Retrieved on November 14, http://en.wikipedia.org/wiki/Supply_chain

Wild, Y., Scharnow, R., & Rühmann, M. (2005). *Container Handbook, 3, Gesamtverband der Deutschen Versicherungswirtschaft e.v.* Berlin, Germany: GDV.

Wisner, J. D., & Fawcett, S. E. (1991). Linking firm strategy to operating decisions through performance measurement. *Production and Inventory Management Journal, 32*(4), 5–11.

World Economic Forum. (2012). *Outlook on the Logistics & Supply Chain Industry 2012.* Author.

World Economic Forum. (2013). *Outlook on the Logistics & Supply Chain Industry.* Global Agenda Council on Logistics & Supply Chain Systems.

Wortmann, J. C. (1998). Evolution of ERP systems. In *Proceedings of International conference of the manufacturing value chain.* Troon: Kluwer Academic Publishing doi:10.1007/978-0-387-35321-0_2

Wouters, M. (2009). A developmental approach to performance measures- results from a longitudinal case study. *European Management Journal, 27*(1), 64–78. doi:10.1016/j.emj.2008.06.006

Wu, C.-R., Lin, C.-T., & Lin, Y.-F. (2009). Selecting the preferable bancassurance alliance strategic by using expert group decision technique. *Expert Systems with Applications, 36*(2), 3623–3629. doi:10.1016/j.eswa.2008.02.016

Wyer, P., & Mason, J. (1998). An organizational learning perspective to enhancing understanding of people management in small businesses. *International Journal of Entrepreneurial Behavior & Research, 4*(2), 112–128. doi:10.1108/13552559810224576

Xu, H. (2010). Agile manufacturing in complex supply networks. In L. Wang & S. C. Lenny Koh (Eds.), *Enterprise networks and logistics for agile manufacturing* (pp. 39–66). London: Springer. doi:10.1007/978-1-84996-244-5_3

Yang, C.-L., Chuang, S.-P., & Huang, R.-H. (2009). Manufacturing evaluation system based on AHP/ANP approach for wafer fabricating industry. *Expert Systems with Applications, 36*(8), 11369–11377. doi:10.1016/j.eswa.2009.03.023

Yayla, A. Y., Özbek, A., & Yildiz, A. (2012). Fuzzy TOPSIS Method in Supplier Selection and Application in the Garment Industry. *FIBRES & TEXTILES in Eastern Europe, 4*(93), 20–23.

Yeh, C. H., & Deng, H. (2004). A practical approach to fuzzy utilities comparison in fuzzy multi criteria analysis. *International Journal of Approximate Reasoning, 35*(2), 179–194. doi:10.1016/j.ijar.2003.09.002

Yeh, C. H., Deng, H., & Chang, Y.-H. (2000). Fuzzy multi criteria analysis for performance evaluation of bus companies. *European Journal of Operational Research, 126*(3), 459–473. doi:10.1016/S0377-2217(99)00315-X

Yiakoumi. (2014). Overcoming Automotive Logistics Challenges. *Supply Chain Nation.*

Yigrem, . (2008). *Dairy production, processing and marketing systems of Shashemene–Dilla area, South Ethiopia.* Nairobi, Kenya: International Livestock Research Institute.

Yokoyama, M. (1921). The Nature of the affective judgment in the method of paired comparison. *The American Journal of Psychology, 32*(3), 357–369. doi:10.2307/1414000

Zahedi, F. (1986). The analytic hierarchy process: A survey of the method and its applications. *Interface, 16*(4), 96–108. doi:10.1287/inte.16.4.96

Zahedirad & Shivaraj. (2011, October-December). Supply Chain: Barriers and Benefits Indian SMEs. *SCMS Journal of Indian Management*, 11-30.

Zayed, T. M., & Chang, L. M. (2002). Prototype model for build-operate-transfer risk assessment. *Journal of Management Engineering*, *18*(1), 7–16. doi:10.1061/(ASCE)0742-597X(2002)18:1(7)

Zeng, D. Z. (2010). *Building engines for growth and competitiveness in China*. New York: World Bank. doi:10.1596/978-0-8213-8432-9

Zeng, J., An, M., & Smith, N. J. (2007). Application of a fuzzy based decision making methodology to construct project risk assessment. *International Journal of Project Management*, *25*(6), 589–600. doi:10.1016/j.ijproman.2007.02.006

Zhang, W., & Lam, J. S. L. (2013). Maritime cluster evolution based on symbiosis theory and Lotka–Volterra model. *Maritime Policy & Management*, *40*(2), 161–176. doi:10.1080/03088839.2012.757375

Zhao, X., Xie, J., & Wei, J. C. (2002). The impact of forecast errors on early order commitment in a supply chain. *Decision Sciences*, *33*(2), 251–280. doi:10.1111/j.1540-5915.2002.tb01644.x

Zhou. (2009). *Study on the Implementation of Green Supply Chain Management in Textile Enterprises*. Academic Press.

Zhu, Q., & Geng, Y. (2001). Integrating environmental issues into supplier selection and management: a study of Large and Medium-sized State-owned Manufacturers in China. *Green. Manag. Int.*, 27–40.

Zhu, Sarkis, & Geng. (2005). Green supply chain management in China: pressures, practices and performance. *International Journal of Operations & Production Management, 25*(5), 449 – 468.

Zhu, Q. H., & Cote, R. (2002). Green supply chain management in China: how and why?*The Fifth International Eco-city Conference*, Shenzhen, China.

Zhu, Q., Geng, Y., Fujita, T., & Hashimoto, S. (2010). Green supply chain management in leading manufacturers: Case studies in Japanese large companies. *Management Research Review*, *33*(4), 380–392. doi:10.1108/01409171011030471

Zhu, Q., & Sarkis, J. (2004). Relationships between operational practices and performance among early adopters of green supply chain management practices in Chinese manufacturing manufacturers. *Journal of Operations Management*, *22*(3), 265–289. doi:10.1016/j.jom.2004.01.005

Zhu, Q., & Sarkis, J. (2007, September15). The moderating effects of institutional pressures on emergent green supply chain practices and performance. *International Journal of Production Research*, *45*(18–19), 4333–4355. doi:10.1080/00207540701440345

Zielinski. (n.d.). *Logistics Challenges in a Changing Auto Industry*. CEVA Logistics.

Zimmermann, H. J. (1996). *Fuzzy Set Theory and its Applications*. Boston: Kluwer Academic Publishers. doi:10.1007/978-94-015-8702-0

Zou, P. X. W., Zhang, G., & Wang, J. (2007). Understanding the key risks in construction projects in china. *International Journal of Project Management*, *25*(6), 601–614. doi:10.1016/j.ijproman.2007.03.001

Zsidisin, G. A., & Hendrick, T. E. (1998). Purchasing's involvement in environmental issues:a multi-country perspective. *Industrial Management & Data Systems*, *7*(7), 313–320. doi:10.1108/02635579810241773

Zsidisin, G., Jun, M., & Adams, L. (2000). The relationship between information technology & service quality in the dual-direction supply chain. *International Journal of Service Industry Management*, *11*(4), 312–328. doi:10.1108/09564230010355359

Z, Yang,., Hu, Z., Liu, J., & Zhao, P. (2007). The Relationships between Manufacturer Justice and Distributor Trust: A Case Study of Channel Relations in the Chinese Cell Phone Industry[In Chinese]. *Journal of Marine Science*, *3*(1), 91–105.

About the Contributors

Ashish Dwivedi is the Deputy Director of the Centre for Systems Studies at Hull University Business School, UK. Previously, he was the Deputy Graduate Research Director at Hull University Business School, and was also associated with the management of the high-tech Management Learning Laboratory. His primary research interests are in knowledge management (in which he obtained his PhD), supply chain management, healthcare management and information, and communication technologies. He has published 4 books, and over 50 journal and conference papers. He has served as an Invited reviewer and Guest Editor for several journals, including the IEEE Transactions on Information Technology in Biomedicine.

* * *

Suresh Garg is Pro-Vice Chancellor of Delhi Technological University, formerly known as Delhi College of Engineering, New Delhi, India. He has published more than 130 research papers in international, national journals, conferences. His areas of interests include supply chain management, six sigma, total quality management and flexibility.

Rajneesh Mahajan has received DAAD's Sandwich Scholarship from June 2012 - Sept. 2013. To conduct part of doctoral research in Gottingen University in Germany. He had been associated as full-time Doctoral Research Scholar in Delhi School of Management, Delhi Technological University, formerly known as Delhi College of Engineering, New Delhi, India. He is presently associated as Associate Professor with Apeejay School of Management, Dwarka, New Delhi, India. His research interests are in the area of cold chain management, global supply chain management, global food Safety, international trade operations, international business and international marketing and system dynamic.

Amit Kumar Marwah is a Ph.D. student in the Department of Industrial and Production Engineering at Shri Govindram Seksaria Institute of Technology and Science, Indore (M.P.), India. His research interests focus on supply chain management, supply chain performance measurement and competitiveness. He has published several research papers of national and international repute like Udyog Pragati, International Journal of Business Research & Management and International Journal of Engineering. He has also reviewed various research papers of journals like Business Process Management Journal and Asia Pacific Journal of Marketing and Logistics.

Ravi Potti has more than 8 years of work experience spanning IT, Automobile and ITES sectors. As part of current assignment, his focus lies in analyzing different sectors and come up with actionable inputs.

Swapnil Saurav loves to write and speak on Supply Chain Management, Retail and emerging technologies. Accomplished and versatile professional with 12 years of experience and solid understanding of a diverse range of business management applications around Retail and Manufacturing Industries, including market analysis, sales and marketing support, project management and quality assurance. He loves to figure out ways to do what others say can't be done. He believes in Sam Goldwyn's words "The harder I work, the luckier I get."

P. B. Sharma is the Founder, Vice-Chancellor of Delhi Technological University, formerly known as Delhi College of Engineering, at New Delhi, India. He is an Academician, Researcher and Administrator. His areas of interests include knowledge and technology management, innovations management, technical education, etc.

Index

R

S

T

U

W

www.ingramcontent.com/pod-product-compliance
Lightning Source LLC
LaVergne TN
LVHW060422300926
849162LV00006B/1504